High-Yield™

Gross Anatomy

FOURTH EDITION

High-Yield™
Gross Anatomy
FOURTH EDITION

Ronald W. Dudek, PhD
Professor
Brody School of Medicine
East Carolina University
Department of Anatomy and Cell Biology
Greenville, North Carolina

Thomas M. Louis, PhD
Professor
Brody School of Medicine
East Carolina University
Department of Anatomy and Cell Biology
Greenville, North Carolina

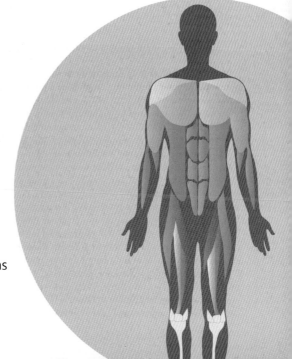

 Wolters Kluwer | Lippincott Williams & Wilkins
Health
Philadelphia · Baltimore · New York · London
Buenos Aires · Hong Kong · Sydney · Tokyo

Acquisitions Editor: Crystal Taylor
Product Manager: Sirkka E. Howes
Manufacturing Coordinator: Margie Orzech
Vendor Manager: Bridgett Dougherty
Designer: Teresa Mallon
Compositor: Aptara, Inc.

Fourth Edition

351 West Camden Street   530 Walnut Street
Baltimore, MD 21201   Philadelphia, PA 19106

Printed in China

9 8 7 6 5 4 3 2 1

Library of Congress Cataloging-in-Publication Data

Dudek, Ronald W., 1950-
  High-yield gross anatomy / Ronald W. Dudek, Thomas M. Louis. — 4th ed.
      p. ; cm.
  Includes index.
  ISBN 978-1-60547-763-3
  1. Human anatomy—Outlines, syllabi,.etc. 2. Anatomy, Surgical and
topographical—Outlines, syllabi, etc. I. Louis, Thomas. II. Title.
  [DNLM: 1. Anatomy—Outlines. QS 18.2 D845h 2011]
  QM31.D83 2011
  612.002'02—dc22

                                                            2009047422

DISCLAIMER

Care has been taken to confirm the accuracy of the information present and to describe generally accepted practices. However, the authors, editors, and publisher are not responsible for errors or omissions or for any consequences from application of the information in this book and make no warranty, expressed or implied, with respect to the currency, completeness, or accuracy of the contents of the publication. Application of this information in a particular situation remains the professional responsibility of the practitioner; the clinical treatments described and recommended may not be considered absolute and universal recommendations.

The authors, editors, and publisher have exerted every effort to ensure that drug selection and dosage set forth in this text are in accordance with the current recommendations and practice at the time of publication. However, in view of ongoing research, changes in government regulations, and the constant flow of information relating to drug therapy and drug reactions, the reader is urged to check the package insert for each drug for any change in indications and dosage and for added warnings and precautions. This is particularly important when the recommended agent is a new or infrequently employed drug.

Some drugs and medical devices presented in this publication have Food and Drug Administration (FDA) clearance for limited use in restricted research settings. It is the responsibility of the health care provider to ascertain the FDA status of each drug or device planned for use in their clinical practice.

To purchase additional copies of this book, call our customer service department at (800) 638-3030 or fax orders to (301) 223-2320. International customers should call (301) 223-2300.

Visit Lippincott Williams & Wilkins on the Internet: http://www.lww.com. Lippincott Williams & Wilkins customer service representatives are available from 8:30 am to 6:00 pm, EST.

Preface

High-Yield Gross Anatomy addresses many of the recurring clinical themes of the USMLE Step 1. The information presented in this text prepares you to handle not only the clinical vignettes found on the USMLE Step 1, but also the questions concerning basic gross anatomy concepts.

Like the USMLE Step 1, the discussions are comprehensively illustrated with a combination of drawings, MRIs, CT scans, radiographs, and cross-sectional anatomy. In addition, *High-Yield Gross Anatomy* directly addresses clinical issues and common clinical techniques (e.g., liver biopsy, tracheostomy, and lumbar puncture) that require knowledge of basic gross anatomy to deduce the correct answer.

For *High-Yield Gross Anatomy*, Fourth Edition, Dr. Thomas Louis has again contributed his considerable gross anatomy teaching experience to improve and narrow the focus of the book. Dr. Louis has taught gross anatomy for about 30 years in both cadaver-dissection and computer-assisted distance-learning gross anatomy courses. He has been a leader in developing computer-assisted distance learning at the Brody School of Medicine and has received national recognition for his efforts.

Dr. Louis used *High-Yield Gross Anatomy* in his physician assistant gross anatomy course for 4 years with excellent success and supplemented the clinical anatomy presented in the book with critical basic anatomy figures and diagrams to assist students in learning the gross anatomy relationships of these clinically relevant areas. Dr. Louis and I have added some of these figures and diagrams to further enhance your understanding. In addition, Dr. Louis has compiled a plethora of clinical vignettes, many of which are included in this edition.

I would like to thank Dr. Edward A. Monaco, III, for his suggestions and contributions to many of the chapters (especially the chapters on the Lymphatic System and Eye). Dr. Monaco has a PhD in Neuroscience from SUNY Upstate Medical University and is currently completing his medical studies at Columbia University College of Physicians and Surgeons in New York City.

I would appreciate your comments or suggestions about this book, especially after you have taken the USMLE Step 1, so that future editions can be improved and made more relevant to the test. You may contact me at dudekr@ecu.edu.

Ronald W. Dudek, PhD

Contents

6 Pleura, Tracheobronchial Tree, and Lungs 50

7 Heart 69

8 Abdominal Wall 89

9 Peritoneal Cavity 94

10 Abdominal Vasculature 97

Chapter 1

Vertebral Column

Ⅰ The Vertebral Column (Figure 1-1)

A. The vertebral column consists of 33 vertebrae (cervical C1–7, thoracic T1–12, lumbar L1–5, sacral S1–5 [sacrum], and coccygeal Co1–4 [coccyx]).

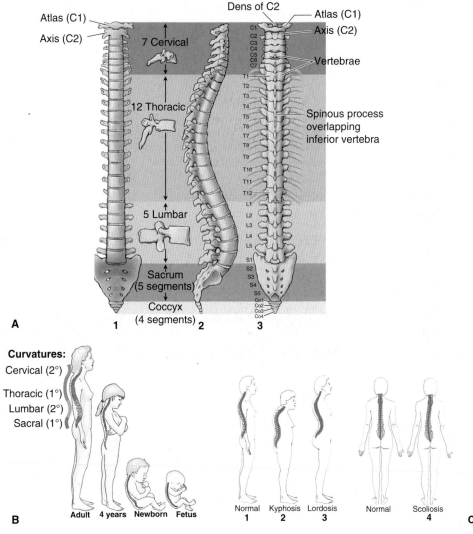

● **Figure 1-1 A:** Vertebral column: *(1)* anterior view; *(2)* right lateral view; *(3)* posterior view with vertebral ends of the ribs. **B:** Curvatures of the vertebral column from fetus to adult. **C:** Curvatures of the vertebral column: *(1)* normal; *(2)* kyphosis; *(3)* lordosis; *(4)* scoliosis. *(continued)*

VERTEBRAL LEVELS FOR REFERENCE	
Anatomic Structure	**Vertebral Level**
Hyoid bone, bifurcation of common carotid artery	C4
Thyroid cartilage, carotid pulse palpated	C5
Cricoid cartilage, start of trachea, start of esophagus	C6
Sternal notch, arch of aorta	T2
Sternal angle, junction of superior and inferior mediastinum, bifurcation of trachea	T4
Pulmonary hilum	T5–7
Inferior vena cava hiatus	T8
Xiphisternal joint	T9
Esophageal hiatus	T10
Upper pole of left kidney	T11
Aortic hiatus	T12
Duodenum	T12–L1
Celiac artery, upper pole of right kidney	T12
Superior mesenteric artery, end of spinal cord in adult (conus medullaris and pia mater)	L1
Renal artery	L2
End of spinal cord in newborn, inferior mesenteric artery, umbilicus	L3
Iliac crest, bifurcation of aorta	L4
Sacral promontory, start of sigmoid colon	S1
End of dural sac, dura, arachnoid, subarachnoid space, and cerebrospinal fluid	S2
End of sigmoid colon	S3

D

● **Figure 1-1** *Continued* **D:** Vertebral levels. Vertebral levels are used to reference the location of important anatomic structures. Knowledge of these vertebral levels will assist in deciphering clinical vignette questions. For example, a clinical vignette question may describe a pulsatile swelling located at vertebral level T2. Knowledge that the arch of the aorta is found at T2 will allow you deduce an aortic arch aneurysm.

   B. The **vertebral canal** contains the spinal cord, dorsal rootlets, ventral rootlets, dorsal nerve root, ventral nerve root, and meninges.

   C. The spinal nerve is located outside the vertebral canal by exiting through the **intervertebral foramen.**

II Curves (Figure 1-1)

   A. PRIMARY CURVES are the thoracic and sacral curvatures, which form during the fetal period.

   B. SECONDARY CURVES are the cervical and lumbar curvatures, which form after birth as a result of lifting the head and walking, respectively.

   C. KYPHOSIS is an exaggeration of the thoracic curvature, which may occur in the aged due to osteoporosis or disc degeneration.

   D. LORDOSIS is an exaggeration of the lumbar curvature, which may occur as a result of pregnancy, spondylolisthesis, or "pot belly."

   E. SCOLIOSIS is a complex lateral deviation/torsion, which may occur due to poliomyelitis, a short leg, or hip disease.

III Joints

   A. ATLANTO-OCCIPITAL JOINTS (FIGURE 1-2)
      1. Atlanto-occipital joints are the articulations between the superior articular surfaces of the atlas (C1) and the occipital condyles.

2. The action of **nodding the head (as in indicating "yes")** and **sideways tilting of the head** occurs at these joints.
3. These are synovial joints and have **no** intervertebral disc.
4. The **anterior and posterior atlanto-occipital membranes** limit excessive movement at this joint.

B. ATLANTOAXIAL JOINTS (FIGURE 1-2)

1. Atlantoaxial joints are the articulations between the atlas (C1) and axis (C2) that include two **lateral atlantoaxial joints** between the inferior facets of C1 and superior facets of C2, and one **median atlantoaxial joint** between the anterior arch of C1 and the dens of C2.
2. The action of **turning the head side to side (as in indicating "no")** occurs at these joints.
3. These are synovial joints and have **no** intervertebral disc.
4. The **alar ligaments**, which extend from the sides of the dens to the lateral margins of the foramen magnum, limit excessive movement at this joint.

C. CLINICAL CONSIDERATION: ATLANTOAXIAL DISLOCATION (SUBLUXATION)

1. Atlantoaxial dislocation (subluxation) is caused by the **rupture of the transverse ligament of the atlas** due to trauma (e.g., Jefferson fracture) or rheumatoid arthritis. This allows mobility of the **dens** (part of C2) within the vertebral canal, which places at risk the cervical spinal cord (leading to quadriplegia) and/or medulla (respiratory paralysis leading to sudden death).
2. The **dens** is secured in its position by the:
   a. **Transverse ligament of the atlas**, which together with the **superior longitudinal band** and **inferior longitudinal band** form the **cruciate ligament**. A widening of the atlantodental interval (distance from the anterior arch of C1 to the dens) suggests tearing of the transverse ligament.
   b. **Alar ligaments**
   c. **Tectorial membrane**, which is a continuation of the posterior longitudinal ligament.

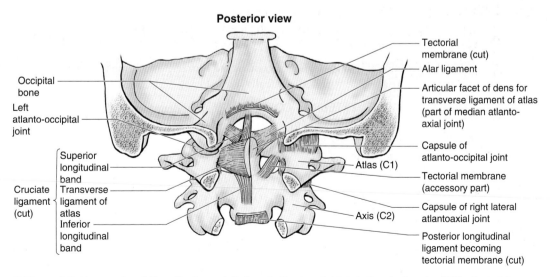

Posterior view

Occipital bone
Left atlanto-occipital joint
Cruciate ligament (cut)
Superior longitudinal band
Transverse ligament of atlas
Inferior longitudinal band

Tectorial membrane (cut)
Alar ligament
Articular facet of dens for transverse ligament of atlas (part of median atlanto-axial joint)
Capsule of atlanto-occipital joint
Tectorial membrane (accessory part)
Capsule of right lateral atlantoaxial joint
Posterior longitudinal ligament becoming tectorial membrane (cut)

Atlas (C1)
Axis (C2)

● **Figure 1-2 Ligaments of the atlanto-occipital and atlantoaxial joints (posterior view).** The tectorial membrane and the right side of the cruciate ligament have been removed to show the attachment of the right alar ligament to the dens of the C2.

Vasculature of the Vertebral Column

A. ARTERIAL SUPPLY

1. The vertebrae are supplied by **periosteal branches**, **equatorial branches**, and **spinal branches** from larger parent arteries that include the vertebral arteries, ascending cervical arteries, segmental arteries of the trunk, posterior intercostal arteries, subcostal and lumbar arteries in the abdomen, and iliolumbar and lateral and medial sacral arteries in the pelvis.

2. The periosteal and equatorial branches arise from these parent arteries as they travel along the anterolateral surface of the vertebrae.

3. The spinal branches enter the intervertebral foramina and divide into the **anterior vertebral canal branch**, which sends **nutrient arteries** into the vertebral bodies, and the **posterior vertebral canal branch**. The spinal branches terminate as the **segmental medullary arteries** or **radicular arteries**, which supply the spinal cord.

B. VENOUS DRAINAGE

1. The vertebrae are drained by **spinal veins**, which form the **internal vertebral venous plexus** and the **external vertebral venous plexus**.

2. The **basivertebral veins** form within the vertebral bodies, exit via foramina on the vertebral surface, and drain into the internal vertebral venous plexus (anterior portion).

3. The **intervertebral veins** receive veins from the spinal cord and the vertebral venous plexuses as they accompany spinal nerves through the intervertebral foramina.

Clinical Considerations

A. DENERVATION OF ZYGAPOPHYSEAL (FACET) JOINTS. The zygapophyseal (facet) joints are synovial joints between **inferior and superior articular processes**. These joints are located near the intervertebral foramen. If these joints are traumatized or diseased (e.g., rheumatoid arthritis), a spinal nerve may be impinged and cause severe pain. To relieve the pain, medial branches of the dorsal primary ramus are severed (i.e., dorsal rhizotomy).

B. DISLOCATIONS WITHOUT FRACTURE occur only in the cervical region because the articular surfaces are inclined horizontally. Cervical dislocations will stretch the posterior longitudinal ligament.

C. DISLOCATIONS WITH FRACTURE occur in the thoracic and lumbar regions because the articular surfaces are inclined vertically.

D. STABILITY OF THE VERTEBRAL COLUMN is mainly determined by four ligaments:

1. Anterior longitudinal ligament
2. Posterior longitudinal ligament
3. Ligamentum flavum
4. Interspinous ligaments

E. A ROUTE OF METASTASIS for breast, lung, and prostate cancer to the brain exists because the **internal vertebral venous plexus, basivertebral veins,** and **external vertebral venous plexus** surrounding the vertebral column communicate with the cranial dural sinuses and veins of the thorax, abdomen, and pelvis.

F. PROTRUSION OF THE NUCLEUS PULPOSUS (FIGURE 1-3). An intervertebral disc consists of the **annulus fibrosus** (fibrocartilage) and **nucleus pulposus** (remnant of the embryonic notochord). The nucleus pulposus generally herniates in a **posterior-lateral direction** and compresses a nerve root.

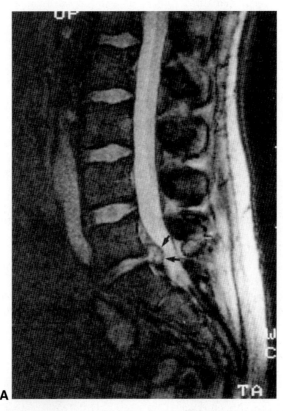

● **Figure 1-3 Herniated disc. A:** Magnetic resonance image (MRI; sagittal view) shows a herniated disc (*arrows*) between L5 and S1. (*continued*)

Herniated Disc Between	Compressed Nerve Root	Dermatome Affected	Muscles Affected	Movement Weakness	Nerve and Reflex Involved
C4 and C5	**C5**	C5 Shoulder Lateral surface of upper limb	Deltoid	Abduction of arm	Axillary nerve ↓ Biceps jerk
C5 and C6	**C6**	C6 Thumb	Biceps Brachialis Brachioradialis	Flexion of forearm Supination/ pronation	Musculocutaneous nerve ↓ Biceps jerk ↓ Brachioradialis jerk
C6 and C7	**C7**	C7 Posterior surface of upper limb Middle and index fingers	Triceps Wrist extensors	Extension of forearm Extension of wrist	Radial nerve ↓ Triceps jerk
L3 and L4	**L4**	L4 Medial surface of leg Big toe	Quadriceps	Extension of knee	Femoral nerve ↓ Knee jerk
L4 and L5	**L5**	L5 Lateral surface of leg Dorsum of foot	Tibialis anterior Extensor hallucis longus Extensor digitorum longus	Dorsiflexion of ankle (patient cannot stand on heels) Extension of toes	Common fibular nerve No reflex loss
L5 and S1 (most common)	**S1**	S1 Posterior surface of lower limb Little toe	Gastrocnemius Soleus	Plantar flexion of ankle (patient cannot stand on toes) Flexion of toes	Tibial nerve ↓ Ankle jerk

B

● **Figure 1-3** *Continued* **B:** Important features of a herniated disc at various vertebral levels are shown. From various clinical signs, you should be able to deduce which nerve root is compressed and then identify the appropriate intervertebral disc on a radiograph or MRI.

G. SPONDYLOLYSIS (FIGURE 1-4) is a stress fracture of the **pars interarticularis** (an area between the pedicle and lamina of a vertebra). It is often seen in adolescent athletes, most commonly at the L4 or L5 vertebra. The oblique radiograph shows a fracture at the pars interarticularis with sclerotic margins (*small arrows*), which appears as a **radiolucent "collar" around the neck of a Scottie dog**. Note that the pars interarticularis at the L4 vertebra is normal (*large arrow*).

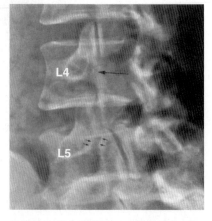

● **Figure 1-4 Oblique radiograph of a spondylolysis with spondylolisthesis.**

H. SPONDYLOLISTHESIS (FIGURE 1-5) (Greek: "spondylo" = vertebra; "olisthesis" = to slide on an incline) is the anterior subluxation of the vertebral body so that the body of the vertebra moves anterior with respect to the vertebrae below it, causing a lordosis. This occurs when the **pedicles** of a lumbar vertebra degenerate or fail to develop properly, or as a sequela of spondylolysis. Consequently, this may result in a **degenerative spondylolisthesis**, which usually occurs at the L4–5 vertebral level, or a **congenital spondylolisthesis**, which usually occurs at the L5–S1 vertebral level. The lateral radiograph shows spondylolysis at L5 (*small arrows*) with a spondylolisthesis where the L5 vertebra is subluxed anteriorly with respect to S1.

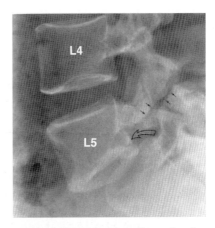

● **Figure 1-5 Lateral radiograph of a spondylolysis with spondylolisthesis.**

I. HANGMAN FRACTURE (TRAUMATIC SPONDYLOLISTHESIS OF C2) (FIGURE 1-6) occurs when a force is applied with the neck *hyperextended* (e.g., extension component of whiplash, car accident when chin or forehead strikes dashboard, head-on collision in football, or hanging) and places the spinal cord at risk. A traumatic spondylolisthesis of C2 includes the following pathology: fracture of the pars interarticularis bilaterally of the C2 vertebra, anterior subluxation of the C2 vertebra, tear of the anterior longitudinal ligament, and posterior fractured portion of C2 remains attached to C3 (in a legal drop hanging). The lateral radiograph shows a traumatic spondylolisthesis. Note the fracture of the pars interarticularis of the C2 vertebra (*solid arrow*) and the anterior subluxation of the C2 vertebra with respect to the C3 vertebra (*open arrow*).

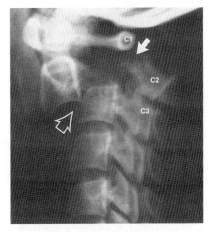

● **Figure 1-6 Lateral radiograph of a traumatic spondylolisthesis.**

J. SPONDYLOSIS (FIGURE 1-7) is a very common degenerative process of the vertebral column that occurs in the cervical region of elderly patients. The extent of degeneration may range from mild disc space narrowing and bone spur formation to severe **spondylosis deformans** (which includes disc space narrowing, facet joint narrowing, and bone spur formation). The lateral radiograph shows narrowing of all the disc spaces below C4, resulting in a severe cervical spondylosis. The bone spurs encroach the vertebral canal, and sclerosis of the facet joints is apparent.

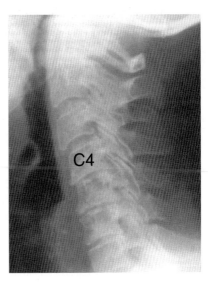

● **Figure 1-7 Severe cervical spondylosis.**

K. TEARDROP FRACTURE (FIGURE 1-8) is caused by **hyperflexion of the cervical region** (e.g., diving into shallow water, rebound flexion component of whiplash from a rear-end car accident, or head-on collision in football) and places the spinal cord at risk. A triangular fragment ("teardrop body") is sheared off of the anterior-inferior corner of the dislocating vertebral body. The result is a complete disruption of the cervical spine, with the upper portion of the vertebra displaced posteriorly and angulated anteriorly. A teardrop fracture includes the following pathology: avulsion fracture of a cervical vertebral body ("teardrop body"), fracture of the spinous process, posterior subluxation of vertebrae, compression of the spinal cord, tear of the anterior longitudinal ligament, and tear/disruption of the posterior longitudinal ligament, ligamentum flavum, interspinous ligament, and supraspinous ligament. The lateral radiograph shows a fracture of the C5 vertebral body ("teardrop body"; *arrow* and *dotted line*) and the posterior subluxation of the C5 vertebra.

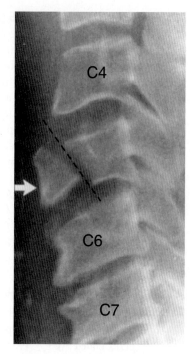

● **Figure 1-8 Lateral radiograph of a teardrop hyperflexion injury.**

L. JEFFERSON FRACTURE (FIGURE 1-9) is caused by **compression of the cervical region** (e.g., force applied to top of head) and places the spinal cord at risk. A Jefferson fracture includes the following pathology: fracture of the C1 vertebra at multiple sites, lateral displacement or C1 vertebra beyond the margins of the C2 vertebra, and tear of the transverse ligament. The computed tomography scan shows a fracture of the C1 vertebra at multiple sites (*arrows*).

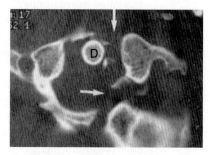

● **Figure 1-9 Computed tomography scan of a Jefferson fracture.** D, dens.

M. HYPEREXTENSION (WHIPLASH) INJURY (FIGURE 1-10) is caused by **hyperextension of the cervical region** (e.g., extension component of whiplash from a rear-end car accident, car accident when chin or forehead strikes dashboard, or head-on collision in football). The usual whiplash injury is a strain of the paravertebral and neck muscles. In more severe injuries, tear of the anterior longitudinal ligament, tear of the anterior attachment of the intervertebral disc, and widening of the intervertebral space may occur (bony fractures and dislocations are uncommon). However, in more violent hyperextension injuries (e.g., head-on collision in football), fracture of the posterior portion of the cervical vertebrae may occur. The lateral radiograph of a hyperextension injury shows the anterior widening of the intervertebral space at C5–6 (*arrow*).

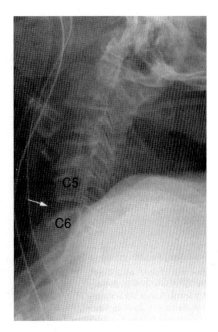

● **Figure 1-10 Lateral radiograph of a hyperextension (whiplash) injury.**

N. CHANCE FRACTURE (FIGURE 1-11) is caused by **hyperflexion of the thoracic or lumbar region** (e.g., "seat-belt injury" most commonly at vertebral level L2 or L3 when car occupant is thrown forward against a restraining seat belt during sudden deceleration and associated with intra-abdominal injuries) and generally does not place the spinal cord at risk. A Chance fracture includes the following pathology: transverse fracture of the vertebral body and arch, rupture of the intervertebral disc, and tear of the posterior longitudinal ligament, ligamentum flavum, interspinous ligament, and supraspinous ligament. The lateral radiograph of a Chance fracture shows the compressed L3 vertebral body (*arrowheads*) due to the transverse fracture (*arrows*). Note the increased distance between the spinous processes due to a tear of the ligamentum flavum, interspinous ligament, and supraspinous ligament (*long double-headed arrow*).

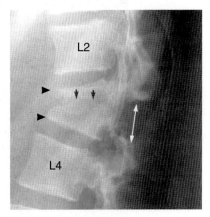

● **Figure 1-11 Lateral radiograph of a Chance fracture.**

VI Normal Radiology

A. CERVICAL AND THORACIC REGIONS (FIGURE 1-12)

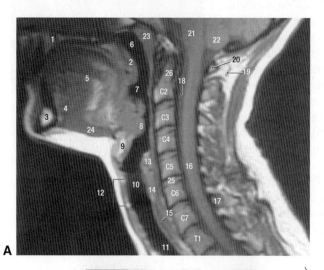

1 Hard palate	14 Cricoid cartilage
2 Soft palate	15 Esophagus
3 Mandible	16 Spinal cord
4 Genioglossus	17 Spinous process
5 Tongue	18 Cerebrospinal fluid in
6 Nasopharynx	subarachnoid space
7 Oropharynx	19 Nuchal ligament
8 Epiglottis	20 Posterior arch of atlas
9 Hyoid bone	21 Medulla oblongata
10 Vestibule	22 Tonsil of cerebellum
11 Trachea	23 Pharyngeal tonsil (adenoid)
12 Thyroid cartilage	24 Geniohyoid
13 Arytenoid cartilage	25 Intervertebral disc
	26 Dens of axis

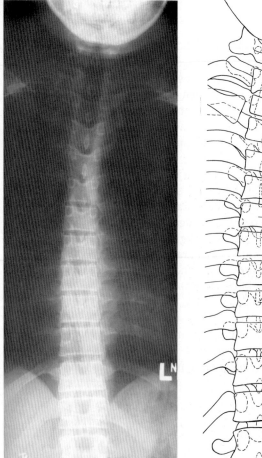

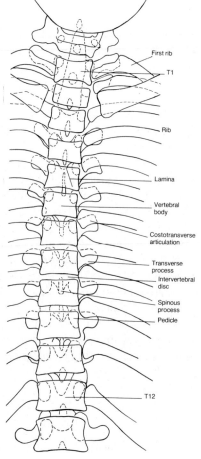

● **Figure 1-12 Normal radiology of the cervical and thoracic regions. A:** Median sagittal magnetic resonance image of the cervical region. Note the superior projection of the dens axis (26) and its relationship to the atlas, spinal cord (16), and medulla (21). The dens axis is secured in its position predominately by the transverse ligament, the rupture of which will place the spinal cord and possibly the medulla in jeopardy. **B:** Anteroposterior radiograph of the thoracic region.

B. LUMBOSACRAL REGION (FIGURE 1-13)

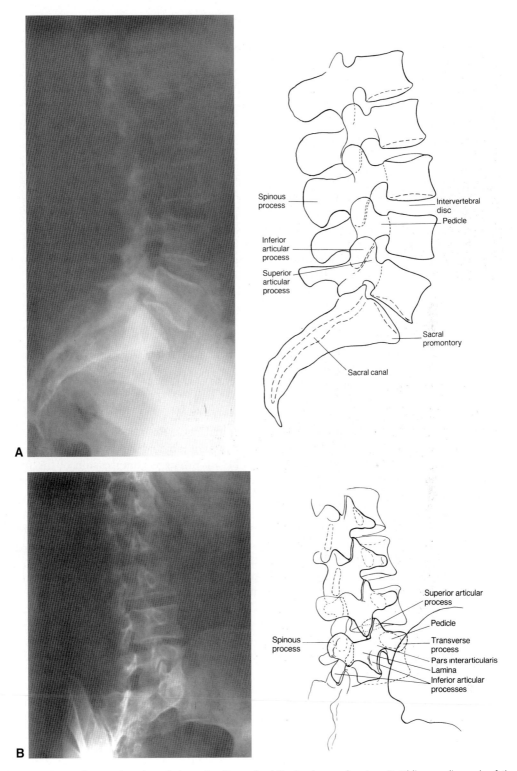

● **Figure 1-13 Lumbosacral region. A:** Lateral radiograph of the lumbosacral region. **B:** Oblique radiograph of the lumbosacral region ("Scottie dog" projection). The anatomic structures of lumbar vertebrae portray a "Scottie dog" appearance in an oblique view. The ears of the Scottie dog are the superior articular processes, the legs are the inferior articular processes, the nose is the transverse process, the neck is the pars interarticularis, and the eye is the pedicle.

Case Study 1-1

A 35-year-old construction worker experienced pain in his lower back while trying to move a beam from one side of the construction site to another site. The man comes into your office and tells you that the pain was "sudden" and "sharp" and that "over the last several days the pain has begun to move from my lower back down to my right leg." After further questioning, he tells you that "there is some numbness and tingling in my right leg, foot, and little toe." A little later in the conversation, he tells you that "when I cough, it really hurts bad." What is the most likely diagnosis?

Relevant Physical Exam Findings

- Analgesic gait
- Uncomfortable as he sits
- Pain upon raising his right extended leg
- Weakness in plantar flexion of his right foot ("cannot stand on his toes")
- Loss of sensation over the dorsal side of the right fourth and fifth toes
- Reduced ankle jerk reflex
- The straight leg raise (SLR) test exacerbates right lower limb pain at 45 degrees' elevation and the crossed SLR test exacerbates the pain at 40 degrees' elevation
- Pain restricts active flexion of the lumbosacral spine to 20 degrees
- Palpation of the lower back shows a flattening of the normal lordosis

Relevant Lab Findings

- Complete blood count (CBC), urinalysis, and urine culture are negative, which rules out deep organ etiology (e.g., cancer metastasis)
- A magnetic resonance imaging (MRI) scan was ordered

Diagnosis

Herniated Disc at L5–S1 Compressing the Right Spinal Nerve Root at S1

- The patient has weakness in plantar flexion of his right foot and loss of sensation over the dorsal side of the right fourth and fifth toes because there is **compression of the tibial nerve (L4–S3 rami)**. The tibial nerve provided the motor function for the posterior compartment muscles of the thigh (except for the short head of the biceps femoris), leg, and sole of the foot.
  - The muscles that produce plantar flexion of the foot are posterior compartment muscles of the leg (i.e., gastrocnemius, soleus, and plantaris). When the tibial nerve is compressed, weakness of plantar flexion occurs.
  - The **medial sural cutaneous nerve** (a branch of the tibial nerve) is usually joined by the **sural communicating branch of the common fibular nerve** to form the **sural nerve**. The sural nerve supplies the skin of the lateral and posterior part of the inferior one third of the leg and the lateral side of the foot.
- Approximately 95% of lumbar disc protrusions occur at either the L4–5 or L5–S1 levels.
- Protrusions of the nucleus pulposus usually occur **posterolaterally.** This is because of the following:
  - The nucleus pulposus is pushed farther posteriorly during flexion.
  - The anulus fibrosus is weaker posteriorly and laterally.
  - The posterior longitudinal ligament does not completely support the discs.

Chapter **2**

Spinal Cord and Spinal Nerves

Ⓘ Components of the Spinal Cord (Figure 2-1)

A. **GRAY MATTER** of the spinal cord consists of neuronal cell bodies and is divided into the **dorsal horn, ventral horn,** and **lateral horn.**

B. **WHITE MATTER** of the spinal cord consists of neuronal fibers and is divided into the **dorsal funiculus, ventral funiculus,** and **lateral funiculus.**

C. **VENTRAL MEDIAN FISSURE** is a distinct surface indentation present at all spinal cord levels and is related to the anterior spinal artery.

D. **DORSAL MEDIAN FISSURE** is a less distinct surface indentation present at all spinal cord levels.

E. **DORSAL INTERMEDIATE SEPTUM** is a surface indentation present only **at and above** T6 that distinguishes ascending fibers within the **gracile fasciculus** (from the lower extremity) from ascending fibers within the **cuneate fasciculus** (from the upper extremity).

F. **CONUS MEDULLARIS** is the end of the spinal cord, which occurs at vertebral level **L1 in the adult** and vertebral level **L3 in the newborn.**

G. **CAUDA EQUINA** consists of the dorsal and ventral nerve roots of L2 through coccygeal 1 spinal nerves traveling in the subarachnoid space below the conus medullaris.

H. **FILUM TERMINALE** is a prolongation of the **pia mater** from the conus medullaris to the end of the dural sac at vertebral level S2 where it blends with the dura. The dura continues caudally as the **filum of the dura mater** (or coccygeal ligament), which attaches to the dorsum of the coccyx bone.

Ⅱ Meninges and Spaces (Figure 2-1)

A. **EPIDURAL SPACE** is located between the vertebra and dura mater. This space contains fat and the **internal vertebral venous plexus.**

B. **DURA MATER** is the tough, outermost layer of the meninges.

C. **SUBDURAL SPACE** is located between the dura mater and arachnoid.

D. **ARACHNOID** is a filmlike, transparent layer connected to the pia mater by **trabeculations.**

E. **SUBARACHNOID SPACE** is located between the arachnoid and pia mater and is filled with **cerebrospinal fluid.**

F. **PIA MATER** is a thin layer that is closely applied to the spinal cord and has lateral extensions called **denticulate ligaments,** which attach to the dura mater and thereby suspend the spinal cord within the dural sac.

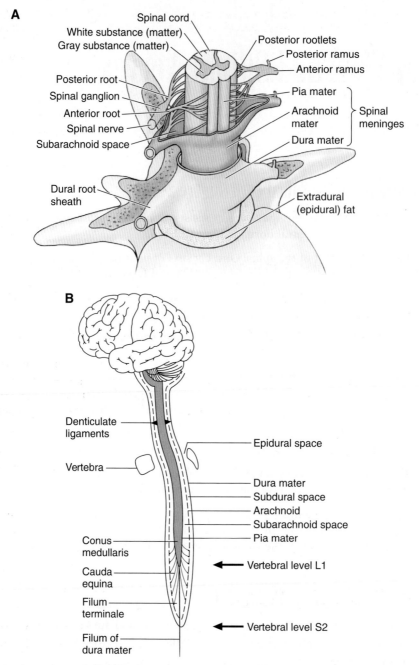

● **Figure 2-1 Spinal cord anatomy. A:** A diagram of the spinal cord, spinal nerves, and meninges. **B:** A diagram indicating craniocaudal extent of the spinal cord and meninges.

III Arterial Supply of the Spinal Cord

A. ANTERIOR SPINAL ARTERY AND POSTERIOR SPINAL ARTERIES

1. There is only **one anterior spinal artery**, which arises from the vertebral arteries and runs in the anterior median fissure. The anterior spinal artery gives rise to **sulcal arteries**, which supply the **ventral two thirds** of the spinal cord.

2. There are **two posterior spinal arteries**, which arise from either the vertebral arteries or the posterior inferior cerebellar arteries. The posterior spinal arteries supply the **dorsal one third** of the spinal cord.

3. The anterior and posterior spinal arteries supply only the short superior part of the spinal cord. The circulation of the rest of the spinal cord depends on the **segmental medullary arteries** and **radicular arteries**.

B. ANTERIOR AND POSTERIOR MEDULLARY SEGMENTAL ARTERIES

1. These arteries arise from the spinal branches of the ascending cervical, deep cervical, vertebral, posterior intercostal, and lumbar arteries.

2. The anterior and posterior medullary segmental arteries occur irregularly in place of radicular arteries and are located mainly in the cervical and lumbosacral spinal enlargements.

3. The medullary segmental arteries are actually "large radicular arteries" that connect with the anterior and posterior spinal arteries, whereas the radicular arteries do not.

C. GREAT ANTERIOR SEGMENTAL MEDULLARY (OF ADAMKIEWICZ)

1. This artery generally arises on the left side from a posterior intercostal artery or a lumbar artery and enters the vertebral canal through the intervertebral foramen at the lower thoracic or upper lumbar level.

2. This artery is clinically important since it makes a major contribution to the anterior spinal artery and the lower part of the spinal cord.

3. If this artery is ligated during resection of an **abdominal aortic aneurysm**, **anterior spinal artery syndrome** may result. Clinical symptoms include paraplegia, impotence, loss of voluntary control of the bladder and bowel (incontinence), and loss of pain and temperature, but vibration and proprioception sensation are preserved.

D. ANTERIOR AND POSTERIOR RADICULAR ARTERIES.
These arteries are small and supply only the dorsal and ventral roots of spinal nerves and superficial parts of the gray matter.

IV Components of a Spinal Nerve (Figure 2-2)

A. There are 31 pairs of spinal nerves: 8 cervical, 12 thoracic, 5 lumbar, 5 sacral, and 1 coccygeal.

B. Small bundles of nerve fibers called the **dorsal (posterior) rootlets** and **ventral (anterior) rootlets** arise from the dorsal and ventral surfaces of the spinal cord, respectively.

C. The dorsal rootlets converge to form the **dorsal (posterior) root** (containing afferent or sensory fibers) and the ventral rootlets converge to form the **ventral (anterior) root** (containing efferent or motor fibers).

D. The dorsal root and ventral root join to form the **mixed spinal nerve** near the intervertebral foramen.

E. Each spinal nerve divides into a **dorsal (posterior) primary ramus** (which innervates the skin and deep muscles of the back) and **ventral (anterior) primary ramus** (which innervates the remainder of the body).

F. Spinal nerves are connected to the paravertebral ganglia (sympathetic chain ganglia) and prevertebral ganglia by the **white communicating rami** (containing **myelinated** preganglionic sympathetic nerve fibers present in spinal nerves T1–L3) and **gray communicating rami** (containing **unmyelinated** postganglionic sympathetic nerve fibers present in all spinal nerves).

A

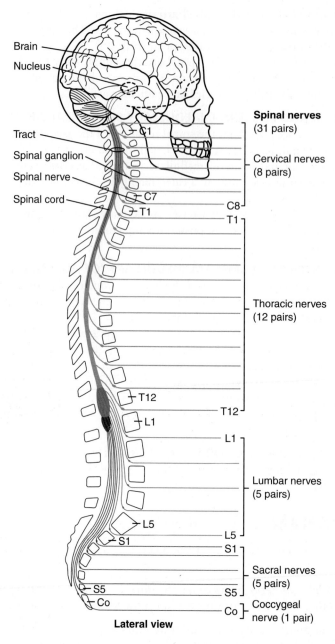

Brain

Nucleus

Tract

Spinal ganglion

Spinal nerve

Spinal cord

C1

C7

T1

T12

L1

L5

S1

S5

Co

Spinal nerves
(31 pairs)

Cervical nerves
(8 pairs)

C8

T1

Thoracic nerves
(12 pairs)

T12

L1

Lumbar nerves
(5 pairs)

L5

S1

Sacral nerves
(5 pairs)

S5

Co

Coccygeal
nerve (1 pair)

Lateral view

B

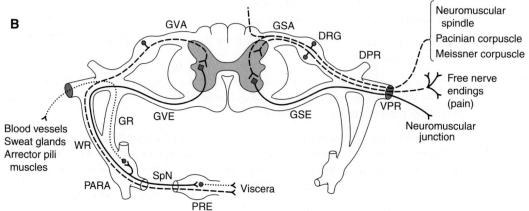

GVA

GSA

DRG

DPR

Neuromuscular
spindle
Pacinian
corpuscle
Meissner corpuscle

Free nerve
endings
(pain)

VPR

Neuromuscular
junction

GR

GVE

GSE

Blood vessels
Sweat glands
Arrector pili
muscles

WR

PARA

SpN

PRE

Viscera

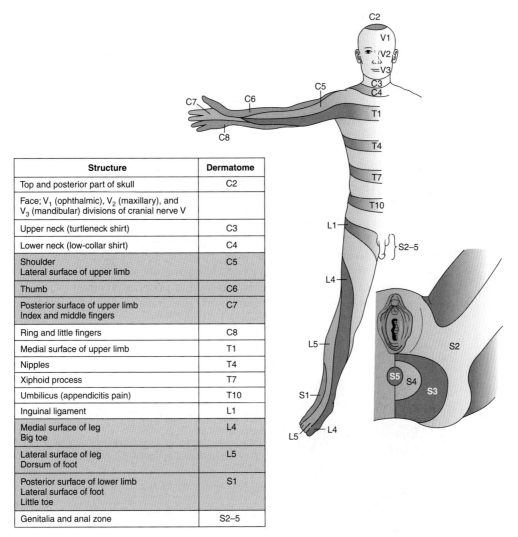

Structure	Dermatome
Top and posterior part of skull	C2
Face; V₁ (ophthalmic), V₂ (maxillary), and V₃ (mandibular) divisions of cranial nerve V	
Upper neck (turtleneck shirt)	C3
Lower neck (low-collar shirt)	C4
Shoulder Lateral surface of upper limb	C5
Thumb	C6
Posterior surface of upper limb Index and middle fingers	C7
Ring and little fingers	C8
Medial surface of upper limb	T1
Nipples	T4
Xiphoid process	T7
Umbilicus (appendicitis pain)	T10
Inguinal ligament	L1
Medial surface of leg Big toe	L4
Lateral surface of leg Dorsum of foot	L5
Posterior surface of lower limb Lateral surface of foot Little toe	S1
Genitalia and anal zone	S2–5

● **Figure 2-3 Dermatomes.** Anterior view of some important dermatomes of the body. Although dermatomes are shown as distinct segments, in reality, there is overlap between any two adjacent dermatomes. Note that the sensory innervation of the face does not involve dermatomes but is carried by cranial nerve (CN) V: V₁ (ophthalmic division), V₂ (maxillary division), and V₃ (mandibular division). Shaded areas in the table indicate dermatomes affected by a herniated disc (see Chapter 1). Knowledge of dermatomes is important because clinical vignette questions will include a description of sensory loss at a specific dermatome level.

 Dermatomes (Figure 2-3) are strips of skin extending from the posterior midline to the anterior midline that are supplied by sensory branches of dorsal and ventral rami of a single spinal nerve. A clinical finding of sensory deficit in a dermatome is important in order to assess what spinal nerve, nerve root, or spinal cord segment may be damaged.

● **Figure 2-2 A:** Basic organization of the spinal nerves. Note that each spinal nerve bears the same letter–numeric designation as the vertebra forming the *superior boundary* of its exit from the vertebral column, except in the cervical region. In the cervical region, each spinal nerve bears the same letter–numeric designation as the vertebra forming the *inferior boundary* of its exit from the vertebral column. Note that spinal nerve C8 exits between vertebrae C7 and T1. **B:** Components of a typical thoracic spinal nerve. The four functional components are indicated: general somatic afferent (GSA), general somatic efferent (GSE), general visceral afferent (GVA), and general visceral efferent (GVE). The muscle stretch (myotactic) reflex includes the neuromuscular spindle, GSA dorsal root ganglion cell, GSE ventral horn motor neuron, and neuromuscular junction. DPR = dorsal primary ramus, DRG = dorsal root ganglion, GR = gray communicating ramus, PARA = paravertebral (sympathetic chain) ganglion, PRE = prevertebral ganglion, SpN = splanchnic nerve, VPR = ventral primary ramus, WR = white communicating ramus.

Ⅵ Clinical Procedures (Figure 2-4)

A. LUMBAR PUNCTURE

1. Lumbar puncture can be done to either withdraw cerebrospinal fluid (CSF) or inject an anesthetic (e.g., spinal anesthesia).
2. A needle is inserted above or below the spinous process of the **L4 vertebra**.
3. The needle will pass through the following structures: skin → superficial fascia → supraspinous ligament → interspinous ligament → ligamentum flavum → epidural space containing the internal vertebral venous plexus → dura mater → arachnoid → subarachnoid space containing CSF. The pia mater is not pierced.

B. SPINAL ANESTHESIA (SPINAL BLOCK OR SADDLE BLOCK)

1. Spinal anesthesia is produced by injecting anesthetic into the **subarachnoid space** and may be used during childbirth.
2. Sensory nerve fibers for pain from the uterus travel with the:
   a. **Pelvic splanchnic nerves** (parasympathetic) to S2–4 spinal levels from the cervix (may be responsible for referred pain to the gluteal region and legs)
   b. **Hypogastric plexus** and **lumbar splanchnic nerves** (sympathetic) to L1–3 spinal levels from the fundus and body of the uterus and oviducts (may be responsible for referred pain to the back)

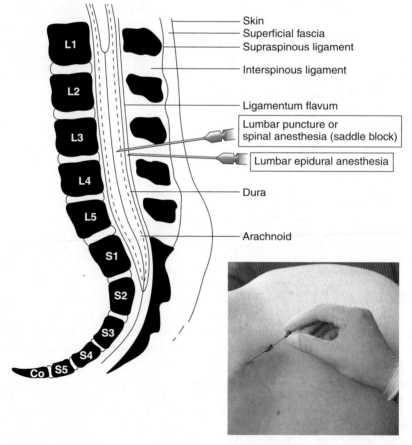

● **Figure 2-4 Lumbar vertebral column and spinal cord.** A needle is shown inserted into the subarachnoid space above the spinous process of L4 (L3–4 interspace) to withdraw cerebrospinal fluid (CSF) as in a lumbar puncture or to administer spinal anesthesia (saddle block). A second needle is shown inserted into the epidural space to administer lumbar epidural anesthesia. Note the sequence of layers (superficial to deep) that the needle must penetrate. Inset: Photograph shows a physician inserting a needle during a lumbar puncture procedure. The dotted line indicates the arachnoid.

3. Spinal anesthesia up to **spinal nerve T10** is necessary to block pain for vaginal childbirth and up to **spinal nerve T4** for cesarean section.

4. Pregnant women require a smaller dose of anesthetic (than nonpregnant patients) because the subarachnoid space is compressed because the **internal vertebral venous plexus** is engorged with blood from the pregnant uterus compressing the inferior vena cava.

5. Complications may include **hypotension** due to sympathetic blockade and vasodilation, **respiratory paralysis** involving the phrenic nerve due to high spinal blockade, and **spinal headache** due to CSF leakage.

C. LUMBAR EPIDURAL ANESTHESIA

1. Lumbar epidural anesthesia is produced by injecting anesthetic into the **epidural space** and may be used during childbirth.

2. Complications may include **respiratory paralysis** due to high spinal blockage if the dura and arachnoid are punctured and the usual amount of anesthetic is injected in the subarachnoid space by mistake, and **central nervous system (CNS) toxicity** (slurred speech, tinnitus, convulsions, cardiac arrest) due to injection of the anesthetic into the **internal vertebral venous plexus** (intravenous injection versus epidural application).

Ⅶ Clinical Considerations

A. ANTERIOR SPINAL ARTERY OCCLUSION results in damage to the lateral corticospinal tracts, lateral spinothalamic tracts, hypothalamospinal tracts, ventral gray horns, and corticospinal tracts to sacral parasympathetic centers at S2–4. **Clinical findings include** bilateral spastic paresis with pyramidal signs below the lesion, bilateral loss of pain and temperature sensation below the lesion, bilateral Horner syndrome, bilateral flaccid paralysis, and loss of voluntary bladder and bowel control.

B. SYRINGOMYELIA is a central cavitation of the cervical spinal cord of unknown etiology and results in damage to ventral white commissure involving the decussating lateral spinothalamic axons and ventral gray horns. **Clinical findings include** bilateral loss of pain and temperature sensation and flaccid paralysis of the intrinsic muscles of the hand.

C. SPINAL CORD INJURY (SCI)

1. **Complete SCI (transection of spinal cord)** results in loss of sensation and motor function below the lesion. There are two types of complete SCI:
   a. **Paraplegia** (i.e., paralysis of lower limbs) occurs if the transection occurs anywhere between the cervical and lumbar enlargements of the spinal cord.
   b. **Quadriplegia** (i.e., paralysis of all four limbs) occurs if the transection occurs above C3. These individuals may die quickly due to respiratory failure if the phrenic nerve is compromised.

2. **Incomplete SCI** can be ameliorated somewhat by rapid surgical intervention. There are three situations that may lead to an incomplete SCI: a concussive blow, anterior spinal artery occlusion, or a penetrating blow (e.g., Brown-Sequard syndrome).

3. **Complications of any SCI** include hypotension in the acute setting, ileus (bowel obstruction due to lack of motility), renal stones, pyelonephritis, renal failure, and deep venous thrombosis. **Methylprednisolone** may be of benefit if administered within 8 hours of injury.

D. CHORDOMAS are malignant, midline, lobulated, mucoid tumors that arise from remnants of the embryonic notochord and usually occur in the sacral (most common site) or clival region. Chordomas have histologic features, which include physaliphorous (bubble-bearing) cells with mucoid droplets in the cytoplasm.

E. **ASTROCYTOMAS (FIGURE 2-5)** (account for 70% of all neuroglial tumors) typically arise from astrocytes and are composed of cells with elongated or irregular, hyperchromatic nuclei and an eosinophilic glial fibrillary acidic protein (GFAP)–positive cytoplasm. **Glioblastoma multiforme (GBM)** is the most common primary brain tumor in adults (men 40 to 70 years of age), are highly malignant, and pursue a rapidly fatal course. A common site of GBMs is the frontal lobe, which commonly crosses the corpus callosum, producing a butterfly appearance on magnetic resonance imaging (MRI). The MRI shows an astrocytoma that is an excellent example of an intramedullary (within the spinal cord) tumor. Note that the astrocytoma (*arrows*) within the substance of the spinal cord has a cystic appearance.

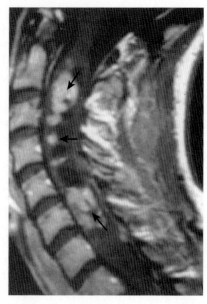

● Figure 2-5 Magnetic resonance image of an astrocytoma.

F. **MENINGIOMAS (FIGURE 2-6)** (90% are benign) arise from arachnoid cap cells of the arachnoid villi of the meninges and are found at the skull vault, sites of dural reflection (e.g., falx cerebri, tentorium cerebelli), optic nerve sheath, and choroid plexus. Meningiomas occur more commonly in women, may increase in size during pregnancy, have an increased incidence in women taking postmenopausal hormones, and are associated with breast cancer, all of which suggest the potential involvement of steroid hormones. The MRI shows a meningioma that is an excellent example of an intradural (within the meninges) tumor. Note the meningioma (*arrow*) outside of the spinal cord causing some compression of the spinal cord.

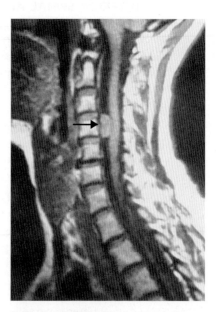

● Figure 2-6 Magnetic resonance image of a meningioma.

G. SCHWANNOMAS (FIGURE 2-7) are benign, well-circumscribed , encapsulated tumors that arise from Schwann cells located on cranial nerves, spinal nerve roots (present as dumb-bell-shaped tumors protruding through the intervertebral foramen), or spinal nerves. The most common intracranial site is the cerebellopontine angle with involvement of cranial nerve VIII (acoustic neuroma), where expansion of the tumor results in tinnitus and sensorineural deafness. Multiple schwannomas may occur associated with neurofibromatosis type II. The MRI shows a schwannoma protruding through the intervertebral foramen (*arrow*), which is a clear characteristic of a schwannoma (or neurofibroma).

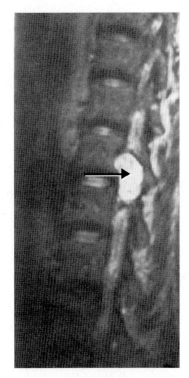

● **Figure 2-7 Magnetic resonance image of a schwannoma.**

Chapter 3

Autonomic Nervous System

ⓘ General Features of the Nervous System

A. The nervous system can be **anatomically** divided into the **central nervous system (CNS)**, which consists of the brain and spinal cord, and the **peripheral nervous system (PNS)**, which consists of 12 pairs of cranial nerves and 31 pairs of spinal nerves along with their associated ganglia.

B. The nervous system can also be **functionally** divided into the **somatic nervous system**, which controls voluntary activities by innervating skeletal muscle, and the **visceral (or autonomic) nervous system**, which controls involuntary activities by innervating smooth muscle, cardiac muscle, and glands.

C. The autonomic nervous system (ANS) is divided into the **sympathetic (thoracolumbar) division** and the **parasympathetic (craniosacral) division**.

D. The **hypothalamus** has central control of the ANS, whereby the hypothalamus coordinates all ANS actions.

E. The ANS has a **visceromotor component** and a **viscerosensory component** (although traditionally only the visceromotor component has been emphasized).

ⓘⓘ Sympathetic Division of the ANS (Thoracolumbar)

A. MOTOR (EFFERENT) COMPONENT (FIGURE 3-1). The motor component of the sympathetic nervous system has a "fight-or-flight" or **catabolic function** that is necessary in **emergency situations** where the body needs a sudden burst of energy. The whole motor component of the sympathetic nervous system tends to "go off at once" in an emergency situation. In a controlled environment, the sympathetic nervous system is not necessary for life but is essential for any stressful situation. The motor component of the sympathetic nervous system is a two-neuron chain that consists of a **preganglionic sympathetic neuron** and a **postganglionic sympathetic neuron** that follows this general pattern: CNS → short preganglionic neuron → ganglion → long postganglionic neuron → smooth muscle, cardiac muscle, and glands.

1. **Preganglionic Sympathetic Neuron.** The preganglionic neuronal cell bodies are located in the gray matter of the **T1–L2/L3 spinal cord** (i.e., **intermediolateral cell column**). Preganglionic axons have a number of fates:

   a. Preganglionic axons enter the paravertebral chain ganglia through white communicating rami, where they synapse with postganglionic neurons at that level.

   b. Preganglionic axons travel up or down the paravertebral chain ganglia, where they synapse with postganglionic neurons at upper or lower levels, respectively.

   c. Preganglionic axons pass through the paravertebral chain ganglia (i.e., no synapse) as **thoracic splanchnic nerves (greater, lesser, and least)**, **lumbar splanchnic nerves (L1–4)**, and **sacral splanchnic nerves (L5 and S1–3)**,

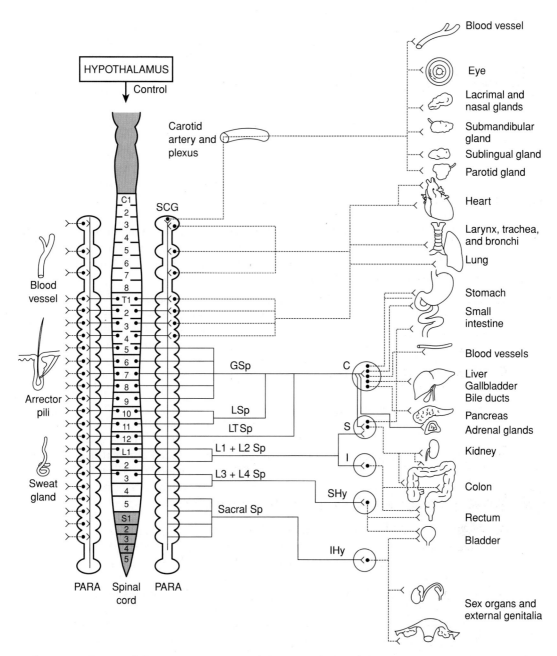

● **Figure 3-1 Diagram of the motor component of the sympathetic nervous system.** Solid lines = preganglionic sympathetic neurons. Dashed lines = postganglionic sympathetic neurons. C = celiac ganglion, I = inferior mesenteric ganglion, GSp = greater thoracic splanchnic nerve, IHy = inferior hypogastric plexus, LSp = lesser thoracic splanchnic nerve, LTSp = least thoracic splanchnic nerve, PARA = paravertebral chain ganglia, S = superior mesenteric ganglion, SHy = superior hypogastric plexus, SCG = superior cervical ganglion.

which synapse with postganglionic neurons in prevertebral ganglia (i.e., celiac ganglion, aorticorenal ganglion, superior mesenteric ganglion, inferior mesenteric ganglion) as well as in the superior hypogastric plexus and inferior hypogastric plexus.

d. Preganglionic axons pass through the paravertebral chain ganglia (i.e., no synapse) as thoracic splanchnic nerves, which synapse with modified postganglionic sympathetic neurons in the **adrenal medulla** called **chromaffin cells**.

2. **Postganglionic Sympathetic Neuron.** The postganglionic neuronal cell bodies are located in the **paravertebral chain ganglia** and the **prevertebral ganglia**. Postganglionic sympathetic neurons use **norepinephrine** as a neurotransmitter (except for those innervating eccrine sweat glands, which use acetylcholine), which binds to α_1-, α_2-, β_1-, β_2-, and β_3-adrenergic receptors located on the cell membrane of **smooth muscle, cardiac muscle, and glands**. Postganglionic axons have a number of fates:

   a. Postganglionic axons leave the paravertebral chain ganglia through gray communicating rami and join all 31 pairs of spinal nerves to innervate **smooth muscle of blood vessels, arrector pili smooth muscle of hair follicles**, and **sweat glands of the skin**.

   b. Postganglionic axons leave the superior cervical ganglion of the prevertebral chain ganglia and follow the **carotid arterial system** into the head and neck to innervate **smooth muscle of blood vessels, the dilator pupillae muscle, the superior tarsal muscle, the lacrimal gland, the submandibular gland, the sublingual gland**, and **the parotid gland**.

   c. Postganglionic axons leave the paravertebral chain ganglia (from the superior cervical ganglion → T4 levels) to enter the **cardiac nerve plexus** and **pulmonary nerve plexus** to innervate the heart and lung, respectively.

   d. Postganglionic axons leave prevertebral ganglia and the superior and inferior hypogastric plexuses to innervate **smooth muscle of various visceral organs**.

   e. Modified postganglionic sympathetic neurons called **chromaffin cells within the adrenal medulla** release epinephrine (the majority product; 90%) and norepinephrine (the minority product; 10%) into the bloodstream, both of which are potent sympathetic neurotransmitters.

B. SENSORY (AFFERENT) COMPONENT (FIGURE 3-2)

1. The sensory component of the sympathetic nervous system carries **visceral pain sensation** from **nociceptors** located in viscera to the CNS.

2. Nociceptors are free nerve endings that respond to pathologic stimuli such as myocardial infarction, appendicitis, and gastrointestinal cramping or bloating.

3. Visceral pain sensation is carried almost exclusively by the sensory (afferent) component of the sympathetic nervous system.

4. Visceral pain sensation is **poorly localized** because nociceptor density is low, nociceptor fields are large, and its projection to higher CNS levels is widespread.

5. The sensory component of the sympathetic nervous system has the following neuronal chain:

   a. The first neuron in the chain has its neuronal cell body located in the **dorsal root ganglia at T1–L2/L3 spinal cord levels**. This neuron sends a peripheral process to the viscera that ends as a free nerve ending (or nociceptor) and sends a central process into the spinal cord, which synapses with a second neuron **within the spinal cord**.

   b. The second neuron in the chain (within the spinal cord) projects axons to the **ventral posterolateral nucleus of the thalamus (VPL)** and the **reticular formation**, where they synapse with a third neuron.

   c. The third neuron in the chain (within the VPL and the reticular formation) projects axons to **diverse areas of the cerebral cortex, hypothalamus**, and **intralaminar nuclei of the thalamus**.

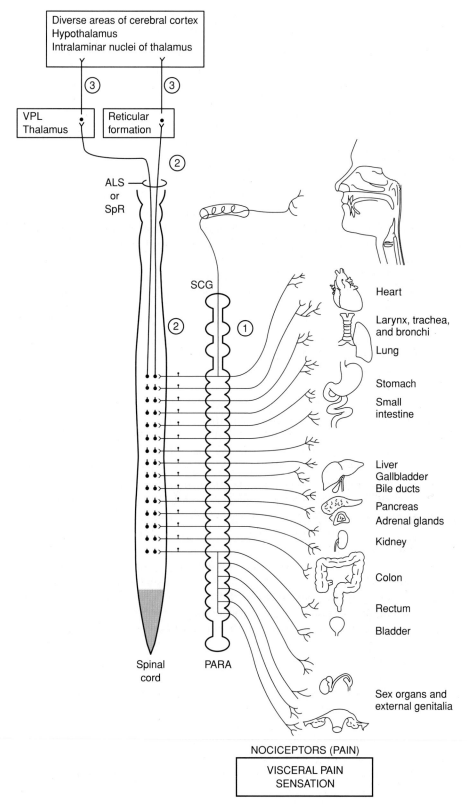

● **Figure 3-2 Diagram of the sensory component of the sympathetic nervous system (visceral pain sensation).**
The circled numbers indicate the three-neuron chain involved in visceral pain sensation. ALS = anterolateral system,
PARA = paravertebral chain ganglia, SpR = spinoreticular tract, VPL = ventral posterolateral nucleus of the thalamus.

ⓘⓘⓘ Parasympathetic Division of the ANS (Craniosacral)

A. MOTOR (EFFERENT) COMPONENT (FIGURE 3-3). The motor component of the parasympathetic nervous system has a **"rest and digest"** or **anabolic function** that is necessary to conserve energy, restore body resources, and get rid of waste. The whole motor component of the parasympathetic nervous system does *not* "go off at once"; instead, specific activities are initiated when appropriate. The motor component of the parasympathetic nervous system is a two-neuron chain that consists of a **preganglionic parasympathetic neuron** and a **postganglionic parasympathetic neuron** that follows this general pattern: CNS → long preganglionic neuron → ganglion → short postganglionic neuron → smooth muscle, cardiac muscle, and glands.

1. **Preganglionic Parasympathetic Neuron.** The preganglionic neuronal cell bodies are located in the **Edinger-Westphal nucleus, lacrimal nucleus, superior salivatory nucleus, inferior salivatory nucleus, dorsal motor nucleus of the vagus nerve,** and **gray matter of the S2–4 spinal cord.** Preganglionic axons have a number of fates:

    a. Preganglionic axons from the Edinger-Westphal nucleus run with cranial nerve (CN) III and enter the ciliary ganglia, where they synapse with postganglionic neurons.

    b. Preganglionic axons from the lacrimal nucleus run with **CN VII** and enter the **pterygopalatine ganglion**, where they synapse with postganglionic neurons.

    c. Preganglionic axons from the superior salivatory nucleus run with **CN VII** and enter the **submandibular ganglion**, where they synapse with postganglionic neurons.

    d. Preganglionic axons from the inferior salivatory nucleus run with **CN IX** and enter the **otic ganglion**, where they synapse with postganglionic neurons.

    e. Preganglionic axons from the dorsal motor nucleus of the vagus nerve run with **CN X** and travel to **various visceral organs (up to the splenic flexure of the transverse colon)**, where they synapse with postganglionic neurons.

    f. Preganglionic axons from the gray matter of the S2–4 spinal cord run as **pelvic splanchnic nerves**, which interact with the inferior hypogastric plexus and travel to **various visceral organs (distal to the splenic flexure of the transverse colon)**, where they synapse with postganglionic neurons.

2. **Postganglionic Parasympathetic Neuron.** The postganglionic neuronal cells bodies are located in the **ciliary ganglion, pterygopalatine ganglion, submandibular ganglion,** and **otic ganglion** and **within various visceral organs.** Postganglionic parasympathetic neurons use **acetylcholine** as a neurotransmitter, which binds to M_1, M_2, and M_3 **muscarinic acetylcholine receptors** located on the cell membrane of **smooth muscle, cardiac muscle,** and **glands.** Postganglionic axons have a number of fates:

    a. Postganglionic axons leave the ciliary ganglion to innervate the sphincter pupillae muscle and ciliary muscle.

    b. Postganglionic axons leave the pterygopalatine ganglion to innervate the **lacrimal glands** and **nasal glands.**

    c. Postganglionic axons leave the submandibular ganglion to innervate the **submandibular glands** and **sublingual glands.**

    d. Postganglionic axons leave the otic ganglion to innervate the **parotid gland.**

    e. Postganglionic axons associated with CN X innervate **various visceral organs** and **cardiac muscle.**

    f. Postganglionic axons associated with the pelvic splanchnic nerves innervate **various visceral organs.**

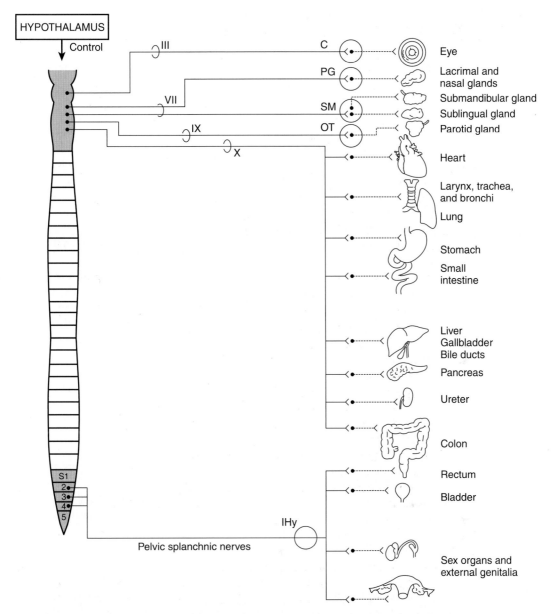

● **Figure 3-3 Diagram of the motor component of the parasympathetic nervous system.** Solid lines = preganglionic parasympathetic neurons. Dashed lines = postganglionic parasympathetic neurons. C = ciliary ganglion, IHy = inferior hypogastric plexus, OT = otic ganglion, PG = pterygopalatine ganglion, SM = submandibular ganglion.

B. SENSORY (AFFERENT) COMPONENT (FIGURE 3-4)

1. The sensory component of the parasympathetic nervous system carries the following:

   a. **Visceral pressure and movement sensation** from **rapidly adapting mechanoreceptors**

   b. **Visceral stretch sensation** from **slowly adapting mechanoreceptors**

   c. **Arterial oxygen tension (PaO$_2$)** and **arterial pH** information from **chemoreceptors** (i.e., carotid bodies located at the bifurcation of the common carotid artery and aortic bodies located in the aortic arch)

   d. **Blood pressure** information from **baroreceptors** (i.e., carotid sinus located in the walls of the common carotid artery and baroreceptors located in the great veins, atria, and aortic arch)

   e. **Osmolarity** information from **osmoreceptors**

   f. **Temperature** from **internal thermal receptors**

2. The sensory component of the parasympathetic nervous system follows this neuronal chain:

   a. The first neuron in the chain has its neuronal cell body located in the **geniculate ganglion of CN VII, inferior (petrosal) ganglion of CN IX, inferior (nodose) ganglion of CN X**, and **dorsal root ganglia of the S2–4 spinal cord**. This neuron sends a peripheral process to the viscera that ends at the rapidly adapting mechanoreceptors, slowly adapting mechanoreceptors, chemoreceptors, baroreceptors, osmoreceptors, and internal thermal receptors. These neurons also send a central process into the brainstem or spinal cord, which synapses with a second neuron either in the **solitary nucleus, dorsal horn of the spinal cord**, or **gray matter of the S2–4 spinal cord**.

   b. The second neuron in the chain (in the solitary nucleus) projects axons to the **dorsal motor nucleus of the vagus nerve (DMN)** and the **rostral ventrolateral medulla (RVLM)**, where they synapse with a third neuron. The second neuron in the chain (in the dorsal horn of the spinal cord) projects axons to the **anterolateral system (ALS)** and the **spinoreticular tract**, which terminate in the reticular formation. The second neuron in the chain (in the gray matter of the S2–4 spinal cord) is actually a **preganglionic parasympathetic motor neuron of a pelvic splanchnic nerve** (forming a sensory–motor reflex arc).

   c. The third neuron in the chain (in the DMN) is actually a **preganglionic parasympathetic motor neuron of CN X** (forming a sensory–motor reflex arc). The third neuron in the chain (in the RVLM) projects axons to the **intermediolateral cell column** of the spinal and thereby controls the activity of preganglionic *sympathetic* motor neurons (forming a sensory– motor reflex arc).

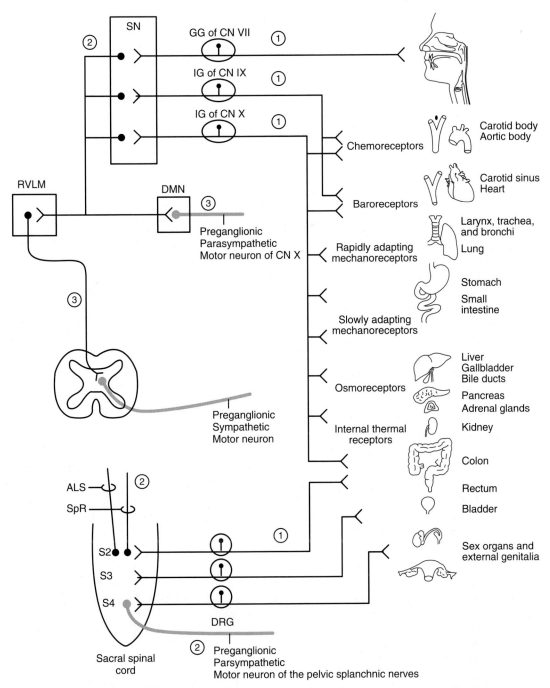

● **Figure 3-4 Diagram of the sensory component of the parasympathetic nervous system (visceral pressure and movement sensation, visceral stretch sensation, arterial oxygen tension and arterial pH, blood pressure, osmolarity, and temperature).** The circled numbers indicate the three-neuron chain involved in visceral sensation. ALS = anterolateral system, DMN = dorsal motor nucleus of the vagus nerve, DRG = dorsal root ganglion, GG = geniculate ganglion of cranial nerve (CN) VII, IG of CN IX = inferior (petrosal) ganglion of the glossopharyngeal nerve, IG of CN X = inferior (nodose) ganglion of the vagus nerve, RVLM = rostral ventrolateral medulla, SN = solitary nucleus, SpR = spinoreticular tract.

Summary Table of Sympathetic and Parasympathetic Motor Actions (Table 3-1)

TABLE 3-1	SUMMARY TABLE OF SYMPATHETIC AND PARASYMPATHETIC MOTOR ACTIONS

Specific Sympathetic Motor Actions

Smooth Muscle
Contracts dilator pupillae muscle, causing dilation of pupil (mydriasis)
Contracts arrector pili muscle in skin
Contracts smooth muscle in skin, skeletal muscle, and visceral blood vessels
Relaxes smooth muscle in skeletal muscle blood vessels
Relaxes bronchial smooth muscle in lung (bronchodilation)
Relaxes smooth muscle in GI tract wall
Contracts smooth muscle in GI tract sphincters
Relaxes smooth muscle in urinary bladder
Contracts smooth muscle in urinary tract sphincter
Contracts smooth muscle of ductus deferens, causing ejaculation (emission); "shoot"
Female reproductive tract[a]

Cardiac Muscle
Accelerates SA node (increases heart rate); positive chronotropism
Increases conduction velocity in AV node; positive dromotropism
Increases contractility of cardiac muscle; positive inotropism

Glands
Increases viscous secretion from salivary glands
Increases eccrine sweat gland secretion (thermoregulation)
Increases apocrine sweat gland secretion (stress)
Stimulates seminal vesicle and prostate secretion during ejaculation (emission)

Other
Stimulates gluconeogenesis and glycogenolysis in hepatocytes (hyperglycemia)
Stimulates lipolysis in adipocytes
Stimulates renin secretion from JG cells in kidney (increases blood pressure)
Inhibits insulin secretion from pancreatic β cells (hyperglycemia)

Specific Parasympathetic Motor Actions

Smooth Muscle
Contracts sphincter pupillae muscle, causing constriction of pupil (miosis)
Contracts ciliary muscle, causing accommodation for near vision
Contracts bronchial smooth muscle in lung (bronchoconstriction)
Contracts smooth muscle in GI tract wall
Relaxes smooth muscle in GI tract sphincters
Contracts smooth muscle in urinary bladder
Relaxes smooth muscle in urinary tract sphincter
Relaxes smooth muscle in penile blood vessels, causing dilation (erection of penis); "point"
Female reproductive tract[a]

Cardiac Muscle
Decelerates SA node (decreases heart rate; vagal arrest); negative chronotropism
Decreases conduction velocity in AV node; negative dromotropism
Decreases contractility of cardiac muscle (atrial); negative inotropism

Glands
Increases watery secretion from salivary glands
Increases secretion from lacrimal gland

AV = atrioventricular, GI = gastrointestinal, JG = juxtaglomerular, SA = sinoatrial.
[a]Despite numerous studies, specific parasympathetic actions on the female reproductive tract remain inconclusive.

Chapter 4

Lymphatic System

I Central Lymphatic Drainage (Figure 4-1)

A. **GENERAL FEATURES.** The lymphatic system is a collection of vessels that function to drain extracellular fluid from tissues of the body and return it to the venous system. All regions of the body possess lymphatic drainage **except for the brain and spinal cord.**

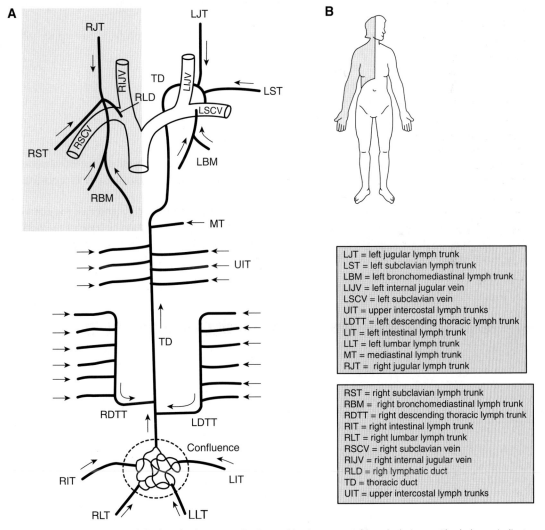

LJT = left jugular lymph trunk	
LST = left subclavian lymph trunk	
LBM = left bronchomediastinal lymph trunk	
LIJV = left internal jugular vein	
LSCV = left subclavian vein	
UIT = upper intercostal lymph trunks	
LDTT = left descending thoracic lymph trunk	
LIT = left intestinal lymph trunk	
LLT = left lumbar lymph trunk	
MT = mediastinal lymph trunk	
RJT = right jugular lymph trunk	

RST = right subclavian lymph trunk
RBM = right bronchomediastinal lymph trunk
RDTT = right descending thoracic lymph trunk
RIT = right intestinal lymph trunk
RLT = right lumbar lymph trunk
RSCV = right subclavian vein
RIJV = right internal jugular vein
RLD = righ lymphatic duct
TD = thoracic duct
UIT = upper intercostal lymph trunks

● **Figure 4-1 A:** Diagram of the lymphatic system. **B:** General body pattern of lymph drainage. Shaded area indicates lymph drainage into the right lymphatic duct. Unshaded area indicates lymph drainage into the thoracic duct. *Arrows* indicate direction of lymph flow).

B. THORACIC DUCT

1. The thoracic duct begins in a majority of individuals as the **abdominal confluence of lymph trunks** at the L1–2 vertebral level.

2. The confluence of lymph trunks receives lymph from four main lymphatic trunks: the **right and left lumbar lymph trunks** and the **right and left intestinal lymph trunks.**

3. In a small percentage of individuals, the abdominal confluence of lymph trunks is represented as a dilated sac (called the **cisterna chyli**).

4. The thoracic duct traverses the **aortic aperture of the diaphragm.**

5. The thoracic duct terminates at the junction of the left internal jugular vein and left subclavian vein (i.e., **left brachiocephalic vein**) at the base of the neck.

6. The thoracic duct drains lymph from the:
    a. **Left side of the head and neck**
    b. **Left breast**
    c. **Left upper limb/superficial thoracoabdominal wall**
    d. **All the body below the diaphragm**

7. Along its course, the thoracic duct receives lymph from the following tributaries:
    a. **Right and left descending thoracic lymph trunks**, which convey lymph from the lower intercostal spaces 6–11
    b. **Upper intercostal lymph trunks**, which convey lymph from the upper inter-costal spaces 1–5
    c. **Mediastinal lymph trunks**
    d. **Left subclavian lymph trunk**
    e. **Left jugular lymph trunk**
    f. **Left bronchomediastinal lymph trunk**

C. RIGHT LYMPHATIC DUCT

1. The right lymphatic duct (a short vessel) begins with a high degree of variability as a convergence of the **right subclavian lymph trunk, right jugular lymph trunk,** and **right bronchomediastinal lymph trunk.**

2. The right lymphatic duct terminates at the junction of the right internal jugular vein and right subclavian vein (i.e., **right brachiocephalic vein**) at the base of the neck.

3. The right lymphatic duct drains lymph from the:
    a. **Right side of the head and neck**
    b. **Right breast**
    c. **Right upper limb/superficial thoracoabdominal wall**

Ⅱ Summary Diagram of Specific Lymphatic Drainage (Figure 4-2)

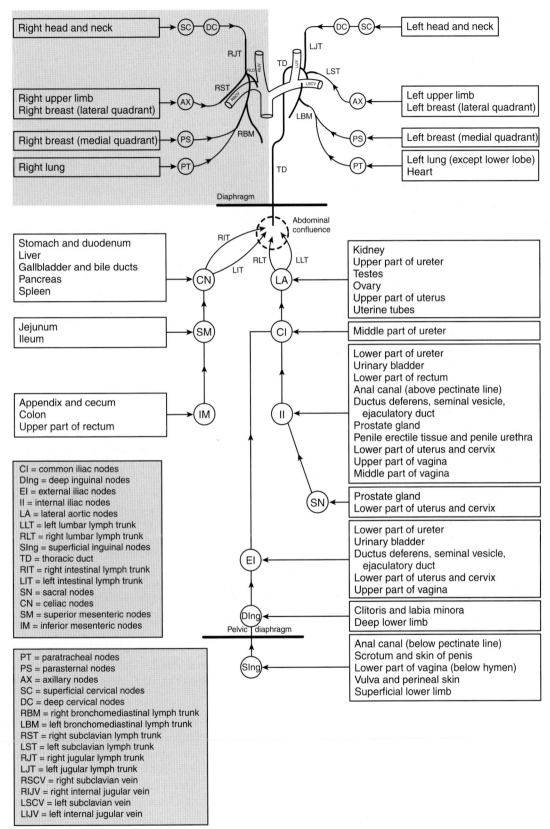

● **Figure 4-2 Summary diagram of specific lymphatic drainage.** *Arrowheads* and *arrows* indicate direction of lymph flow.

Chapter 5

Chest Wall

I General Features of the Thorax

A. The thorax extends from the top of the sternum to the diaphragm.

B. The thorax is bounded by the sternum, ribs, and thoracic vertebrae.

C. The entrance to the thorax (called the **thoracic inlet**) is small and kidney shaped. The boundaries of the thoracic inlet are the manubrium anteriorly, rib 1 laterally, and thoracic vertebrae posteriorly.

D. The outlet from the thorax (called the **thoracic outlet**) is large and is separated from the abdomen by the diaphragm. The boundaries of the thoracic outlet are the xiphoid process anteriorly, costal cartilages 7–10 and rib 12 laterally, and T12 vertebra posteriorly.

II Bones of the Thorax (Figure 5-1)

A. THORACIC VERTEBRAE
1. There are 12 thoracic vertebrae that have facets on their bodies (**costal facets**) for articulation with the heads of ribs, facets on their transverse processes for articulation with the tubercles of rib 9 (except for ribs 11 and 12), and long spinous processes.

B. RIBS
1. There are 12 pairs of ribs that articulate with the thoracic vertebrae.
2. A rib consists of a **head, neck, tubercle,** and **body.**
3. The head articulates with the body of adjacent thoracic vertebrae and the intervertebral disc at the **costovertebral joint.**
4. The tubercle articulates with the transverse process of a thoracic vertebra at the **costotransverse joint.**
5. True (vertebrosternal) ribs are **ribs 1–7**, which articulate individually with the sternum by their costal cartilages.
6. False (vertebrochondral) ribs are **ribs 8–12**. Ribs 8–10 articulate with more superior costal cartilage and form the **anterior costal margin**. Ribs 11 and 12 (often called **floating ribs**) articulate with vertebral bodies but do not articulate with the sternum.

C. STERNUM consists of the following:
1. The **manubrium** forms the **jugular notch** at its superior margin; has a **clavicular notch**, which articulates with the clavicle at the **sternoclavicular joint**; and articulates with the costal cartilages of ribs 1 and 2.
2. The **body** articulates with the manubrium at the **sternal angle of Louis**, articulates with the costal cartilages of ribs 2–7, and articulates with the **xiphoid process** at the **xiphosternal joint**.

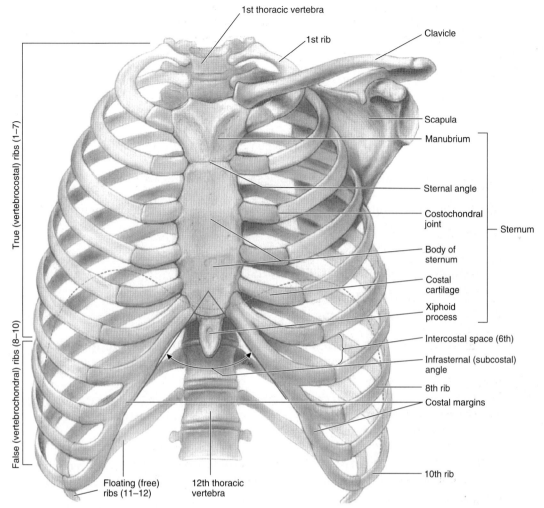

True (vertebrocostal) ribs (1–7)

False (vertebrochondral) ribs (8–10)

1st thoracic vertebra

1st rib

Clavicle

Scapula

Manubrium

Sternal angle

Costochondral joint

Body of sternum

Costal cartilage

Xiphoid process

Intercostal space (6th)

Infrasternal (subcostal) angle

8th rib

Costal margins

10th rib

Sternum

Floating (free) ribs (11–12)

12th thoracic vertebra

● **Figure 5-1 The thoracic skeleton (anterior view).** The osteocartilaginous thoracic cage includes the sternum, 12 pairs of ribs and costal cartilages, and 12 thoracic vertebrae with their intervertebral discs. The clavicles and scapulae form the pectoral (shoulder) girdle. *The dotted line* indicates the position of the diaphragm separating the thoracic cavity from the abdominal cavity.

    3.   The **xiphoid process** articulates with the body of the sternum and attaches to the diaphragm and abdominal musculature via the **linea alba.**

    4.   The **sternal angle of Louis** marks the junction between the manubrium and body of the sternum at vertebral level T4. This is the site where rib 2 articulates with the sternum, the aortic arch begins and ends, the trachea bifurcates, and the superior mediastinum ends.

Ⅲ Muscles of the Thorax

A. DIAPHRAGM

    1.   The diaphragm is the **most important muscle of inspiration.**

    2.   The diaphragm elevates the ribs and increases the vertical, transverse ("bucket handle" movement), and anteroposterior ("pump handle" movement) diameters of the thorax.

    3.   The diaphragm is innervated by the **phrenic nerves (ventral primary rami of C3–5),** which provide motor and sensory innervation. Sensory innervation to the periphery of the diaphragm is provided by the **intercostal nerves.**

4. A lesion of the phrenic nerve may result in **paralysis** and **paradoxical movement** of the diaphragm. The paralyzed dome of the diaphragm does not descend during inspiration and is consequently forced upward due to increased abdominal pressure.

B. INTERCOSTAL MUSCLES
1. The intercostal muscles are thin multiple layers of muscle that occupy the **intercostal spaces (1–11)** and keep the intercostal space rigid during inspiration or expiration.
2. The **external intercostal muscles** elevate the ribs and play a role in inspiration during exercise or lung disease.
3. The **internal intercostal muscles** play a role in expiration during exercise or lung disease.
4. The **innermost intercostal muscles** are presumed to act with the internal intercostal muscles.
5. The **intercostal vein, artery, and nerve** run between the internal intercostal muscles and innermost intercostal muscles.

C. SERRATUS POSTERIOR SUPERIOR MUSCLE

D. SERRATUS POSTERIOR INFERIOR MUSCLE

E. LEVATOR COSTARUM MUSCLE

F. TRANSVERSE THORACIC MUSCLE

G. STERNOCLEIDOMASTOID, PECTORALIS MAJOR AND MINOR, AND SCALENE MUSCLES. These muscles attach to the ribs and play a role in inspiration during exercise or lung disease.

H. EXTERNAL OBLIQUE, INTERNAL OBLIQUE, TRANSVERSE ABDOMINAL, AND RECTUS ABDOMINIS MUSCLES. These abdominal muscles play a role in expiration during exercise, lung disease, or the Valsalva maneuver.

 Movement of the Thoracic Wall is concerned with increasing or decreasing the **intrathoracic pressure**. The act of breathing involves changes in intrathoracic pressure and is called **inspiration** and **expiration**.

A. INSPIRATION
1. The inferiorly sloped **ribs 1–6 elevate** by rotating on their tubercles within the vertebral facets. This causes an increase in the anteroposterior diameter of the thorax (the sternum is pushed forward and upward), which has been compared to a "pump handle" movement.
2. The **lower ribs elevate** by swinging upward and laterally due to their shape and the limited movement of their tubercles. This causes an increase in the lateral diameter of the thorax, which has been compared to a "**bucket handle**" movement.

B. EXPIRATION
1. Expiration is predominately passive and depends on the elasticity of the lungs.
2. In contrast, forced expiration involves contraction of abdominal wall muscles, primarily the **external oblique, internal oblique,** and **transverse abdominal muscles**. This contraction of abdominal muscles pushes against the diaphragm.

Arteries of the Thorax

A. INTERNAL THORACIC ARTERY. The internal thoracic artery is a branch of the **subclavian artery** that descends just lateral to the sternum and terminates at intercostal space 6 by dividing into the **superior epigastric artery** and **musculophrenic artery**.

B. ANTERIOR INTERCOSTAL ARTERIES
1. The anterior intercostal arteries that supply intercostal spaces 1–6 are branches of the **internal thoracic artery.**
2. The anterior intercostal arteries that supply intercostal spaces 7–9 are branches of the **musculophrenic artery.**
3. There are two anterior intercostal arteries within each intercostal space that anastomose with the posterior intercostal arteries.

C. POSTERIOR INTERCOSTAL ARTERIES
1. The posterior intercostal arteries that supply intercostal spaces 1 and 2 are branches of the **superior intercostal artery** that arises from the **costocervical trunk** of the subclavian artery.
2. The posterior intercostal arteries that supply intercostal spaces 3–11 are branches of the **thoracic aorta.**
3. All posterior intercostal arteries give off a posterior branch, which travels with the dorsal primary ramus of a spinal nerve to supply the spinal cord, vertebral column, back muscles, and skin.
4. The posterior intercostal arteries anastomose anteriorly with the anterior intercostal arteries.

VI Veins of the Thorax

A. ANTERIOR INTERCOSTAL VEINS. The anterior intercostal veins drain the anterior thorax and empty into the **internal thoracic veins,** which then empty into the **brachiocephalic veins.**

B. POSTERIOR INTERCOSTAL VEINS. The posterior intercostal veins drain the lateral and posterior thorax and empty into the **hemiazygos veins** on the left side and the **azygos vein** on the right side. The hemiazygos veins empty into the azygos vein, which empties into the superior vena cava.

VII Nerves of the Thorax

A. The **intercostal nerves** are the ventral primary rami of T1–11 and run in the **costal groove** between the internal intercostal muscles and innermost intercostal muscles.

B. The **subcostal nerve** is the ventral primary ramus of T12.

C. Intercostal nerve injury is evidenced by a sucking in (upon inspiration) and bulging out (upon expiration) of the affected intercostal space.

VIII Breast (Figure 5-2)

A. The breast lies in the superficial fascia of the anterior chest wall overlying the **pectoralis major** and **serratus anterior muscles** and extends into the **superior lateral quadrant** of the axilla as the **axillary tail,** where a high percentage of tumors occur.

B. In a well-developed female, the breast extends vertically from **rib 2 to rib 6** and laterally from the **sternum to the midaxillary line.**

C. At the greatest prominence of the breast is the **nipple,** which is surrounded by a circular pigmented area of skin called the **areola.**

D. The **retromammary space** lies between the breast and the **pectoral (deep) fascia** and allows free movement of the breast. If breast carcinoma invades the retromammary space and pectoral fascia, contraction of the pectoralis major may cause the **whole breast to move superiorly.**

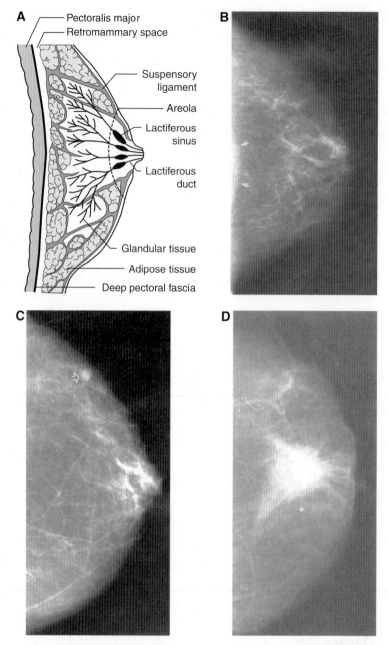

● **Figure 5-2 Diagram and mammograms of the breast. A:** A sagittal diagram of the breast. **B:** A craniocaudal (CC) mammogram of a normal left breast. The pectoralis major muscle (*arrows*) is seen. **C:** A CC mammogram of a benign mass (*arrow*). A benign mass has the following characteristics: **shape** is round/oval, **margins** are well circumscribed, **density** is low-medium contrast, it becomes smaller over time, and **calcifications** are large, smooth, and uniform. **D:** A CC mammogram of a malignant mass. A malignant mass has the following characteristics: **shape** is irregular with many lobulations, **margins** are irregular or spiculated, **density** is medium-high, breast architecture may be distorted, it becomes larger over time, and **calcifications** (not shown) are small, irregular, variable, and found within ducts (ductal casts).

E. Suspensory ligaments (Cooper's) extend from the dermis of the skin to the pectoral fascia and provide support for the breast. If breast carcinoma invades the suspensory ligaments, the ligaments may shorten and cause **dimpling of the skin** or **inversion of the nipple.**

F. **Adipose tissue** within the breast contributes largely to the contour and size of the breast.

G. ARTERIAL SUPPLY. The arterial supply of the breast is from the medial mammary branches of the **internal thoracic artery**, lateral mammary branches of the **lateral thoracic artery**, pectoral branches of the **thoracoacromial artery**, perforating branches of the **anterior intercostal arteries**, and **posterior intercostal arteries.**

H. VENOUS DRAINAGE
  1. The venous drainage from the breast is mainly to the **axillary vein** via lateral mammary veins and the lateral thoracic vein.
  2. Additional venous drainage from the breast is to the **internal thoracic vein** via medial mammary veins, **anterior intercostal veins**, and **posterior intercostal veins** (drain into the azygos system).
  3. Metastasis of breast carcinoma to the brain may occur by the following route: cancer cells enter an intercostal vein → external vertebral venous plexuses → internal vertebral venous plexus → cranial dural sinuses.

I. INNERVATION. The nerves of the breast are derived from anterior and lateral cutaneous branches of **intercostal nerves 4–6** (i.e., T4, T5, T6 dermatomes).

J. LYMPH DRAINAGE. The breast has lymphatic plexuses that communicate freely, called the **circumareolar plexus, perilobular plexus,** and **interlobular plexus,** all of which drain into the deep **subareolar plexus.** From the subareolar plexus, lymph flows as follows:
  1. Lymph Drainage from the Lateral Quadrant
    a. A majority of the lymph (>75%) from the lateral quadrant of the right and left breast drains as follows: **axillary nodes (humeral, subscapular, pectoral, central, and apical)** → **infraclavicular and supraclavicular nodes** → **right subclavian lymph trunk** (for the right breast) or **left subclavian lymph trunk** (for the left breast).
    b. The remaining ~25% of lymph drainage occurs via the **interpectoral, deltopectoral, supraclavicular,** and **inferior deep cervical nodes.**
  2. Lymph Drainage from the Medial Quadrant
    a. The lymph from the medial quadrant of the right and left breast drains as follows: **parasternal nodes** → **right bronchomediastinal lymph trunk** (for the right breast) or **left bronchomediastinal lymph trunk** (for the left breast).
    b. Lymph from the medial quadrant may also **drain into the opposite breast.**
  3. Lymph Drainage from the Inferior Quadrant. The lymph from the inferior quadrant of the right and left breast drains into the nodes of the upper abdomen (e.g., inferior phrenic lymph nodes).

K. CLINICAL CONSIDERATIONS
  1. Fibroadenoma is a benign proliferation of connective tissue such that the mammary glands are compressed into cords of epithelium. A fibroadenoma presents clinically as a sharply circumscribed, spherical nodule that is freely movable. It is the most common benign neoplasm of the breast.
  2. Infiltrating duct carcinoma is a malignant proliferation of duct epithelium where the tumor cells are arranged in cell nests, cords, anastomosing masses, or a mixture of these. It is the most common type of breast cancer, accounting for 65% to 80% of all breast cancers. An infiltrating duct carcinoma presents clinically as a jagged density, fixed in position; dimpling of skin; inversion of the nipple; and thick, leathery skin.

IX Anterior Chest Wall (Figure 5-3)

A. INSERTION OF A CENTRAL VENOUS CATHETER. In clinical practice, access to the superior vena cava (SVC) and right side of the heart is required for monitoring blood pressure, long-term feeding, or administration of drugs. The internal jugular vein and subclavian vein are generally used.

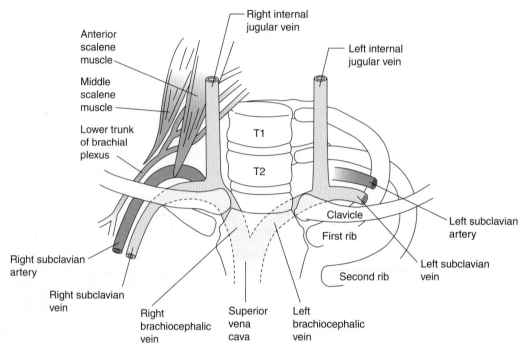

● **Figure 5-3 Anterior chest wall.** The first pair of ribs is shown with their articulation with the T1 vertebra and manubrium of the sternum. On the right, structures crossing rib 1 are shown (subclavian vein, subclavian artery, and brachial plexus). Note the relationship of these structures to the clavicle. Note also the arrangement of the large veins in this area and their use in a placing a central venous catheter (inferior jugular venous approach or subclavian approach).

1. Internal Jugular Vein (Central or Anterior Approach) (Figure 5-4). The needle is inserted at the apex of a triangle formed by the two heads of the sternocleidomastoid muscle and the clavicle of the right side. The diagram shows the correct central approach when inserting a catheter into the internal jugular vein.

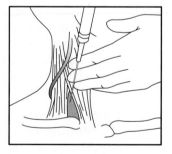

● **Figure 5-4 Insertion of catheter (central approach).**

2. **Subclavian Vein (Infraclavicular Approach) (Figure 5-5).** Place index finger at the sternal notch and thumb at the intersection of the clavicle and first rib as anatomic landmarks. The needle is inserted below the clavicle and lateral to your thumb on the right side. The diagram shows the correct infraclavicular approach when inserting a catheter into the right subclavian vein.

3. **Complications of a central venous catheter** may include the following: puncture of the subclavian artery or subclavian vein, pneumothorax, hemothorax, trauma to trunks of the brachial plexus, arrhythmias, venous thrombosis, erosion of the catheter through the SVC, damage to the tricuspid valve, and infections.

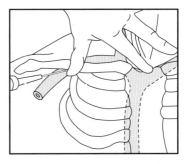

● **Figure 5-5 Insertion of catheter (infraclavicular approach).**

B. **POSTDUCTAL COARCTATION OF THE AORTA (FIGURE 5-6)** is a congenital malformation associated with increased blood pressure to the upper extremities, diminished and delayed femoral artery pulse, and high risk of cerebral hemorrhage and bacterial endocarditis. A postductal arctation of the aorta is generally located distal to the left subclavian artery and the ligamentum arteriosum. The **internal thoracic artery** → **intercostal arteries** → **superior epigastric artery** → **inferior epigastric artery** → **external iliac arteries** are involved in the collateral circulation to bypass the constriction and become dilated. Dilation of the intercostal arteries causes erosion of the lower border of the ribs, termed "**rib notching.**" A **preductal coarctation** is less common and occurs proximal to the ductus arteriosus; blood reaches the lower part of the body via a patent ductus arteriosus. The angiogram shows a narrowing (*arrow*) just distal to the prominent left subclavian artery. The aortic arch is hypoplastic. Note the tortuous internal thoracic artery (*arrowhead*).

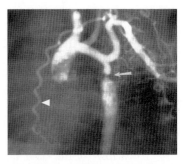

● **Figure 5-6 Postductal coarctation of the aorta.**

C. ANEURYSM OF THE AORTA (FIGURE 5-7) may compress the trachea or tug on the trachea with each cardiac systole such that it can be felt by palpating the trachea at the sternal notch (T2). The angiogram shows an atherosclerotic aneurysm (*curved arrows*) protruding from the ascending aorta.

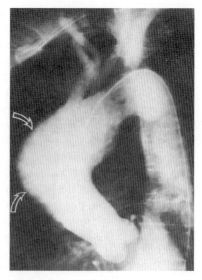

● Figure 5-7 Aortic aneurysm.

D. AORTIC DISSECTION (FIGURE 5-8) is a result of a deceleration injury where the aorta tears just distal to the left subclavian artery. The tear is through the tunica intima and tunica media. The computed tomography (CT) scan shows a tunica intima flap within the ascending (*closed arrow*) and descending (*open arrow*) aorta. The larger false lumen compresses the true lumen.

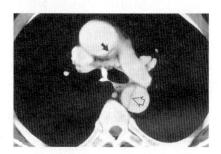

● Figure 5-8 Aortic dissection.

E. THORACIC OUTLET SYNDROME (FIGURE 5-9) may be the result of an anomalous cervical rib that compresses the lower trunk of the brachial plexus, subclavian artery, or both. Clinical findings include atrophy of thenar and hypothenar eminences, atrophy of interosseous muscles, sensory deficits on the medial side of the forearm and hand, diminished radial artery pulse upon moving the head to the opposite side, and a bruit over the subclavian artery. The angiogram, taken with abduction of both arms, shows that blood flow is partially occluded in the subclavian arteries (*arrows*).

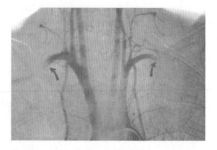

● Figure 5-9 Thoracic outlet syndrome.

F. KNIFE WOUND TO CHEST WALL ABOVE THE CLAVICLE may damage structures at the root of the neck. The **subclavian artery** may be cut. The **lower trunk of the brachial plexus** may be cut, causing loss of hand movements (ulnar nerve involvement) and loss of sensation over the medial aspect of the arm, forearm, and last two digits (C8 and T1 dermatomes). The **cervical pleura** and **apex of the lung** may be cut, causing an open pneumothorax and collapse of the lung. These structures project superiorly into the neck through the thoracic inlet and posterior to the sternocleidomastoid muscle.

G. PROJECTIONS OF THE DIAPHRAGM ON THE CHEST WALL. The central tendon of the diaphragm lies directly posterior to the xiphosternal joint. The **right dome** of the diaphragm arches superiorly to the *upper* border of rib 5 in the midclavicular line. The **left dome** of the diaphragm arches superiorly to the *lower* border of rib 5 in the midclavicular line.

H. SCALENE LYMPH NODE BIOPSY. Scalene lymph nodes are located behind the clavicle surrounded by pleura, lymph ducts, and the phrenic nerve. Inadvertent damage to these structures will cause the following clinical findings: pneumothorax (pleura), lymph leakage (lymph ducts), and diaphragm paralysis (phrenic nerve).

X Lateral Chest Wall (Figure 5-10)

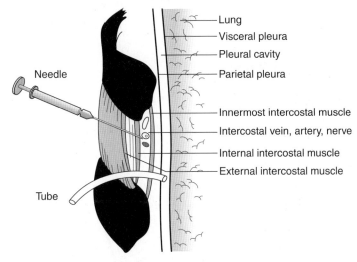

● **Figure 5-10 Lateral chest wall.** A schematic diagram of an intercostal space and layers. Note their relationship to pleura and lung. The needle indicates the positioning for an intercostal nerve block. The tube indicates the positioning for thoracocentesis.

A. TUBE THORACOSTOMY (FIGURE 5-11) is performed to evacuate ongoing production of air/fluid into the pleural cavity. A tube is inserted through intercostal space 5 in the anterior axillary line (i.e., posterior approach) close to the *upper* border of the rib to avoid the **intercostal vein, artery, and nerve**, which run in the costal groove between the internal intercostal muscle and innermost intercostal muscle. An incision is made at intercostal space 6 lateral to the nipple, but medial to the latissimus dorsi muscle. The tube will penetrate **skin → superficial fascia → serratus anterior muscle → external intercostal muscle → internal intercostal muscle → innermost intercostal muscle → parietal pleura.** The diagram shows the surgical approach for a tube thoracostomy.

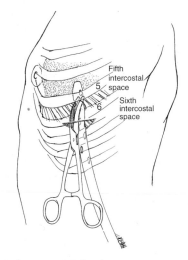

● **Figure 5-11 Tube thoracostomy.**

B. **INTERCOSTAL NERVE BLOCK (FIGURE 5-12)**
may be necessary to relieve pain associated
with a rib fracture or herpes zoster (shingles).
A needle is inserted at the posterior angle of
the rib along the *lower* border of the rib in order to bathe the nerve in anesthetic. The needle penetrates the following structures: **skin
→ superficial fascia → serratus anterior muscle → external intercostal muscle → internal
intercostal muscle.** Several intercostal nerves
must be blocked to achieve pain relief because
of the presence of nerve collaterals (i.e., over-
lapping of contiguous dermatomes). The photograph shows the approach for an intercostal
nerve block.

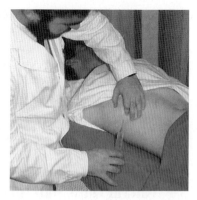

● **Figure 5-12 Intercostal nerve block.**

XI Posterior Chest Wall (Figure 5-13)

A. **FRACTURES OF THE LOWER RIBS.** A rib fracture on the right side may damage the
right kidney and **liver.** A rib fracture on the left side may damage the **left kidney** and
spleen. A rib fracture on either side may damage the **pleura** as it crosses rib 12.

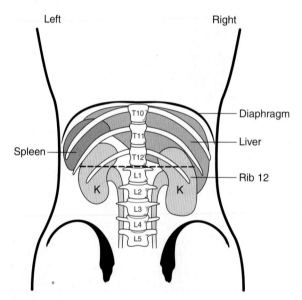

● **Figure 5-13 Posterior chest wall.** Note that the kidneys are located from T12 to L3 vertebrae and that the right
kidney is lower than the left. The pleura extends across rib 12 (*dotted line*). Note the structures that may be injured by
fractures to the lower ribs. During a splenectomy, the left kidney may be damaged due to its close anatomic relationship and connection via the splenorenal ligament. K, kidney.

XII Mediastinum (Figure 5-14). The mediastinum is defined as the space between the
pleural cavities in the thorax. It is bounded laterally by the pleural cavities, anteriorly by
the sternum, and posteriorly by the vertebral column and is divided artificially into a **superior division** and an **inferior division** by a line from the sternal angle of Louis to the T4–5
intervertebral disc. The inferior division is then further divided into the **anterior, middle,**
and **posterior** divisions.

A. SUPERIOR MEDIASTINUM
1. The contents of the superior mediastinum include the trachea, esophagus, thymus, phrenic nerves, azygous vein, SVC, brachiocephalic artery and veins, aortic arch, left common carotid artery, left subclavian artery, and thoracic duct.
2. Common pathologies found in this area include **aortic arch aneurysm, esophageal perforation either from endoscopy or invading malignancy,** and **traumatic rupture of the trachea.**

B. ANTERIOR MEDIASTINUM
1. The contents of the anterior mediastinum include the thymus, fat, lymph nodes, and connective tissue.
2. Common pathologies found in this area include **thymoma associated with myasthenia gravis and red blood cell aplasia, thyroid mass, germinal cell neoplasm,** and **lymphomas (Hodgkin or non-Hodgkin).**

C. MIDDLE MEDIASTINUM
1. The contents of the middle mediastinum include the heart, pericardium, phrenic nerves, ascending aorta, SVC, and coronary arteries and veins.
2. Common pathologies found in this area include **pericardial cysts, bronchiogenic cysts,** and **sarcoidosis.**

D. POSTERIOR MEDIASTINUM
1. The contents of the posterior mediastinum include the descending aorta, esophagus, thoracic duct, azygous vein, splanchnic nerves, and vagus nerves (cranial nerve X).
2. Common pathologies found in this area include **ganglioneuromas, neuroblastomas,** and **esophageal diverticula or neoplasms.**

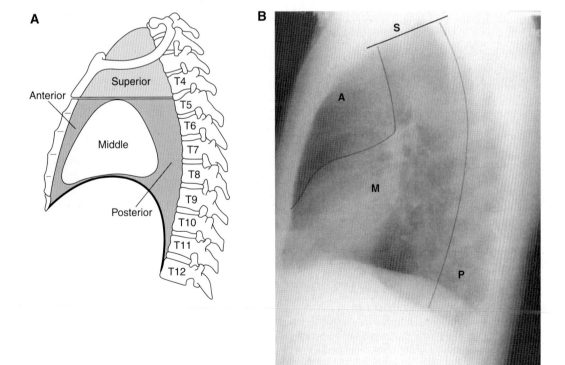

● **Figure 5-14 Mediastinum. A:** A diagram indicating the superior, anterior, middle, and posterior divisions of the mediastinum. **B:** A lateral radiograph demarcating the superior (S), anterior (A), middle (M), and posterior (P) divisions of the mediastinum.

XIII Radiology

A. POSTERIOR-ANTERIOR (PA) CHEST RADIOGRAPH (FIGURE 5-15)

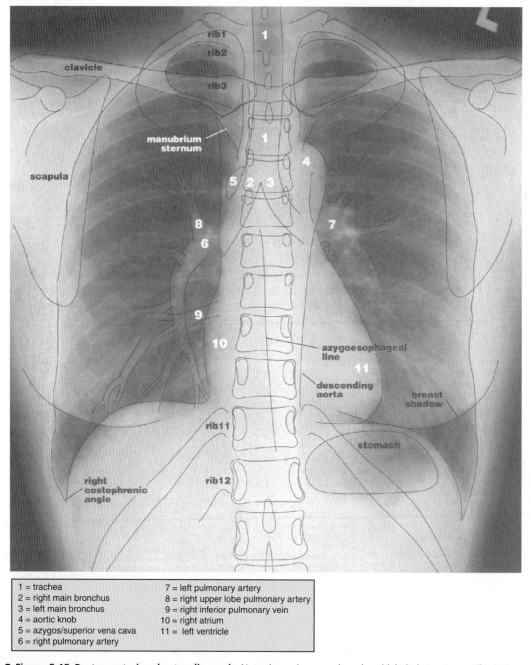

1 = trachea	7 = left pulmonary artery
2 = right main bronchus	8 = right upper lobe pulmonary artery
3 = left main bronchus	9 = right inferior pulmonary vein
4 = aortic knob	10 = right atrium
5 = azygos/superior vena cava	11 = left ventricle
6 = right pulmonary artery	

● **Figure 5-15 Posteroanterior chest radiograph.** Note the various numbered and labeled structures. Ribs 1–8 can generally be traced from their articulation with the vertebral column to the union of the rib with the costal cartilage. The liver and right dome of the diaphragm cast a domed water-density shadow at the base of the right lung. The stomach, spleen, and left dome of the diaphragm cast a domed water-density shadow at the base of the left lung. Both domes generally lie just below vertebra T10. The left dome is lower than the right dome due to the downward thrust of the heart. The right border of the cardiovascular shadow includes the brachiocephalic artery and right brachiocephalic vein, superior vena cava and ascending aorta, right atrium, and inferior vena cava. The left border of the cardiovascular shadow includes the left subclavian artery and left brachiocephalic vein, aortic arch (or aortic knob), pulmonary trunk, auricle of left atrium, and left ventricle. The angle between the right and left main bronchi at the carina is generally 60 to 75 degrees. The left hilum is generally higher than the right hilum.

B. LATERAL CHEST RADIOGRAPH (FIGURE 5-16)

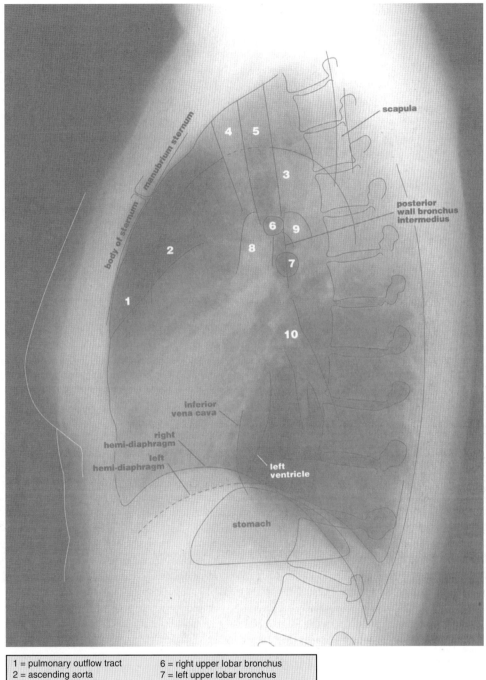

1 = pulmonary outflow tract	6 = right upper lobar bronchus
2 = ascending aorta	7 = left upper lobar bronchus
3 = aortic arch	8 = right pulmonary artery
4 = brachiocephalic vessels	9 = left pulmonary artery
5 = trachea	10 = confluence of pulmonary veins

● **Figure 5-16 Lateral chest radiograph.** Note the various numbered and labeled structures.

C. AORTIC ANGIOGRAM (LEFT ANTERIOR OBLIQUE VIEW) (FIGURE 5-17)

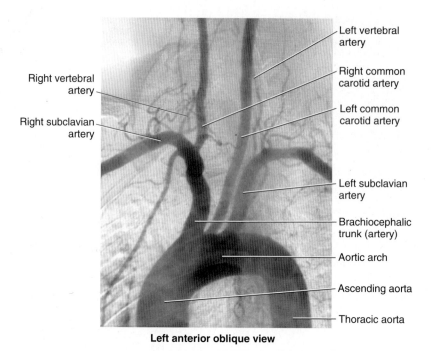

Left anterior oblique view

● **Figure 5-17 Aortic angiogram (left anterior oblique view).** Note that injection of contrast dye into the right subclavian artery will visualize the entire circle of Willis since the dye will enter both the right common carotid artery and the right vertebral artery (not shown). However, injection of contrast dye into the left subclavian artery will visualize only the posterior part of the circle of Willis since the dye will enter only the left vertebral artery (not shown).

Case Study 5-1

A 27-year-old man comes to the emergency room complaining that "my stomach hurts real bad and it has been going on for about 3 days. I've been vomiting a lot also." He also tells you that "I don't know if this means anything but my legs cramp up often and sometimes my feet are cold." After some discussion, you learn that he has had no signs of blood in the vomitus, no fever, no chills, and no history of prior surgeries or other serious medical problems. However, the man admits to drinking five beers and a shot of Jack Daniels whiskey as a daily routine to relax before dinner. You admit the man to the hospital with a diagnosis of alcohol-induced pancreatitis for bowel rest and intravenous therapy. Despite some improvement, the patient is noted to have a persistently elevated systolic blood pressure in the 190 to 200 range and a diastolic blood pressure in the 90 to 110 range. A posteroanterior (PA) chest radiograph and magnetic resonance angiograph (MRA) are ordered.

Relevant Physical Exam Findings

- Normal body temperature
- Blood pressure 198/88 mm Hg
- A 1/6 soft systolic ejection murmur
- S_1 and S_2 heart sounds are normal
- Abdomen is soft but tender to deep palpation in the epigastric region
- Normal rectal exam

Relevant Lab Findings

- Normal complete blood count (CBC) and basic metabolic panel
- Elevated serum amylase (240 U/L; normal range 30–110 U/L)
- Elevated serum lipase (2120 U/L; normal range 46–218 U/L)

Relevant Radiographic Findings

- Cardiomegaly
- Enlarged collateral intercostal arterial circulation
- Rib notching of the inferior-posterior margin
- Classic "figure 3" is observed

Diagnosis

Postductal Coarctation of the Aorta

- Coarctation of the aorta most frequently occurs in association with other congenital heart defects. However, it may also occur as an isolated condition, as in this case.
- Patients presenting beyond infancy often have vague symptoms (e.g., headaches, a propensity for nose bleeds, leg cramps, and cold feet).
- Hypertension is usually present and leads to further testing.
- The hallmark of postductal coarctation of the aorta is that the blood pressure in the arms is at least 20 mm Hg higher than in the legs.

Chapter 6

Pleura, Tracheobronchial Tree, and Lungs

❶ Pleura

A. TYPES OF PLEURA
1. **Visceral Pleura**
   a. The visceral pleura adheres to the lung on all its surfaces.
   b. It is reflected at the root of the lung and continues as parietal pleura.
2. **Parietal pleura** adheres to the chest wall, diaphragm, and pericardial sac.

B. PLEURAL RECESSES
1. **Right and Left Costodiaphragmatic Recesses**
   a. These are slitlike spaces between the costal and diaphragmatic parietal pleura.
   b. **During inspiration**, the lungs descend into the right and left costodiaphragmatic recesses, causing the recesses to appear radiolucent (dark) on radiographs.
   c. **During expiration**, the lungs ascend so that the costal and diaphragmatic parietal pleura come together and the radiolucency disappears on radiographs.
   d. The costodiaphragmatic angle should appear sharp in a posteroanterior (PA) radiograph. If the angle is blunted, pathology of the pleural space may be suspected, such as excess fluid, blood, tumor, or scar tissue.
   e. With a patient in the standing position, excess fluid within the pleural cavity will accumulate in the costodiaphragmatic recesses.
2. **Right and Left Costomediastinal Recesses**
   a. These are slitlike spaces between the costal and mediastinal parietal pleura.
   b. During inspiration, the anterior borders of both lungs expand and enter the right and left costomediastinal recesses. In addition, the **lingula of the left lung** expands and enters a portion of the *left* costomediastinal recess, causing that portion of the recess to appear radiolucent (dark) on radiographs.
   c. During expiration, the anterior borders of both lungs recede and exit the right and left costomediastinal recesses.

C. CLINICAL CONSIDERATIONS
1. **Pleuritis** is inflammation of the pleura. Pleuritis involving only visceral pleura will be associated with **no pain** since the visceral pleura receives no nerve fibers of general sensation. Pleuritis involving the parietal pleura will be associated with **sharp local pain** and **referred pain**. Since parietal pleura is innervated by intercostal nerves and the phrenic nerve (C3, C4, and C5), pain may be referred to the **thoracic wall** and **root of the neck**, respectively.
2. **Inadvertent damage to the pleura** may occur during a:
   a. **Surgical posterior approach to the kidney.** If rib 12 is very short, rib 11 may be mistaken for rib 12. An incision prolonged to the level of rib 11 will damage the pleura.

  b. **Abdominal incision at the right infrasternal angle.** The pleura extends beyond the rib cage in this area.

  c. Stellate ganglion nerve block

  d. Brachial plexus nerve block

  e. **Knife wounds to the chest wall above the clavicle**

  f. **Fracture of lower ribs**

3. **Chylothorax** occurs when lymph accumulates in the pleural cavity due to surgery or trauma that injures the thoracic duct.

4. **Hemothorax** occurs when blood enters the pleura cavity as a result of trauma or rupture of a blood vessel (e.g., a dissecting aneurysm of the aorta).

5. **Empyema** occurs when a thick pus accumulates in the pleural cavity. Empyema is a variant of **pyothorax** whereby a turbid effusion containing many neutrophils accumulates in the pleural cavity, usually as a result of bacterial pneumonia that extends into the pleural surface.

6. **Open pneumothorax** occurs when the parietal pleura is pierced and the pleural cavity is opened to the outside atmosphere. This causes a loss of the negative intrapleural pressure (P_{IP}) because the P_{IP} now equals atmospheric pressure (P_{atm}). This results in an expanded chest wall (its natural tendency) and a collapsed lung (its natural tendency). Upon inspiration, air is sucked into the pleural cavity and results in a **collapsed lung.** The most common causes include chest trauma (e.g., knife wound) and iatrogenic etiology (e.g., thoracocentesis, transthoracic lung biopsy, mechanical ventilation, central line insertion)

7. **Spontaneous pneumothorax (Figure 6-1)** occurs when air enters the pleural cavity usually due to a ruptured bleb (bulla) of a diseased lung. The most common site is in the visceral pleura of the upper lobe of the lung. This results in a loss of negative intrapleural pressure and a **collapsed lung.** Clinical findings include chest pain, cough, and mild to severe dyspnea; spontaneous pneumothorax most commonly occurs in young, tall males. This PA radiograph shows a left apical (*arrows*) and subpulmonic (*curved arrow*) pneumothorax in a 41-year-old woman with adult respiratory distress syndrome.

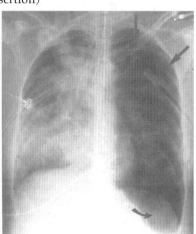

● **Figure 6-1 Spontaneous pneumothorax.**

8. **Tension pneumothorax (Figure 6-2)** may occur as a sequela to an open pneumothorax if the inspired air cannot leave the pleural cavity through the wound upon expiration (check-valve mechanism). This results in a **collapsed lung** on the wounded side and a **compressed lung** on the opposite side due to a deflected mediastinum. Clinical findings include chest pain, shortness of breath, deviated trachea, absent breath sounds on the affected side, and hypotension since the mediastinal shift compresses the superior vena cava (SVC) and inferior vena cava (IVC), thereby obstructing venous return. Tension pneumothorax may cause sudden death. This anteroposterior (AP) radiograph shows a tension pneumothorax as a result of a penetrating chest trauma to the right side.

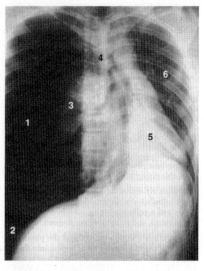

● **Figure 6-2 Tension pneumothorax.** 1 = hyperlucent lung field, 2 = hyperexpansion lower right diaphragm, 3 = collapsed right lung, 4 = deviation of trachea, 5 = mediastinal shift, 6 = compressed left lung.

Ⅱ Tracheobronchial Tree (Figure 6-3)

A. GENERAL CHARACTERISTICS

1. The trachea is a tube composed of **16 to 20 U-shaped hyaline cartilages** and the **trachealis** muscle.
2. The trachea begins just inferior to the cricoid cartilage (C6 vertebral level) and ends at the sternal angle (T4 vertebral level), where it bifurcates into the **right main bronchus and the left main bronchus.**
3. At the bifurcation of the trachea, the last tracheal cartilage forms the **carina,** which can be observed by bronchoscopy as a raised ridge of tissue in the sagittal plane.
4. The right main bronchus is shorter and wider and turns to the right at a shallower angle than the left main bronchus. The right main bronchus branches into **3 lobar bronchi** (upper, middle, and lower) and finally into **10 segmental bronchi.**
5. The left main bronchus branches into **2 lobar bronchi** (upper and lower) and finally into **8 to 10 segmental bronchi.** The branching of segmental bronchi corresponds to the **bronchopulmonary segments** of the lung.

B. CLINICAL CONSIDERATIONS

1. **Compression of the trachea** may be due to an **enlargement of the thyroid gland** or to an **aortic arch aneurysm.** The aortic arch aneurysm may tug on the trachea with each cardiac systole such that it can be felt by palpating the trachea at the sternal notch.
2. **Distortions in the position of the carina** may indicate **metastasis of bronchogenic carcinoma** into the tracheobronchial lymph nodes that surround the tracheal bifurcation or may indicate **enlargement of the left atrium.** The mucous membrane covering the carina is very sensitive in eliciting the cough reflex.
3. **Aspiration of Foreign Objects**
   a. **When a person is sitting or standing** aspirated material most commonly enters the **right lower lobar bronchus** and lodges within the **posterior basal bronchopulmonary segment (#10) of the right lower lobe.**

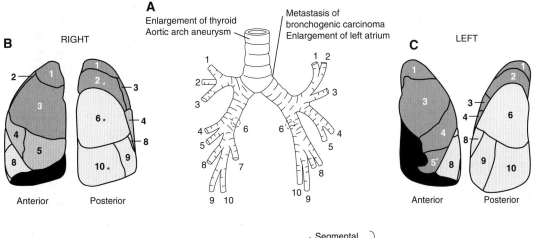

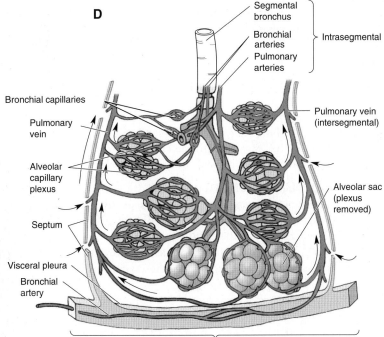

● **Figure 6-3 A:** Diagram of the tracheobronchial tree. **B:** Right lung. The bronchopulmonary segments on the anterior and posterior aspects of the right lung are indicated. 1 = apical, 2 = posterior, 3 = anterior, 4 = lateral, 5 = medial, 6 = superior, 8 = anterior basal, 9 = lateral basal, 10 = posterior basal. Note that #7 (medial basal) is located on the inner mediastinal surface (not shown). **C:** Left lung. The bronchopulmonary segments on the anterior and posterior aspects of the left lung are indicated. 1 = apical, 2, posterior, 3 = anterior, 4 = superior lingular, 5 = inferior lingular, 6 = superior, 8 = anterior basal, 9 = lateral basal, 10 = posterior basal. Note that there is no #7 segmental bronchus or bronchopulmonary segment in the left lung. Asterisk indicates the bronchopulmonary segments involved in aspiration of foreign objects. **D:** Bronchopulmonary segment. Diagram of bronchopulmonary segment #3 shows the centrally located segmental bronchus #3 (SB), branch of the pulmonary artery, and branch of the bronchial artery. Note the location of the pulmonary veins at the periphery of the bronchopulmonary segment.

  b.   **When a person is supine** aspirated material most commonly enters the **right lower lobar bronchus** and lodges within the **superior bronchopulmonary segment (#6) of the right lower lobe.**

  c.   **When a person is lying on the right side** aspirated material most commonly enters the **right upper lobar bronchus** and lodges within the **posterior bronchopulmonary segment (#2) of the right upper lobe.**

  d.   **When a person is lying on the left side** aspirated material most commonly enters the **left upper lobar bronchus** and lodges within the **inferior lingular (#5) bronchopulmonary segment of the left upper lobe.**

III   Lungs (Figure 6-3)

A. RIGHT LUNG

1. The right lung consists of **three lobes (upper, middle, and lower)** separated by a **horizontal fissure** and an **oblique fissure.**

2. The horizontal fissure runs at the level of costal cartilage 4 and meets the oblique fissure at the midaxillary line.

3. The **diaphragmatic surface** consists of the middle lobe and lower lobe.

B. LEFT LUNG

1. The left lung consists of **two lobes (upper and lower)** separated by an **oblique fissure.**

2. The left upper lobe contains the **cardiac notch,** where the left ventricle and pericardial sac abut the lung. The **lingula** (which is the embryologic counterpart to the right middle lobe) lies just beneath the cardiac notch.

3. The **diaphragmatic surface** consists of the lower lobe.

C. BRONCHOPULMONARY SEGMENT

1. The bronchopulmonary segment contains a **segmental bronchus,** a **branch of the pulmonary artery,** and a **branch of the bronchial artery,** which run together through the **central** part of the segment.

2. The **tributaries of the pulmonary vein** are found at the **periphery** between two adjacent bronchopulmonary segments. These veins form surgical landmarks during segmental resection of the lung.

3. The bronchopulmonary segments are both named and numbered as follows:
    a. Right Lung
        i.   **Upper lobe:** apical (#1), **posterior (#2),*** anterior (#3)
        ii. **Middle lobe:** lateral (#4), medial (#5)
        iii. **Lower lobe: superior (#6),** medial basal (#7), anterior basal (#8), lateral basal (#9), **posterior basal (#10)**
    b. Left Lung
        i.   **Upper lobe:** apical (#1), posterior (#2), anterior (#3), superior lingular (#4), **inferior lingular (#5)**
        ii. **Lower lobe:** superior (#6), anterior basal (#8), lateral basal (#9), posterior basal (#10); note that #7 is absent.

D. BREATH SOUNDS

1. Breath sounds from the upper lobe of each lung can be auscultated on the **anterior-superior aspect of the thorax.**

2. Breath sounds from the lower lobe of each lung can be auscultated on the **posterior-inferior aspect of the back.**

3. Breath sounds from the middle lobe of the right lung can be auscultated on the **anterior thorax near the sternum just inferior to intercostal space #4.**

---

*Bronchopulmonary segments in bold are most frequently involved in aspiration of foreign objects.

E. ARTERIAL COMPONENTS OF THE LUNG. The adult lung is supplied by two arterial systems.

   1. Pulmonary System of Arteries
   a. The **pulmonary trunk** is anterior to the ascending aorta and travels in a superior-posterior direction to the left side for about 5 cm and then bifurcates into the **right pulmonary artery** and **left pulmonary artery**, which carry deoxygenated blood to the lung for aeration.
   b. The **right pulmonary** artery runs horizontally toward the hilus beneath the arch of the aorta; posterior to the ascending aorta and superior vena cava; and anterior to the right main bronchus.
   c. The **left pulmonary artery** is shorter and narrower than the right pulmonary artery and is connected to the arch of the aorta by the **ligamentum arteriosum**.
   d. The pulmonary arteries branch to follow the airways to the level of the terminal bronchioles, at which point they form a pulmonary capillary plexus.

   2. Bronchial System of Arteries
   a. The **bronchial arteries** carry oxygenated blood to the parenchyma of the lung.
   b. The **right bronchial artery** is a branch of a posterior intercostal artery.
   c. The two **left bronchial arteries** are branches of the thoracic aorta. The bronchial arteries branch to follow the airways to the level of the terminal bronchioles, at which point they drain into the **pulmonary capillary plexus** (i.e., 70% of bronchial blood drains into the pulmonary capillary plexus).
   d. Bronchial arteries that supply large bronchi drain into **bronchial veins** (i.e., 30% of bronchial blood drains into the bronchial veins).

F. VENOUS COMPONENTS OF THE LUNG. The adult lung is supplied by two venous systems.

   1. Pulmonary System of Veins
   a. The **pulmonary veins** carry oxygenated blood from the pulmonary capillary plexus and deoxygenated bronchial blood to the left atrium.
   b. There are five pulmonary veins that drain each lobe of the lungs. However, the pulmonary veins from the right upper and middle lobes generally join so that only **four pulmonary veins** open into the posterior aspect of the **left atrium**.
   c. Within the lung, small branches of the pulmonary veins run **solo** (i.e., do not run with the airways, pulmonary arteries, or bronchial arteries).
   d. Larger branches of the pulmonary veins are found at the periphery of the bronchopulmonary segments (i.e., **intersegmental location**).

   2. Bronchial System of Veins
   a. The **bronchial veins** carry deoxygenated blood from the bronchial arteries that supply large bronchi.
   b. **Right bronchial veins** drain into the **azygos vein**.
   c. **Left bronchial veins** drain into the **accessory hemiazygos vein**.

G. INNERVATION OF THE LUNG. The lungs are innervated by the **anterior pulmonary plexus** and **posterior pulmonary plexus**, which are located anterior and posterior to the root of the lung at the hilus, respectively. These plexuses contain both **parasympathetic (vagus; cranial nerve [CN] X)** and **sympathetic components**.

   1. Parasympathetic
   a. Motor
      i. Preganglionic neuronal cell bodies are located in the **dorsal nucleus of the vagus** and **nucleus ambiguus** of the medulla. Preganglionic axons run in the **vagus nerve (CN X)**.
      ii. Postganglionic neuronal cells bodies are located in the **pulmonary plexuses** and **within the lung** along the bronchial airways.

      iii. Postganglionic parasympathetic axons terminate on smooth muscle of the bronchial tree, causing **bronchoconstriction**, and seromucous glands, causing **increased glandular secretion**.

    b. **Sensory**. The neuronal cell bodies are located in the inferior (nodose) ganglia of CN X. These neurons send a peripheral process to the lung via CN X and a central process to the solitary nucleus in the brain. These neurons transmit **touch and stretch sensation**.

  2. **Sympathetic**

    a. **Motor**

      i. Preganglionic neuronal cell bodies are located in the **intermediolateral cell column** of the spinal cord. Preganglionic axons enter the **paravertebral ganglion**.

      ii. Postganglionic neuronal cell bodies are located in the paravertebral ganglion at the cervical (superior, middle, and inferior ganglia) and thoracic (T1–4) levels.

      iii. Postganglionic sympathetic axons terminate on smooth muscle of blood vessels within the lung, causing **vasoconstriction**.

      iv. Postganglionic sympathetic axons also terminate on postganglionic *parasympathetic* neurons and modulate their bronchoconstriction activity (thereby causing **bronchodilation**).

      v. Circulating epinephrine from the adrenal medulla acts directly on bronchial smooth muscle to cause **bronchodilation**.

    b. **Sensory**. The neuronal cell bodies are located in the dorsal root ganglia at about the C7–8 and T1–4 spinal cord levels. These neurons send a peripheral process to the lung via the sympathetics and a central process to the spinal cord. These neurons transmit **pain sensation**.

Ⅳ Clinical Considerations

A. ATELECTASIS is the incomplete expansion of alveoli (in neonates) or collapse of alveoli (in adults). **Microatelectasis** is the generalized inability of the lung to expand due to the loss of surfactant usually seen in the following conditions:

  1. **Neonatal respiratory distress syndrome (NRDS)** is caused by a deficiency of surfactant, which may occur due to prolonged intrauterine asphyxia, in premature infants, or in infants of diabetic mothers. Lung maturation is assessed by the **lecithin-to-sphingomyelin ratio** in amniotic fluid (a ratio >2:1 = maturity). Pathologic findings include hemorrhagic edema within the lung, atelectasis, and **hyaline membrane disease** characterized by eosinophilic material consisting of proteinaceous fluid (fibrin, plasma) and necrotic cells. Clinical findings include hypoxemia, which causes pulmonary vasoconstriction, pulmonary hypoperfusion, and capillary endothelium damage.

  2. **Adult respiratory distress syndrome (ARDS) (Figure 6-4)** is defined as a secondary surfactant deficiency due to other primary pathologies that damage either alveolar cells or capillary endothelial cells in the lung. ARDS is a clinical term for diffuse alveolar damage leading to respiratory failure. ARDS may be caused by the following: inhalation of toxic gases (e.g., 9/11 rescue workers), water (as in a near drowning), or extremely hot air; left ventricular failure resulting in cardiogenic

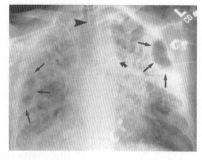

● **Figure 6-4 Adult respiratory distress syndrome (ARDS).**

pulmonary edema; illicit drugs (e.g., heroin); metabolic disorders (e.g., uremia, acidosis, acute pancreatitis); severe trauma (e.g., car accident with multiple fractures); or shock (e.g., endotoxins or ischemia can damage cells). The AP recumbent radiograph shows an endotracheal tube (*arrowhead*), oval collections of air at the periphery of the lung representing pneumatoceles (*arrows*), and right subclavian Swan-Ganz catheter (*curved arrow*).

B. **PULMONARY EMBOLISM (PE; FIGURE 6-5)** is the occlusion of the pulmonary arteries or their branches by an embolic blood clot originating from a deep vein thrombosis (DVT) in the leg or pelvic area. A **large embolus** may occlude the main pulmonary artery or lodge at the bifurcation as a "**saddle embolus**," which may cause sudden death with symptoms easily confused with myocardial infarction (i.e., chest pain, severe dyspnea, shock, increased serum lactate dehydrogenase [LDH] levels). A **medium-sized embolism** may occlude segmental arteries and may produce a **pulmonary infarction**, which is wedge shaped and usually occurs in the lower lobes. A group of **small emboli** ("**emboli showers**") may occlude smaller peripheral branches of the pulmonary artery and cause pulmonary hypertension over time. Risk factors include obesity, cancer, pregnancy, oral contraceptives, hypercoagulability, multiple bone fractures, burns, and prior DVT. A typical clinical scenario involves a postsurgical, bedridden patient who develops sudden shortness of breath. The pulmonary arteriogram shows a large "saddle embolus" (*arrow*). Note the poor perfusion of the right middle and lower lobes compared to the upper lobe.

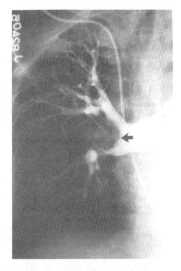

● **Figure 6-5 Pulmonary embolism.**

C. **BRONCHIECTASIS (FIGURE 6-6)** is the abnormal, permanent dilatation of bronchi due to chronic necrotizing infection (e.g., *Staphylococcus, Streptococcus, Haemophilus influenzae*), bronchial obstruction (e.g., foreign body, mucous plugs, or tumors), or congenital conditions (e.g., Kartagener syndrome, cystic fibrosis, immunodeficiency disorders). The **lower lobes** of the lung are predominately affected and the affected bronchi have a **saccular** appearance. Clinical findings include cough, fever, and expectoration of large amounts of foul-smelling purulent sputum. The bronchogram shows dilated bronchi that

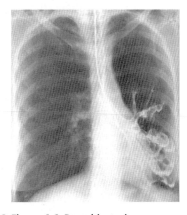

● **Figure 6-6 Bronchiectasis.**

have a saccular appearance and are clearly seen within the left lower lobe.

D. **OBSTRUCTIVE LUNG DISEASES (FIGURE 6-7)** are characterized by an **increase in airway resistance (particularly expiratory airflow).** **Obstructive ventilatory impairment** is the impairment of airflow during expiration with concomitant air trapping and hyperinflation. The increase in airway resistance (due to narrowing of the airway lumen) can be caused by conditions **in the wall of the airway,** where smooth muscle hypertrophy may cause airway narrowing (e.g., asthma); **outside the airway,** where destruction of lung parenchyma may cause airway narrowing upon expiration due to loss of radial traction (e.g., emphysema); and **in the lumen of the airway,** where increased mucus production may cause airway narrowing (e.g., chronic bronchitis). The radiograph shows key features of obstructive lung disease. Note that the hyperinflation of the lung and destruction of the lung interstitium (i.e., bulla formation) cause the lung to appear hyperlucent. Note that the diaphragm is flat and depressed (i.e., lower than rib 11). Specific examples of obstructive lung disease include the following:

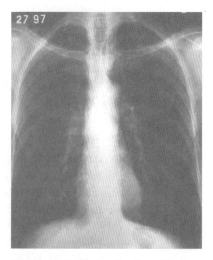

● **Figure 6-7 Obstructive lung disease.**

1. **Asthma.** Asthma is associated with **smooth muscle hyperactivity within bronchi and bronchioles, increased mucus production, and edema of the bronchial wall.** Pathologic findings include inflammatory cell infiltrates containing numerous **eosinophils** within the bronchial wall, hyperplasia of bronchial smooth muscle, hypertrophy of seromucous glands, **Curschmann spirals** (formed from shed epithelium), and **Charcot-Leyden crystals** (formed from eosinophil granules) within the mucous plugs.

2. **Emphysema.** Emphysema is a type of chronic obstructive pulmonary disease (COPD). Patients are referred to as **"pink puffers"** with the following characteristics: a thin, barrel-shaped chest; increased breathing rate (tachypnea); a mildly decreased partial pressure of arterial oxygen (PaO_2; mild hypoxemia); a mildly decreased or normal partial pressure of arterial carbon dioxide ($PaCO_2$; hypocapnia or normocapnia); and decreased diffusion- limited carbon monoxide (DLCO).

    a. **Centriacinar emphysema** (related to **smoking**). Pathologic findings include a widening of the air spaces within the **respiratory bronchioles** only while the surrounding alveoli remain fairly well preserved.

    b. **Panacinar emphysema** (related to α_1-**antitrypsin deficiency**). Pathologic findings include a widening of the air spaces within the **alveolar ducts, alveolar sacs,** and **alveoli** due to destruction of the alveolar walls by enzymes.

3. **Chronic Bronchitis.** Chronic bronchitis is a type of COPD that is related to smoking. Patients are referred to as **"blue bloaters"** with the following characteristics: a muscular, barrel-shaped chest; a severely decreased PaO_2 (severe hypoxemia with cyanosis); an increased $PaCO_2$ (hypercapnia), which leads to chronic respiratory acidosis; an increased HCO_3^- reabsorption by the kidney to buffer the acidemia; right ventricular failure; and systemic edema. Pathologic findings include inflammatory cell infiltrates within the bronchial wall; hypertrophy of seromucous glands

(increase in Reid index); an excessive mucus production leading to copious, purulent sputum production; and recurrent inflammation, infection, and scarring in terminal airways, which results in a decrease in average small airway diameter.

E. RESTRICTIVE LUNG DISEASES (FIGURE 6-8)
are characterized by a **decrease in compliance** (i.e., the distensibility of the lung is restricted). The lungs are said to be "**stiff.**" **Restrictive ventilatory impairment** is the inability to fully expand the lung (**inspiratory airflow**), which results in a decrease in total lung capacity (TLC). The radiograph shows key features of restrictive lung disease. Note a reticular pattern of lung opacities, due to an abnormal lung interstitium, that are interspersed between clear areas (lung cysts or "honeycomb lung"); small, contracted lung; and raised diaphragm. Specific examples of obstructive lung disease include the following:

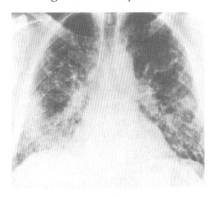
● **Figure 6-8 Restrictive lung disease.**

1. **Idiopathic Pulmonary Fibrosis.** Patients with idiopathic pulmonary fibrosis have the following characteristics: a decreased PaO_2 (hypoxemia) and a mildly decreased or normal $PaCO_2$ (hypocapnia or normocapnia). During exercise, the hypoxemia worsens without hypercapnia. As the condition worsens, hypoventilation leads to a further decreased PaO_2, a severely increased $PaCO_2$ (hypercapnia), respiratory acidosis, a decreased DLCO, and death. Pathologic findings include inflammatory cell infiltrates; thickening of the blood–air barrier (i.e., alveolar wall) due to collagen production by fibroblasts; and destruction of the alveolar architecture, leading to formation of air-filled cystic spaces surrounded by thickened scar tissue (i.e., "honeycomb lung").

2. **Coal Worker Pneumoconiosis.** Coal worker pneumoconiosis results from the inhalation of **coal dust** and is generally benign with little if any reduction in lung function. The appearance of large nodular lesions suggests a change caused by silica in the inhaled dust such that the disease is now called **anthracosilicosis. Anthracosis** is the most innocuous lesion observed, whereby carbon pigment is phagocytosed by alveolar macrophages that accumulate along the lymphatics.

3. **Silicosis.** Silicosis results from the inhalation of **silicon dioxide** (silica; SiO_2) and is associated with an increased disposition to **tuberculosis** (i.e., silicotuberculosis). Silicosis is characterized by small, dense, collagenous nodules that contain birefringent silica crystals.

4. **Asbestosis.** Asbestosis results from the inhalation of asbestos fibers. Asbestosis is characterized by **diffuse pulmonary interstitial fibrosis** and **asbestos bodies.** Asbestos bodies are golden-brown, beaded rods that consist of asbestos fibers coated with iron-containing protein material. These bodies arise when alveolar macrophages attempt to phagocytose asbestos fibers. **Malignant mesothelioma** is the most serious pleural neoplasm and is associated with history of asbestos exposure.

5. **Sarcoidosis.** Sarcoidosis is a **type IV hypersensitivity** reaction to an unknown antigen. Sarcoidosis is characterized by a **noncaseating granuloma** distributed along the lymphatics in the lung. The noncaseating granuloma is an aggregation of epithelioid cells with Langerhans cells and foreign body–type giant cells; **asteroid bodies**, which are stellate inclusions bodies found within giant cells; and **Schaumann bodies**, which are laminated concretions of calcium and proteins.

F. CYSTIC FIBROSIS (CF) (FIGURE 6-9) is caused by production of abnormally thick mucus by epithelial cells lining the respiratory (and gastrointestinal) tract. This results clinically in obstruction of airways and recurrent bacterial infections (e.g., *Staphylococcus aureus, Pseudomonas aeruginosa*). CF is an autosomal recessive genetic disorder caused by more than 1,000 different mutations in the *CFTR* gene on chromosome 7q31.2 for the cystic fibrosis transmembrane conductance regulator, which functions as a chloride ion (Cl^-) channel protein. In North America, 70% of CF cases are due to a three-base deletion that codes for the amino acid **phenylalanine at position #508** such that phenylalanine is missing from *CFTR*. Clinical signs include meconium ileus (i.e., obstruction of the bowel) in the neonate, steatorrhea (fatty stool) or obstruction of the bowel in childhood, and **cor pulmonale** (manifesting as right-sided heart failure) developing secondary to pulmonary hypertension. The PA radiograph of cystic fibrosis shows hyperinflation of both lungs, reduced size of the heart due to pulmonary compression, cyst formation, and atelectasis (collapse of alveoli) in both lungs.

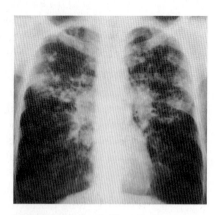

● Figure 6-9 Cystic fibrosis.

G. BRONCHOGENIC CARCINOMA (FIGURE 6-10) begins as hyperplasia of the bronchial epithelium with continued progression occurring through intraluminal growth, infiltrative peribronchial growth, and intraparenchymal growth. Intrathoracic spread of bronchogenic carcinoma may lead to **Horner syndrome** (miosis, ptosis, hemi-anhidrosis, and apparent enophthalmos) due to cervical sympathetic chain involvement; **SVC syndrome** causing dilatation of head and neck veins, facial swelling, and cyanosis; **dysphagia** due to esophageal obstruction; **hoarseness of voice** due to recurrent laryngeal nerve involvement; **paralysis of the diaphragm** due to phrenic nerve involvement; and **Pancoast syndrome** causing ulnar nerve pain and Horner syndrome. Tracheobronchial, parasternal, and supraclavicular lymph nodes are involved in the lymphatic metastasis of bronchogenic carcinoma. Enlargement of the tracheobronchial nodes may **indent the esophagus,** which can be observed radiologically during a barium swallow, or **distort the position of the carina.** Metastasis to the brain via arterial blood may occur by the following route: **cancer cells enter a lung capillary** → **pulmonary**

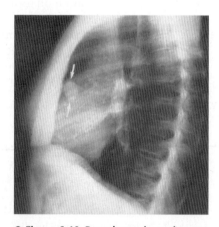

● Figure 6-10 Bronchogenic carcinoma.

vein → left atrium and ventricle → aorta → internal carotid and vertebral arteries. Metastasis to the brain via venous blood may occur by the following route: **cancer cells enter a bronchial vein → azygous vein → external and internal vertebral venous plexuses → cranial dural sinuses.** The lateral radiograph shows the nodule (*arrows*) anterior within the left upper lobe consistent with bronchogenic carcinoma. The most important issue in primary lung cancer is the histologic subclassifications, which include the following:

1. **Adenocarcinoma (AD).** AD has a 35% incidence and is the most common lung cancer in a **nonsmoking history**. AD is peripherally located within the lung as it arises from distal airways and alveoli and forms a well-circumscribed gray-white mass. There are four major histologic subtypes of AD, although it is common to find mixtures of subtypes.

2. **Squamous Cell Carcinoma (SQ).** SQ has a 35% incidence and is *most closely* associated with a **smoking history**. SQ is centrally located as it arises from larger bronchi due to injury of the bronchial epithelium followed by regeneration from the basal layer in the form of squamous metaplasia. SQ begins as a small red granular plaque and progresses to a large intrabronchial mass. SQ may secrete **parathyroid hormone (PTH)**, causing hypercalcemia.

3. **Small Cell Carcinoma (SC).** SC has a 20% incidence and is associated with a **smoking history**. SC is centrally located as it arises from larger bronchi. SC forms large, soft, gray-white masses and contains small, oval-shaped cells ("oat cells") derived from **Kulchitsky cells** (neural crest origin) that may produce **adrenocorticotropic hormone (ACTH)** or **antidiuretic hormone (ADH)**, causing Cushing syndrome or syndrome of inappropriate secretion of ADH (SIADH), respectively. SC is a highly malignant and aggressive tumor (median survival time <3 months), but it does respond favorably to chemotherapy.

4. **Carcinoid Tumor.** Carcinoid tumor has a 2% incidence and is associated with a **nonsmoking history**. Carcinoid tumor is a neuroendocrine neoplasm similar to Kulchitsky cells derived from the pluripotential basal layer of the respiratory epithelium. Carcinoid tumor may be located centrally, peripherally, or in the midportion of the lung. Centrally located carcinoid tumor has a large endobronchial component with a smooth, fleshy mass protruding into the lumen of the bronchus. Patients with carcinoid tumor may develop **carcinoid syndrome**, which is characterized by facial flushing (due to vasomotor disturbances), diarrhea (due to intestinal hypermotility), or wheezing (due to bronchoconstriction).

Ⓥ Cross-sectional Anatomy

A. Computed tomography (CT) scan at the level of origin of the three branches of the aortic arch (about vertebral level T2–3) (Figure 6-11)

B. CT scan and magnetic resonance image (MRI) at the level of the aortic arch (Figure 6-12)

C. CT scan and MRI at the level of the aortic-pulmonary window (at about vertebral level T4) (Figure 6-13)

D. CT scan at the level of origin of the left main pulmonary artery (Figure 6-14)

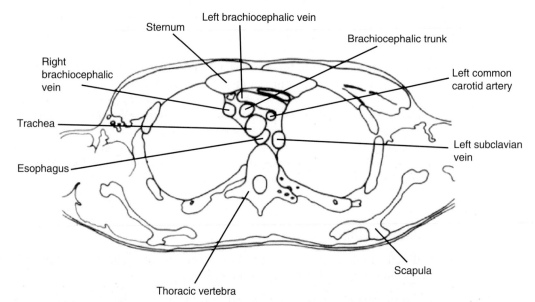

● **Figure 6-11 Computed tomography scan at the level of origin of the three branches of the aortic arch (about vertebral level T2–3).** The esophagus is anterior and to the left of the body of the thoracic vertebra. The trachea is anterior and to the right of the esophagus. The brachiocephalic trunk is anterior and to the right of the trachea. The left common carotid artery is anterior and to the left of the trachea. The left subclavian artery is to the left of the posterior border of the trachea. The right brachiocephalic vein is to the right of the brachiocephalic trunk. The left brachiocephalic vein appears in oblique section as it travels to the right side.

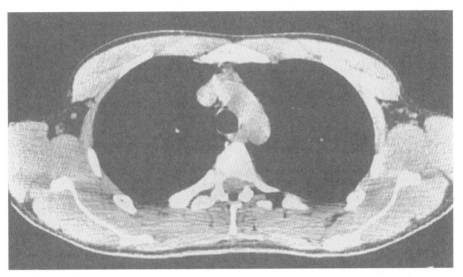

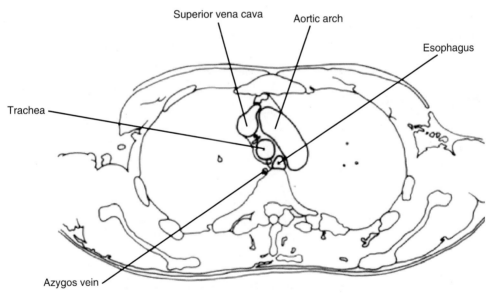

● **Figure 6-12 Computed tomography scan at the level of the aortic arch.** The esophagus is anterior and to the left of the body of the thoracic vertebra. The trachea is anterior and to the right of the esophagus. The azygos vein is posterior to the trachea and to the right of the esophagus. The aortic arch is a curved image that begins to the left of the superior vena cava (or right brachiocephalic vein), curves around the trachea, and ends to the left of the esophagus. The left brachiocephalic vein appears in oblique section at its union with the right brachiocephalic vein emptying into the superior vena cava.

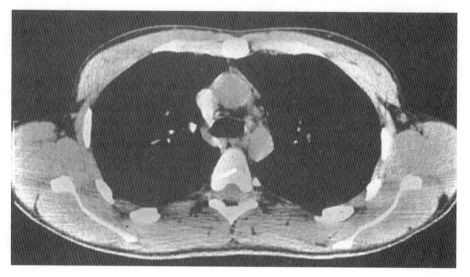

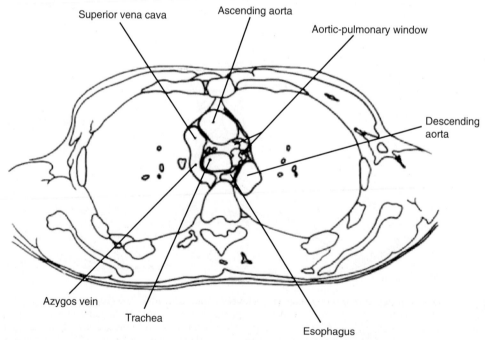

Superior vena cava

Ascending aorta

Aortic-pulmonary window

Descending aorta

Azygos vein

Trachea

Esophagus

● **Figure 6-13 Computed tomography scan at the level of the aortic-pulmonary window (at about vertebral level T4).** The aortic-pulmonary window is the space in the superior mediastinum from the bifurcation of the pulmonary trunk to the undersurface of the aortic arch. The esophagus is anterior and to the left of the body of the thoracic vertebra. At this level, the trachea bifurcates into the right main bronchus and left main bronchus. The azygos vein appears in longitudinal section as it arches over the right main bronchus and empties into the superior vena cava. The ascending aorta is anterior to the right main bronchus and anterior and to the left of the superior vena cava.

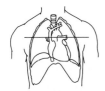

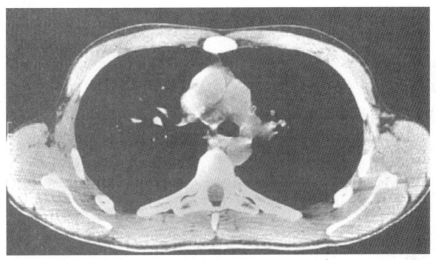

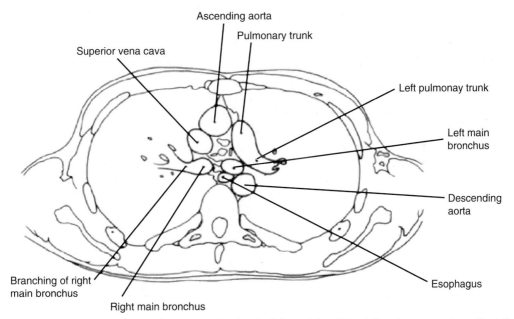

Ascending aorta

Pulmonary trunk

Superior vena cava

Left pulmonay trunk

Left main
bronchus

Descending
aorta

Branching of right
main bronchus

Esophagus

Right main bronchus

● **Figure 6-14 Computed tomography scan at the level of the origin of the left pulmonary artery.** The left main bronchus appears in cross section anterior to the esophagus and descending aorta. The right main bronchus branches into the right upper lobar bronchus and right middle lobar bronchus. The pulmonary trunk is anterior and to the left of the left main bronchus. The left pulmonary artery appears in longitudinal section as it curves posterior-lateral toward the hilum of the left lung. The superior vena cava is anterior to the right main bronchus. The ascending aorta is anterior and between the right main bronchus and left main bronchus.

Case Study 6-1

A mother brings her 4-year-old son to your office complaining that "he coughs constantly and spits up a green mucus." After some discussion, she informs you that her son also has pungent, fatty stools; has had several bouts of severe coughing and fever within the past few months; and had meconium ileus (obstruction of the bowel with a thick meconium) as a baby. Finally, she asks you this question: "Why is my son so small?" What is the most likely diagnosis?

Differentials

- Asthma, foreign body aspiration, pneumonia, selective IgA deficiency

Relevant Physical Exam Findings

- The boy is in the 10th percentile for height and 5th percentile for weight for children 4 years of age.
- Fever = 101°F, tachypnea = 40 breaths/min, tachycardia = 155 bpm.
- Auscultation of the lungs reveals diffuse, bilateral wheezing; rales; and coarse breath sounds.

Relevant Lab Findings

- Sweat test: Na^+ = high; Cl^- = high
- Stool sample showed steatorrhea
- Pulmonary function test: residual volume–to–total lung volume ratio = high; forced expiratory volume in 1 second (FEV_1)–to–forced vital capacity (FVC) ratio = low
- Sputum culture: *Staphylococcus aureus* and *Pseudomonas aeruginosa* positive
- Radiograph: hyperinflation of lungs, early signs of bronchiectasis

Diagnosis

Cystic Fibrosis (CF)

- CF is caused by production of abnormally thick mucus by epithelial cells lining the respiratory (and gastrointestinal) tract. This results in obstruction of airways and recurrent bacterial infections (e.g., *S. aureus, P. aeruginosa*).
- CF is the most common lethal genetic disorder among whites.
- The primary manifestation of CF occurs in the lungs and pancreas. In the lungs, the excessively viscous mucus leads to plugging of the lungs. In the pancreas, the release of pancreatic digestive enzymes is deficient, leading to malabsorption and steatorrhea.
- One of the earliest signs of CF is failure to thrive in early childhood.
- Differentials:
  - Asthmatic patients have a low FEV_1-to-FVC ratio and hyperinflation (similar to CF), but asthma is characterized by the reversibility of attacks and predisposes the patient to infections.
  - Foreign body aspiration is usually unilateral and occurs most often on the right side.
  - This 4-year-old boy has pneumonia, but it is secondary to CF and the sputum culture is classic for pneumonia secondary to CF.
  - Selective IgA deficiency is a congenital disorder in which class-switching in the heavy chain fails to occur; it is the most common congenital B-cell defect. IgA is found in bodily secretions and plays an important role in preventing bacterial colonization of mucosal surfaces, which makes patients with CF susceptible to recurrent sinopulmonary infections.

Case Study 6-2

A 70-year-old man comes to your office complaining that "I'm starting to cough up blood. It's bright red and it scares me. I know I cough a lot because I'm a heavy smoker and when I get a cold in the winter it gets real bad. But, it's always been a green mucus, never blood." He also tells you that "I'm short of breath and I've lost 15 pounds this past month because I have no appetite. I can't describe it, but this cough seems different." After some discussion, he informs you that he has smoked three packs of cigarettes a day for about 50 years and drinks a lot. "You know, Doc, when I smoke, I like to drink, and when I drink, I like to smoke. It's a vicious cycle. But, like I tell my wife, I only have two bad habits in life. That's pretty good, isn't it?" What is the most likely diagnosis?

Differentials

- Bronchiectasis, Goodpasture syndrome, pneumonia, tuberculosis

Relevant Physical Exam Findings

- Tachypnea, positive end-expiratory wheezing, clubbing of the digits
- Lung auscultation: decreased breath sounds on the right side
- Patient appears frail

Relevant Lab Findings

- Blood chemistry: white blood cells (WBCs) = 8,500/μL (normal); hematocrit (Hct) = 30% (low); mean corpuscular volume = 100 μm^3 (high); serum Ca^{2+} = 14.1 mg/dL (high)
- Sputum culture: normal respiratory flora; no acid-fast bacilli
- Urinalysis: ketones = none; leukocyte esterase = none; bacteria = none
- Chest radiograph: large mass in the lower right hilum; atelectasis in the right lower lobe
- Biopsy: keratin pearls and intercellular bridging observed

Diagnosis

Squamous Cell Carcinoma (SQ)

- SQ has a 35% incidence and is most closely associated with smoking history.
- SQ may secrete parathyroid hormone (PTH), causing hypercalcemia.
- SQ appears histologically as polygonal-shaped cells arranged in solid cell nests and bright eosinophilic aggregates of extracellular keratin ("pearls").
- Intracellular keratinization may also be apparent such that the cytoplasm appears glassy and eosinophilic.
- In well-differentiated squamous cell carcinomas, intercellular bridges may be observed, which are cytoplasmic extensions between adjacent cells.
- Another important histologic characteristic of squamous cell carcinoma is the in situ replacement of the bronchial epithelium.
- Differentials:
  - Bronchiectasis is the abnormal, permanent dilatation of bronchi due to chronic necrotizing infection (e.g., *Staphylococcus, Streptococcus, Haemophilus influenzae*), bronchial obstruction (e.g., foreign body, mucous plugs, or tumors), or congenital conditions (e.g., Kartagener syndrome, cystic fibrosis, immunodeficiency disorders). The lower lobes of the lung are predominately affected and the affected bronchi have a saccular appearance.
  - Antiglomerular basement membrane glomerulonephritis (ABMG; Goodpasture syndrome) is caused by deposition of IgG immune complexes (i.e., autoantibodies to a globular noncollagenous domain of type IV collagen) within the glomerular basement membrane. The autoantibodies generally cross-react with pulmonary basement membranes; therefore, when both the lungs and kidneys are involved, the term *Goodpasture syndrome* is used.

- Pneumococcal pneumonia is generally a consequence of altered immunity within the respiratory tract most frequently following a viral infection (e.g., influenza), which damages the mucociliary elevator; chronic obstructive pulmonary disease (COPD); or alcoholism.
- *Mycobacterium tuberculosis* causes tuberculosis (TB), which is the classic mycobacterial disease. Aerosolized infectious particles travel to terminal airways, where *M. tuberculosis* penetrates inactivated alveolar macrophages and inhibits acidification of endolysosomes so that alveolar macrophages cannot kill the bacteria. The Ghon complex is the first lesion of primary TB and consists of a parenchymal granuloma (location is subpleural and in lower lobes of the lung) and prominent, infected mediastinal lymph nodes.

Case Study 6-3

A 25-year-old man is involved in a high-speed automobile accident and is brought into the emergency room. The patient is ill-appearing, combative, and in no condition to offer any assistance as to the extent of his injuries. The emergency medical technicians inform you that he was not wearing a seat belt, that the air bags deployed, and that he was hit broadside on the driver side ("T-boned"). The man had to be removed from the auto by the "jaws of life" and was placed on a backboard with a cervical collar.

Relevant Physical Exam Findings

- Heart rate = 115 bpm
- Blood pressure = 85/50 mm Hg
- Respiratory rate = 32 breaths/min
- 90% O_2 saturation on the nonrebreather mask
- Weakly palpable carotid pulse
- Elevated jugular venous pulse
- Oropharynx is clear
- Trachea is shifted to the right of midline
- Breath sounds are decreased over the left chest
- Percussion of left chest shows hyperresonance
- Standard trauma radiographs including anteroposterior chest radiograph and pelvic scan were ordered

Diagnosis

Tension Pneumothorax

- Tension pneumothorax is a life-threatening condition that occurs when the air in the pleural space is under pressure, thereby displacing mediastinal structures and compromising cardiopulmonary function (reducing venous return and cardiac output).
- Patients may rapidly progress to cardiorespiratory collapse and death.
- In a tension pneumothorax, patients display respiratory distress, tachypnea, tachycardia, cyanosis, jugular vein distention, tracheal deviation away from the injured lung, and pulsus paradoxus (a pulse that markedly decreases in size during inhalation).
- A tension pneumothorax causing hemodynamic compromise should be diagnosed clinically, and treatment should never be delayed in favor of radiographic imaging.
- The emergency room physician placed a 14-gauge Angiocath in the second intercostal space (midclavicular line) of the left chest. A rush of air was appreciated and the blood pressure improved to 95/60 mm Hg.
- The emergency room physician then placed a chest tube in the left fifth intercostal space (midaxillary line).
- The patient was intubated and transported to the operating room.

Chapter 7

Heart

① The Pericardium

A. GENERAL FEATURES. The pericardium consists of three layers: visceral layer of serous pericardium, parietal layer of serous pericardium, and fibrous pericardium.

1. **Visceral Layer of Serous Pericardium**
   a. The visceral layer of serous pericardium is known histologically as the **epi-cardium** and consists of a layer of simple squamous epithelium called **mesothelium.**
   b. Beneath the mesothelium, coronary arteries, coronary veins, and nerves travel along the surface of the heart in a thin collagen bed; adipose tissue is also present.

2. **Parietal Layer of Serous Pericardium.** At the base of the aorta and pulmonary trunk, the mesothelium of the visceral layer of serous pericardium is reflected and becomes continuous with the parietal layer of serous pericardium such that the parietal layer of serous pericardium also consists of a layer of simple squamous epithelium (called **mesothelium**).

3. **Pericardial Cavity**
   a. The pericardial cavity lies between the visceral layer and parietal layer of serous pericardium and normally contains a small amount of pericardial fluid (20 mL), which allows friction-free movement of the heart during diastole and systole.
   b. The **transverse sinus** is a recess of the pericardial cavity. After the pericardial sac is opened, a surgeon can pass a finger or ligature (posterior to the aorta and pulmonary trunk and anterior to the superior vena cava) from one side of the heart to the other through the transverse sinus.
   c. The **oblique sinus** is a recess of the pericardial cavity that ends in a cul de sac surrounded by the pulmonary veins.

4. **Fibrous Pericardium**
   a. Fibrous pericardium is a thick (~1 mm) **collagen layer** (no elastic fibers) with **little ability to distend acutely.**
   b. The fibrous pericardium fuses superiorly to the tunica adventitia of the great vessels, inferiorly to the central tendon of the diaphragm, and anteriorly to the sternum.
   c. The **phrenic nerve** and **pericardiacophrenic artery** descend through the mediastinum lateral to the fibrous pericardium and are in jeopardy during surgery to the heart.

5. The **thoracic portion of the inferior vena cava (IVC)** lies within the pericardium so that to expose this portion of the IVC, the pericardium must be opened.

6. The innervation of the pericardium is supplied by the phrenic nerve (C3–5). Pain sensation carried by the **phrenic nerves** is often referred to the skin (C3–5 dermatomes) of the **ipsilateral supraclavicular region.**

B. CLINICAL CONSIDERATIONS

1. **Cardiac tamponade (heart compression)** is the accumulation of fluid within the pericardial cavity resulting in compression of the heart since the fibrous pericardium is inelastic. Clinical findings include hypotension (blood pressure [BP] 90/40 mm Hg) that does not respond to rehydration; compression of the superior vena cava (SVC), which may cause the veins of the face and neck to engorge with blood; a distention of veins of the neck on inspiration (**Kussmaul sign**); paradoxical pulse (inspiratory lowering of BP by >10 mm Hg); syringe filling spontaneously when blood is drawn due to increased venous pressure; and distant heart sounds. Cardiac tamponade can quickly progress to cardiogenic shock and death.

2. **Pericardiocentesis** is the removal of fluid from the pericardial cavity, which can be approached in two ways.
   a. **Sternal approach**
      i. A needle is inserted at intercostal space 5 or 6 on the left side near the sternum. The cardiac notch of the left lung leaves the fibrous pericardium exposed at this site.
      ii. The needle penetrates the following structures: skin → superficial fascia → pectoralis major muscle → external intercostal membrane → internal intercostal muscle → transverse thoracic muscle → fibrous pericardium → parietal layer of serous pericardium.
      iii. The internal thoracic artery, coronary arteries, and pleura may be damaged during this approach.
   b. **Subxiphoid approach**
      i. A needle is inserted at the left infrasternal angle angled in a superior and posterior position.
      ii. The needle penetrates the following structures: skin → superficial fascia → anterior rectus sheath → rectus abdominus muscle → transverse abdominus muscle → fibrous pericardium → parietal layer of serous pericardium.
      iii. The diaphragm and liver may be damaged during this approach.

❚❚ Heart Surfaces (Figure 7-1A,B). The heart has six surfaces.

A. **POSTERIOR SURFACE (BASE).** The posterior surface consists mainly of the **left atrium**, which receives the pulmonary veins and is related to vertebral bodies T6–9.

B. **Apex**
   1. The apex consists of the inferior lateral portion of the **left ventricle** at intercostal space 5 along the midclavicular line.
   2. The maximal pulsation of the heart (apex beat) occurs at the apex.

C. **ANTERIOR SURFACE (STERNOCOSTAL SURFACE).** The anterior surface consists mainly of the **right ventricle**.

D. **INFERIOR SURFACE (DIAPHRAGMATIC SURFACE).** The inferior surface consists mainly of the **left ventricle** and is related to the central tendon of the diaphragm.

E. **LEFT SURFACE (PULMONARY SURFACE).** The left surface consists mainly of the left ventricle and occupies the cardiac impression of the left lung.

F. **RIGHT SURFACE.** The right surface consists mainly of the right atrium located between the SVC and IVC.

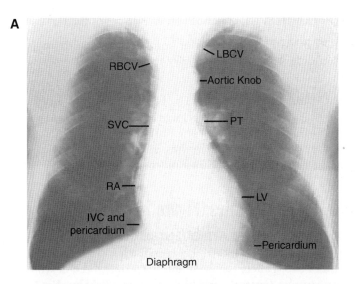

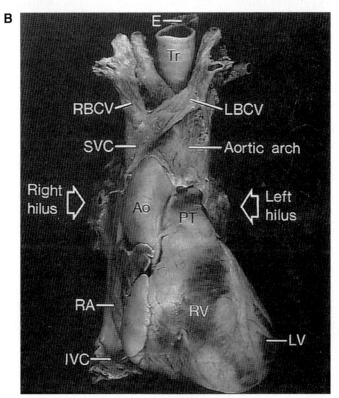

● **Figure 7-1 A:** Radiograph shows the various components of the heart and great vessels. **B:** Photograph of the anatomic heart and great vessels for comparison to the radiograph in A. Ao = aorta, E = esophagus, IVC = inferior vena cava, LBCV = left brachiocephalic vein; LV = left ventricle, PT = pulmonary trunk, RA = right atrium, RBCV = right brachiocephalic vein, RV = right ventricle, SVC = superior vena cava, Tr = trachea.

Ⅲ **Heart Borders (Figure 7-1A,B).** The heart has four borders.

   A. RIGHT BORDER. The right border consists of the **right atrium, SVC, and IVC.**

   B. LEFT BORDER. The left border consists of the **left ventricle, left atrium, pulmonary trunk,** and **aortic arch.**

C. **INFERIOR BORDER.** The inferior border consists of the **right ventricle.**

D. **SUPERIOR BORDER.** The superior border consists of the **right atrium, left atrium, SVC, ascending aorta,** and **pulmonary trunk.**

Ⅳ **Fibrous Skeleton of the Heart.** The fibrous skeleton is a dense framework of collagen within the heart that keeps the orifices of the atrioventricular valves and semilunar valve patent, provides an attachment site of the valve leaflets and cusps, serves as the origin and insertion sites of cardiac myocytes, and forms an electrical "barrier" between the atria and ventricles so that they contract independently.

Ⅴ **Valves and Auscultation Sites (Figure 7-2)**

A. **BICUSPID (MITRAL; LEFT ATRIOVENTRICULAR) VALVE**

   1. The bicuspid valve is located between the left atrium and left ventricle and is composed of **two leaflets (anterior and posterior),** both of which are tethered to **papillary muscles (anterolateral and posteromedial)** by chorda tendineae.

   2. The auscultation site is at the **cardiac apex at left intercostal space 5.**

B. **TRICUSPID (RIGHT ATRIOVENTRICULAR) VALVE**

   1. The tricuspid valve is located between the right atrium and right ventricle and is composed of **three leaflets (anterior, posterior, and septal),** all of which are tethered to **papillary muscles (anterior, posterior, and septal)** by chorda tendineae.

   2. The auscultation site is **over the sternum at intercostal space 5.**

C. **PULMONARY SEMILUNAR VALVE (PULMONIC VALVE)**

   1. The pulmonary semilunar valve is the outflow valve of the right ventricle and is composed of **three cusps (anterior, right, and left)** that fit closely together when closed.

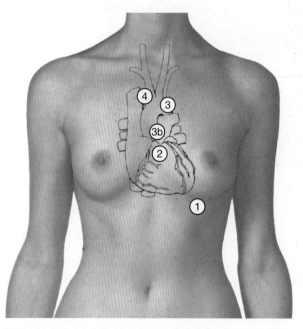

1 = bicuspid (mitral) valve area at the cardiac apex at the left intercostal space 5
2 = tricuspid valve area over the sternum at intercostal space 5
3 = pulmonary valve area lateral to the sternum at left intercostal space 2
3b = secondary pulmonic valve area over the sternum at intercostal space 4
4 = aortic valve area lateral to the sternum at right intercostal space 2

● **Figure 7-2 Auscultatory areas of the chest.** The positions of the auscultatory areas are indicated on the body surface of a woman.

2. The orifice of the pulmonary semilunar valve is directed to the left shoulder.

3. The auscultation site is just **lateral to the sternum at left intercostal space 2.**

4. A secondary pulmonic auscultation site is **over the sternum at about intercostal space 4.**

D. AORTIC SEMILUNAR VALVE

1. The aortic semilunar valve is the outflow valve of the left ventricle and is composed of **three cusps (posterior, right, and left)** that fit closely together when closed.

2. The orifice of the aortic semilunar valve is directed to the right shoulder.

3. The auscultation site is just **lateral to the sternum at right intercostal space 2.**

E. HEART SOUNDS

1. S_1 heart sound (first sound; "lub" sound) is caused by closure of the tricuspid and bicuspid valves.

2. S_2 heart sound (second sound; "dub" sound) is cause by closure of the pulmonary and aortic valves.

Ⓥ Arterial Supply of the Heart (Figure 7-3). The right coronary artery and left coronary artery supply oxygenated arterial blood to the heart. **The coronary arteries fill with blood during diastole.** The coronary arteries have maximal blood flow during diastole and minimal blood flow during systole.

A. RIGHT CORONARY ARTERY (RCA)

1. The RCA arises from the right aortic sinus (of Valsalva) of the ascending aorta and courses in the coronary sulcus.

2. The blood supply of the heart is considered **right-side dominant** (most common) if the posterior interventricular artery arises from the RCA.

3. The RCA branches into the:
   a. **Sinoatrial (SA) nodal artery**
   b. **Conus branch**
   c. **Right marginal artery**
   d. **Atrioventricular (AV) nodal artery**
   e. **Posterior interventricular artery**
   f. **Septal branches**

B. Left Main Coronary Artery (LMCA). The LMCA arises from the left aortic sinus (of Valsalva) of the ascending aorta. The blood supply of the heart is considered **left-side dominant** (less common) if the posterior interventricular artery arises from the LMCA. The LMCA branches into the:

1. **Left circumflex artery (LCx),** which further branches into the:
   a. **Anterior marginal artery**
   b. **Obtuse marginal artery**
   c. **Atrial branches**
   d. **Posterior marginal artery**

2. **Intermediate ramus** (a variable branch)

3. **Anterior interventricular artery** (also called left anterior descending artery [**LAD**]), which further branches into the:
   a. **Anterior diagonal artery**
   b. **Septal branches**

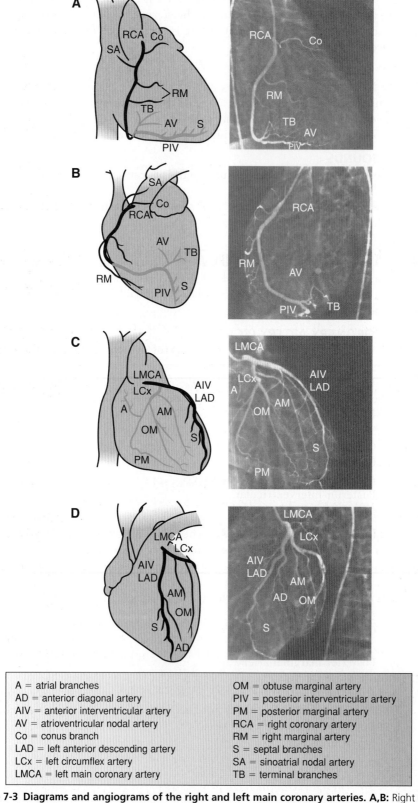

A = atrial branches
AD = anterior diagonal artery
AIV = anterior interventricular artery
AV = atrioventricular nodal artery
Co = conus branch
LAD = left anterior descending artery
LCx = left circumflex artery
LMCA = left main coronary artery

OM = obtuse marginal artery
PIV = posterior interventricular artery
PM = posterior marginal artery
RCA = right coronary artery
RM = right marginal artery
S = septal branches
SA = sinoatrial nodal artery
TB = terminal branches

● **Figure 7-3 Diagrams and angiograms of the right and left main coronary arteries. A,B:** Right coronary artery (RCA). **A:** Right anterior oblique (RAO) angiogram shows the various branches of the RCA that can be observed in this view. **B:** Left anterior oblique (LAO) angiogram shows the various branches of the RCA that can be observed in this view. **C,D:** Left main coronary artery (LMCA). **C:** RAO angiogram shows the various branches of the LMCA that can be observed in this view. **D:** LAO angiogram shows the various branches of the LMCA that can be observed in this view.

 Venous Drainage of the Heart

A. CORONARY SINUS

1. The coronary sinus is the largest vein draining the heart and drains directly into the right atrium.

2. At the opening of the coronary sinus, a crescent-shaped valve remnant (called the **Thebesian valve**) is present.

B. GREAT CARDIAC VEIN. The great cardiac vein follows the **anterior interventricular artery** and drains into the coronary sinus.

C. MIDDLE CARDIAC VEIN. The middle cardiac vein follows the **posterior interventricular artery** and drains into the coronary sinus.

D. SMALL CARDIAC VEIN. The small cardiac vein follows the **right marginal artery** and drains into the coronary sinus.

E. OBLIQUE VEIN OF THE LEFT ATRIUM. The oblique vein of the left atrium is a remnant of the embryonic left superior vena cava and drains into the coronary sinus.

F. LEFT POSTERIOR VENTRICULAR VEIN. The left posterior ventricular vein drains into the coronary sinus.

G. LEFT MARGINAL VEIN. The left marginal vein drains into the coronary sinus.

H. ANTERIOR CARDIAC VEINS. The anterior cardiac veins are found on the anterior aspect of the right ventricle and drain directly into the right atrium.

I. SMALLEST CARDIAC VEINS. The smallest cardiac veins begin within the wall of the heart and drain directly into the nearest heart chamber.

The Conduction System (Figure 7-4)

A. SINOATRIAL NODE

1. The SA node is the **pacemaker** of the heart and is located at the junction of the superior vena cava and right atrium just beneath the epicardium.

2. From the SA node, the impulse spreads throughout the right atrium and to the AV node via the **anterior, middle, and posterior internodal tracts** and to the left atrium via the **Bachmann bundle.**

3. If all SA node activity is destroyed, the AV node will assume the pacemaker role.

B. ATRIOVENTRICULAR NODE

1. The AV node is located on the right side of the AV portion of the atrial septum near the ostium of the coronary sinus in the subendocardial space.

2. The AV septum corresponds to the **triangle of Koch,** an important anatomic landmark because it contains the AV node and the proximal penetrated portion of the bundle of His.

C. BUNDLE OF HIS, BUNDLE BRANCHES, AND PURKINJE MYOCYTES

1. The **bundle of His** travels in the subendocardial space on the right side of the interventricular septum and divides into the **right and left bundle branches.**

2. The left bundle branch is thicker than the right bundle branch.

3. A portion of the right bundle branch enters the septomarginal trabecula (moderator band) to supply the anterior papillary muscle.

4. The left bundle branch further divides into an **anterior segment** and **posterior segment.**

5. The right and left bundle branches both terminate in a complex network of intramural **Purkinje myocytes.**

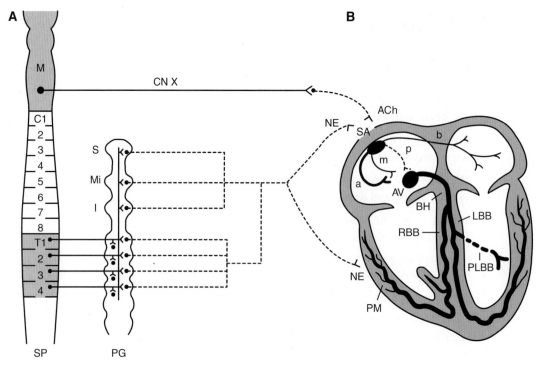

● Figure 7-4 A: Innervation of the heart. Parasympathetic innervation involving the vagus nerve (cranial nerve [CN] X) is shown. Sympathetic innervation is also shown. All preganglionic neurons are indicated by *solid lines*. All postganglionic neurons are indicated by *dotted lines*. Ach = acetylcholine, I = inferior cervical ganglion, M = medulla, Mi = middle cervical ganglion, NE = norepinephrine, PG = paravertebral ganglia, S = superior cervical ganglion, SP = spinal cord. **B:** Diagram of the conduction system and innervation of the heart. AV = atrioventricular node, BH = bundle of His, LBB = left bundle branch, PLBB = posterior segment of the left bundle branch, PM = Purkinje myocytes, RBB = right bundle branch, SA = sinoatrial node, a = anterior internodal tract; m = middle internodal tract, p = posterior internodal tract, b = Bachman bundle.

 Innervation of the Heart (Figure 7-4). The heart is innervated by the **superficial cardiac plexus**, which is located inferior to the aortic arch and anterior to the right pulmonary artery, and the **deep cardiac plexus**, which is located posterior to the aortic arch and anterior to the tracheal bifurcation. These plexuses contain **both parasympathetic (vagus; cranial nerve [CN] X) and sympathetic components.**

A. PARASYMPATHETIC

   1. Motor

      a. Preganglionic neuronal cell bodies are located in the **dorsal nucleus of the vagus** and **nucleus ambiguus** of the medulla. Preganglionic axons run in the **vagus (CN X) nerve.**

      b. Postganglionic neuronal cell bodies are located in the cardiac plexus and atrial wall.

      c. Postganglionic axons are distributed to the **SA node, AV node, atrial myocytes (not ventricular myocytes),** and **smooth muscle of coronary arteries,** causing a:

        i. **Decrease in heart rate**

        ii. **Decrease in conduction velocity through the AV node**

        iii. **Decrease in contractility of atrial myocytes**

      d. The SA node and AV node contain high levels of **acetylcholinesterase** (degrades acetylcholine rapidly) such that any given vagal stimulation is **short-lived.**

      e. **Vasovagal syncope** is a brief period of lightheadedness or loss of consciousness due to an intense burst of CN X activity.

2. **Sensory**
   a. The neuronal cell bodies are located in the inferior (nodose) ganglia of CN X. These neurons send a peripheral process to **baroreceptors** in the great veins, atria, and aortic arch via CN X and a central process to the solitary nucleus in the brain. These neurons transmit **changes in blood pressure.**
   b. The neuronal cell bodies are located in the inferior (nodose) ganglia of CN X. These neurons send a peripheral process to **chemoreceptors** (specifically the **aortic bodies**) via CN X and a central process to the solitary nucleus in the brain. These neurons transmit **changes in the partial pressure of arterial oxygen (PaO_2).**

B. **SYMPATHETIC**
   1. **Motor**
      a. Preganglionic neuronal cell bodies are located in the **intermediolateral columns** of the spinal cord. Preganglionic axons enter the paravertebral ganglion and travel to the stellate/middle cervical ganglia.
      b. Postganglionic neuronal cell bodies are located in the **stellate and middle cervical ganglia.**
      c. Postganglionic axons are distributed to the **SA node, AV node, atrial myocytes, ventricular myocytes**, and **smooth muscle of coronary arteries**, causing an:
         i. **Increase in heart rate**
         ii. **Increase in conduction velocity through the AV node**
         iii. **Increase in contractility of atrial and ventricular myocytes**
      d. Released norepinephrine is either carried away by the bloodstream or taken up by the nerve terminals so that sympathetic stimulation is relatively **long-lived.**
   2. **Sensory**
      a. The neuronal cells bodies are located in the dorsal root ganglia at T1–5 spinal cord levels. These neurons send a peripheral process to the heart via the sympathetics and a central process to the spinal cord. These neurons transmit **pain sensation.**
      b. The pain associated with angina pectoris or a "heart attack" may be referred over the T1–5 dermatomes (i.e., **the classic referred pain down the left arm**).

Ⓧ Gross Anatomy of the Heart

A. **RIGHT ATRIUM.** The right atrium receives venous blood from the SVC, IVC, and coronary sinus. The right atrium consists of the:
   1. **Right auricle,** which is a conical, muscular pouch
   2. **Pectinate muscles,** which form the trabeculated part of the right atrium (2 to 4 mm thick) and develop embryologically from the primitive atrium
   3. **Sinus venarum,** which is the smooth part of the right atrium and develops embryologically from the sinus venous
   4. **Crista terminalis** (an internal muscular ridge 3 to 6 mm thick), which marks the junction between the trabeculated part and smooth part of the right atrium
   5. **Sulcus terminalis** (an external shallow groove), which also marks the junction between the trabeculated part and smooth part of the right atrium
   6. **Openings of the SVC, IVC, coronary sinus, and anterior cardiac vein**
   7. **Atrial septum,** which consists of an interatrial portion and an AV portion
   8. **Fossa ovalis,** which is an oval depression on the interatrial portion consisting of the **valve of the fossa ovalis** (a central sheet of thin fibrous tissue), which is a remnant of septum primum, and the **limbus of the fossa ovalis** (a horseshoe-shaped muscular rim), which is a remnant of the septum secundum

B. RIGHT VENTRICLE. The trabeculated inflow tract of the right ventricle receives venous blood from the right atrium posteriorly through the tricuspid valve while the smooth outflow tract of the right ventricle expels blood superiorly and to the left into the pulmonary trunk. The right ventricle consists of the:

1. **Trabeculae carneae** (irregular muscular ridges), which form the trabeculated part of the right ventricle (inflow tract) and develop embryologically from the primitive ventricle

2. **Conus arteriosus (infundibulum),** which is the smooth part of the right ventricle (outflow tract) and develops embryologically from the bulbus cordis

3. **Supraventricular crest** (a C-shaped internal muscular ridge), which marks the junction between the trabeculated part and smooth part of the right ventricle

4. **Tricuspid valve (anterior, posterior, and septal cusps),** which attaches at its base to the fibrous skeleton

5. **Chordae tendineae,** which are cords that extend from the free edge of the tricuspid valve to the papillary muscles and prevent eversion of the tricuspid valve into the right atrium, thereby preventing regurgitation of ventricular blood into the right atrium during systole

6. **Papillary muscles (anterior, posterior, and septal),** which are conical muscular projections from the ventricular wall and are attached to the chordae tendineae

7. **Interventricular septum,** which consists of a **membranous part** (located in a superior-posterior position and continuous with the fibrous skeleton) and a **muscular part**

8. **Septomarginal trabecula (moderator band),** which is a curved muscular bundle that extends from the interventricular septum to the anterior papillary muscle and contains part of the right bundle branch of the bundle of His to the anterior papillary muscle

9. **Right AV orifice**

10. **Opening of the pulmonary trunk**

11. **Pulmonary semilunar valve (anterior, right, and left cusps),** which lies at the apex of the conus arteriosus and prevents blood from returning to the right ventricle

12. In fetal and neonatal life, the thickness of the right ventricular wall is similar to the thickness of the left ventricular wall due to the equalization of pulmonary and aortic pressures by the ductus arteriosus. By 3 months of age, the infant heart shows regression of the right ventricular wall thickness.

C. LEFT ATRIUM. The left atrium receives oxygenated blood from the lungs through the pulmonary veins. The left atrium consists of the:

1. **Left auricle,** which is a tubular muscular pouch

2. **Pectinate muscles,** which form the trabeculated part of the left atrium and develop embryologically from the primitive atrium

3. **Smooth part of the left atrium,** which develops embryologically by incorporation of the transient common pulmonary vein into its wall

4. **Openings of the valveless pulmonary veins**

5. **Atrial septum,** which consists only of an interatrial portion

6. **Semilunar depression,** which indicates the valve of the fossa ovalis. The limbus of the fossa ovalis and the AV septum are not visible from the left atrium.

D. LEFT VENTRICLE. The trabeculate inflow tract of the left ventricle receives oxygenated blood from the left atrium through the mitral valve while the smooth outflow tract of the left ventricle expels blood superoanteriorly into the ascending aorta. The left ventricle consists of the:

1. **Trabeculae carneae** (irregular muscular ridges), which form the trabeculated part of the left ventricle (inflow tract) and develop embryologically from the primitive ventricle

2. **Aortic vestibule,** which is the smooth part of the left ventricle (outflow tract) and develops embryologically from the bulbus cordis
3. **Mitral valve (anterior and posterior cusps),** which attaches at its base to the fibrous skeleton
4. **Chordae tendineae,** which are cords that extend from the free edge of the mitral valve to the papillary muscles and prevent eversion of the mitral valve into the left atrium, thereby preventing regurgitation of ventricular blood into the left atrium during systole
5. **Papillary muscles (anterior and posterior),** which are conical muscular projections from the ventricular wall and are attached to the chordae tendineae
6. **Left AV orifice**
7. **Opening of the ascending aorta**
8. **Aortic semilunar valve (posterior, right, and left cusps),** which lies at the apex of the aortic vestibule and prevents blood from returning to the left ventricle

XI Clinical Considerations

A. **ATHEROSCLEROSIS.** The characteristic lesion of atherosclerosis is an **atheromatous plaque (fibrofatty plaque; atheroma)** within the **tunica intima** of blood vessels. An early stage in the formation of an atheromatous plaque is the subendothelial **fatty streak.** Fatty streaks are elevated, pale yellow, smooth surfaced, focal in distribution, and irregular in shape with well-defined borders. Fatty streaks can be seen as early as the second decade of life.

B. **ISCHEMIC HEART DISEASE.** Coronary artery atherosclerosis leads to three major clinical conditions:
1. **Angina pectoris** is the sudden onset of precordial (anterior surface of the body over the heart and stomach) pain.
2. **Myocardial Infarction (MI) (Figure 7-5).** An MI is the ischemic necrosis of the myocardium of the heart. Complications of an MI include **hemopericardium** caused by rupture of the free ventricular wall; **arterial emboli; pericarditis** (only in transmural infarcts); **ventricular aneurysm,** which is a bulge in the heart during systole at the postinfarction scar; and **postmyocardial infarction syndrome (Dressler syndrome),** which is an autoimmune pericarditis. There are two types of infarcts:
   a. **Transmural infarct** is unifocal and solid, follows the distribution of a specific coronary artery, often causes shock, and is caused by an occlusive thrombus; pericarditis is common with a transmural infarct. The volume of **collateral arterial blood flow** is the chief factor that affects the progression of a transmural infarct. In chronic cardiac ischemia, extensive collateral blood vessels develop over time that supply the subepicardial portion of the myocardium and thereby limit the infarct to the subendocardial portion of the myocardium.
   b. **Subendocardial infarct** is multifocal and patchy, follows a circumferential distribution, and is caused by hypoperfusion of the heart (e.g., aortic stenosis, hemorrhagic shock, or hypoperfusion during cardiopulmonary bypass); pericarditis is uncommon with subendocardial infarct.
3. **Congestive Heart Failure (CHF).** CHF is the inability of the heart to pump blood at a rate commensurate with the requirements of the body tissues, or it can do so only from elevated filling pressures. Most instances of CHF are due to the progressive deterioration of myocardial contractile function (i.e., systolic dysfunction) as occurs in ischemic heart disease or hypertension (i.e., the hypertensive left heart). CHF is characterized by reduced cardiac output (i.e., forward failure) or damming back of blood into the venous system (i.e., backward failure), or both.

A

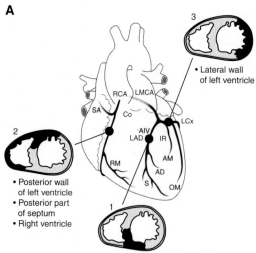

3
• Lateral wall of left ventricle

2
• Posterior wall of left ventricle
• Posterior part of septum
• Right ventricle

1
• Anterior wall of left ventricle
• Anterior part of septum

B

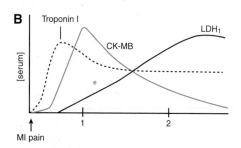

Troponin I

LDH₁

CK-MB

[serum]

↑ MI pain

C Early Anterior MI (2–24 hours)
• ST segment elevated in V₃ and V₄

Recent Anterior MI (24–72 hours)
• Q waves in V₃ and V₄
• T waves inverted in V₃ and V₄

Old Anterior MI
• Q waves persist in V₃ and V₄
• No T-wave inversion

D

Day 1

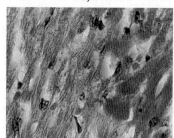

• Coagulation necrosis
• Wavy myocytes
• Pyknotic nuclei
• Eosinophilic cytoplasm
• Contraction bands

Days 2–4

• Total coagulation necrosis
• Loss of nuclei
• Loss of striations
• Dilated vessels (hyperemia)
• Neutrophil infiltration

Days 5–10

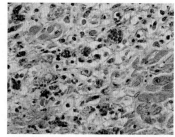

• Macrophage infiltration
• Phagocytosis of necrotic myocytes

Week 7

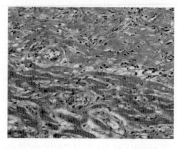

• Collagenous scar

C. RIGHT VENTRICLE (RV) FAILURE (FIGURE 7-6A,B)

1. **General Features.** The RV is susceptible to failure in situations that cause an increase in afterload on the RV. Pure RV failure most often occurs with **cor pulmonale**, which can be induced by intrinsic diseases of the lung or **pulmonary arterial hypertension (PAH)**. **Acute cor pulmonale** is RV dilation caused by a large thrombopulmonary embolism. **Chronic cor pulmonale** is RV hypertrophy followed by RV enlargement and RV failure caused by PAH. PAH is defined as pulmonary artery pressures above the normal systolic value of 30 mm Hg. There are numerous causes of PAH, including vasculitis, idiopathic ("primary PAH"), chronic pulmonary emboli, chronic lung disease, emphysema, and Eisenmenger syndrome (mnemonic: "VICE").

2. **Clinical findings include** right hypogastric quadrant discomfort due to hepatomegaly, a cut section of the liver demonstrating a "nutmeg" pattern of chronic passive congestion, peripheral edema (e.g., hallmark of RV failure is ankle swelling), pulmonary edema absent, jugular vein and portal vein distention, enlarged spleen, peritoneal cavity ascites, pleural effusion, palpable parasternal "heave," presence of S_4 heart sound ("atrial gallop"), and tricuspid valve murmur. Ascent to high altitudes is contraindicated due to hypoxic pulmonary vasoconstriction, which will exacerbate the condition.

D. LEFT VENTRICLE (LV) FAILURE (FIGURE 7-6C,D)

1. **General Features.** LV failure most often occurs due to impaired left ventricle function caused by MI. The left ventricle is usually hypertrophied and quite massively dilated. In LV failure, there is progressive damming of blood within the pulmonary circulation such that pulmonary vein pressure mounts and pulmonary edema with wet, heavy lungs is apparent. Coughing is a common feature of LV failure. Transferrin and hemoglobin, which leak from the congested capillaries, are phagocytosed by macrophages in the alveoli (called "heart failure cells"). In LV failure, the decreased cardiac output causes a reduction in kidney perfusion, which may lead to acute tubular necrosis and also activates the renin-angiotensinogen system.

2. **Clinical findings include** patient is overweight, has a poor diet, and has occasional episodes of angina; crushing pressure on the chest with pain radiating down the left arm ("referred pain"); nausea; profuse sweating and cold, clammy skin due to stress-induced release of catecholamines (epinephrine and norepinephrine) from

---

● **Figure 7-5 Myocardial infarction (MI).** **A:** Transmural MIs are caused by thrombotic occlusion of a coronary artery. Infarction is localized to the anatomic area supplied by the occluded artery. Coronary artery occlusion occurs most commonly in the anterior interventricular artery (AIV; also called the left anterior descending [LAD]), followed by the right coronary artery (RCA) and then the left circumflex artery (LCx). This is indicated by the numbers 1, 2, and 3. AD = anterior diagonal artery, AM = anterior marginal, IR = intermediate ramus, LMCA = left main coronary artery, OM = obtuse marginal artery, RM = right marginal artery, S = septal branches, SA = sinoatrial artery. **B:** Serum markers of MI. Troponin I is a highly specific cardiac marker that can be detected within 4 hours to 7 to 10 days after MI pain. Creatine kinase (CK) consists of M and B subunits. CK-MM is found in skeletal muscle and cardiac muscle. CK-MB is found mainly in cardiac muscle. CK-MB is the test of choice in the first 24 hours after MI pain. CK-MB begins to rise 4 to 8 hours after MI pain, peaks at 24 hours, and returns to normal within 48 to 72 hours. This sequence is important because skeletal muscle injury or non-MI conditions may raise serum CK-MB but do not show this pattern. It is common to calculate the ratio of CK-MB to total CK. A CK-MB–to–total CK ratio greater than 2.5% indicates MI. Lactate dehydrogenase (LDH) consists of H and M subunits. LDH-HHHH (or LDH_1) and LDH-HHHM (or LDH_2) are found in cardiac muscle. LDH_1 is the test of choice 2 to 3 days after MI pain since CK-MB levels have already returned to normal at this time. It is common to calculate the ratio of LDH_1 to LDH_2. A LDH_1-to-LDH_2 ratio greater than 1.0 indicates MI. **C:** Electrocardiograms (ECGs). An acute MI is associated with ST elevation. A recent MI (within 1 to 2 days) is associated with deep Q waves and inverted T waves. An old MI (weeks later) is associated with persistence of deep Q waves but no T-wave inversion. **D:** Evolution of a MI. The histologic changes of an MI are indicated.

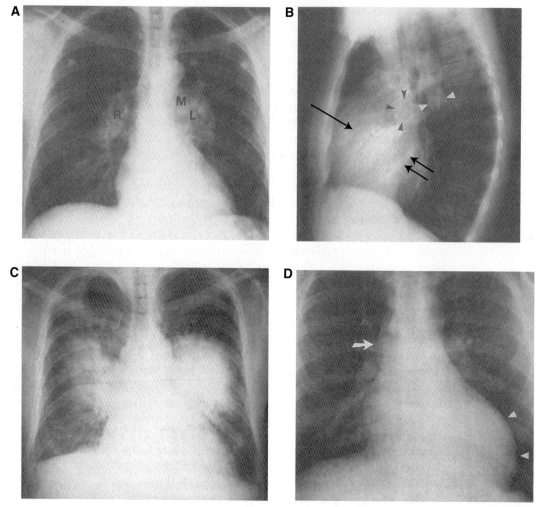

● **Figure 7-6 A,B:** Right ventricle failure. **A:** Posteroanterior radiograph of pulmonary arterial hypertension (PAH) shows enormously dilated pulmonary trunk (M) and right (R) and left (L) pulmonary arteries with diminutive peripheral pulmonary vessels. **B:** Lateral radiograph of PAH shows the enlarged right ventricle (RV hypertrophy) and atrium extending anteriorly into the anterior mediastinum (*arrow*). Note that the posterior border of the heart (left ventricle) is flat (*double arrows*). Fine curvilinear calcifications can be seen outlining the enlarged right (*black arrowheads*) and left (*white arrowheads*) pulmonary arteries. **C,D:** Left ventricle failure. **C:** Anteroposterior (AP) radiograph shows alveolar (air-space) pulmonary edema at the central, parahilar regions of the lung in the classic "bat's wing" appearance. **D:** AP radiograph shows left ventricle (LV) enlargement. Note the prominence of the LV with rounding along the inferior heart border and a downward pointing apex (*arrowheads*).

adrenal medulla, which stimulates sweat glands and causes peripheral vasoconstriction; dyspnea; orthopnea; auscultation of pulmonary rales due to "popping open" of small airways that were closed off due to pulmonary edema; noisy breathing ("cardiac asthma"); pulmonary wedge pressure (indicator of left atrial pressure) increased versus normal (30 vs. 5 mm Hg, respectively); and ejection fraction decreased versus normal (0.35 vs. 0.55, respectively).

XII Radiology

A. MAGNETIC RESONANCE IMAGE (MRI) AT ABOUT T2–3 (FIGURE 7-7)

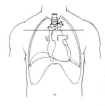

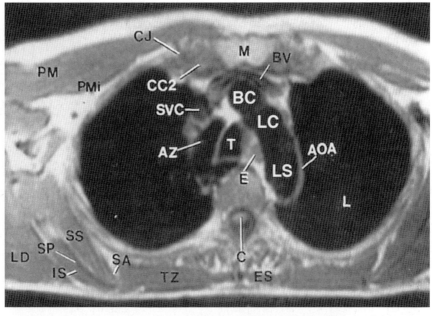

AOA = aortic arch	LC = left common carotid artery
AZ = azygos vein	LS = left subclavian artery
BC = brachiocephalic artery	M = manubrium
BV = left brachiocephalic vein	PM = pectoralis major
C = spinal cord	PMi = pectoralis minor
CC2 = second costal cartilage	SA = serratus anterior muscle
CJ = costochondral junction	SP = scapula
E = esophagus	SS = subscapularis muscle
ES = erector spinae muscle	SVC = superior vena cava
IS = infraspinatus muscle	T = trachea
L = lung	TZ = trapezius muscle

● **Figure 7-7 Magnetic resonance imaging scan at about T2–3.** The line diagram shows the level of the cross section.

B. MRI AT ABOUT T5–6 (FIGURE 7-8)

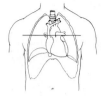

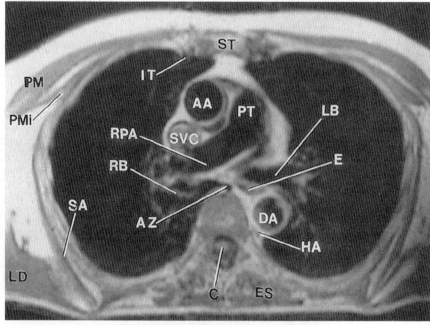

AA = ascending aorta	LD = latissimus dorsi muscle
AZ = azygos vein	PM = pectoralis major muscle
C = spinal cord	PMi = pectoralis minor muscle
DA = descending aorta	PT = pulmonary trunk
E = esophagus	RB = right main bronchus
ES = erector spinae muscle	RPA = right pulmonary artery
HA = hemiazygos vein	SA = serratus anterior muscle
IT = internal thoracic vein	ST = sternum
LB = left main bronchus	SVC = superior vena cava

● Figure 7-8 **Magnetic resonance imaging scan at about T5–6 (at the level of the origin of the right pulmonary artery).** The line diagram shows the level of the cross section.

C. MRI AT ABOUT T7–8 (FIGURE 7-9)

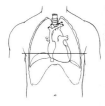

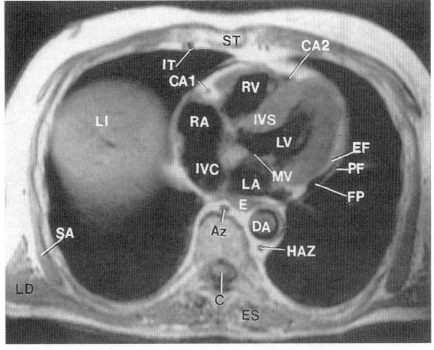

AZ = azygos vein	IVS = interventricular septum
C = spinal cord	LA = left atrium
CA1 = right coronary artery	LD = latissimus dorsi muscle
CA2 = left anterior descending artery	LI = liver
DA = descending aorta	LV = left ventricle
E = esophagus	MV = mitral valve
EF = epicardial artery	PF = pericardial fat
ES = erector spinae muscle	RA = right atrium
FP = fibrous pericardium	RV = right ventricle
HAZ = hemiazygos vein	SA = serratus anterior muscle
IT = internal thoracic artery	ST = sternum
IVC = inferior vena cava	

● **Figure 7-9 Magnetic resonance imaging scan at about T7–8.** The line diagram shows the level of the cross section. It is important to know the arrangement of the heart chambers in the anteroposterior direction.

Case Study 7-1

Karen, a 33-year-old fashion buyer, comes to your office complaining that "sometimes I get these sudden pains in my chest right behind the sternum and sometimes I feel pain around the left shoulder and my jaw. They seem to come and go." She also tells you that "yesterday I ran to catch a bus and the pain was real bad but stopped after I sat down on the bus, and I can't even climb one flight of stairs anymore." After some discussion, she informs you that she is a nonsmoker and does not take oral contraceptives, her father died at age 40 due to a myocardial infarction, and her older sister has just undergone coronary bypass surgery.

Relevant Physical Exam Findings

- Blood pressure = 125/82 mm Hg
- Pulse = 70 and regular
- A soft systolic murmur in the aortic area
- No evidence of left ventricular hypertrophy
- Carotid pulses normal
- Prominent corneal arcus lipoides
- Swelling in the tendons of the hands
- Thickening of both Achilles tendons

Relevant Lab Findings

- Cholesterol = >450 mg/dL
- Low-density lipoprotein (LDL) = 250 mg/dL
- Angiogram shows occlusion of the anterior interventricular artery (or the left anterior descending artery [LAD])

Diagnosis

Angina Pectoris

- Angina pectoris results from a mismatch between the supply and demand of myocardial oxygen.
- The patient's familial history of premature coronary artery disease suggests familial hyper-cholesterolemia (FH). The findings of prominent corneal arcus lipoides, swelling in the tendons of the hands, and thickening of both Achilles tendons suggest that a lipid disorder is present in this patient.
- FH is an autosomal dominant genetic disorder caused by more than 400 different mutations in the **LDLR gene** on chromosome 19p13.1-13.3 for the **low-density lipoprotein receptor**, which binds LDL and delivers LDL into the cell cytoplasm.
- In FH, **3-hydroxy-3-methylglutaryl-CoA (HMG-CoA) reductase** activity and de novo cholesterol biosynthesis are unchecked.
- Cardiac referred pain is a phenomenon whereby pain originating in the heart is sensed by the person as originating from a superficial part of the body (e.g., upper left limb). The afferent (sensory) pain fibers from the heart run centrally in the thoracic cardiac branches of the sympathetic trunk and enter spinal cord segments T1–5 (dermatomes T1–5), especially on the left side. Upon entering the spinal cord, afferent neurons travel to higher brain centers in the central nervous system and some afferent neurons synapse on interneurons that cross to the contralateral side.

Case Study 7-2

A 58-year-old man comes to the emergency room in obvious distress complaining that "I've been having pain in my chest for the last 8 hours or so and I can't take it anymore. The pain feels heavy and very tight. I've never felt anything like this before." He also tells you that "sometimes it feels like the pain goes into my left arm and shoulder. And, I don't know if this is related, but I feel nauseated and I'm vomiting like I ate some bad food." After some discussion, you learn that the man is a heavy smoker (two packs of cigarettes per day for 40 years), does not exercise at all, and is obviously overweight.

Relevant Physical Exam Findings

- Diaphoretic (sweating)
- Tachycardia
- Acute distress

Relevant Lab Findings

- Troponin I = 4 ng/mL (high)
- CK-MB = 5 U/L (high)
- Amylase = 50 U/L (normal)
- Lipase = 70 mIU/mL (normal)
- Electrocardiogram (ECG) = ST-segment elevation in leads V_4, V_5, and V_6

Diagnosis

Acute Myocardial Infarction (MI)

- MI is characterized by two distinct patterns of myocardial necrosis: transmural and subendocardial.
- Transmural infarction is necrosis that traverses the entire ventricular wall and is associated with coronary artery occlusion. Immediate ECG changes show elevation of the ST segment. Q waves are permanent evidence of a transmural MI and rarely disappear over time.
- Subendocardial infarction is necrosis limited to the interior one third of the ventricular wall. ECG changes show ST-segment depression. Q waves are absent.
- Some differentials include angina pectoris, aortic dissection, esophageal spasm, pancreatitis, pericarditis, and pulmonary embolism.

Case Study 7-3

A 17-year-old young man is involved in a sports-related collision and is brought into the emergency room. The emergency medical technicians inform you that the patient was hit by a fastball in the middle of the chest while he was at bat. The patient immediately dropped to the ground and was unresponsive. CPR was initiated by the coach after no pulses were palpated. The emergency medical technicians arrived on the scene 15 minutes later and noted the patient to have ventricular fibrillation on a rhythm strip. A 200-joule countershock was administered, which converted the ventricular fibrillation to a normal sinus rhythm and the patient regained consciousness. The patient was transported to the emergency room.

Relevant Physical Findings

- Mild ecchymosis on the anterior chest wall
- Blood pressure = 130/72 mm Hg
- Heart rate = 106 bpm
- Normal cardiac rhythm
- Survey of airway, breathing, and circulation is unremarkable
- Respirations = 30 breaths/min
- Oxygen saturation is 82% breathing room air and corrects to 98% on a nonrebreather mask
- Patient is alert
- Lungs are clear to auscultation bilaterally

Relevant Lab Findings

- Initial troponin I = 0.04 ng/mL (normal)
- Metabolic panel is normal
- 12-lead electrocardiogram (ECG) shows sinus tachycardia = 110 bpm
- QRS complex, QT interval, ST/T waves, and P waves are normal

Diagnosis

Commotio Cordis (Concussion of the Heart)

- Commotio cordis is an instantaneous cardiac arrest produced by a witnessed, nonpenetrating blow to the chest in the absence of any preexisting heart disease.
- The threshold speed of impact at which a standard baseball can cause ventricular fibrillation is 25 to 30 mph.
- The impact must be directly over the cardiac silhouette near or just to the left of the sternum in order to cause ventricular fibrillation.
- The impact must be delivered 10 to 30 milliseconds before the peak of the T wave in the cardiac cycle.
- In documented sports-related cases of commotio cordis, if CPR is delayed by more than 3 minutes of impact, only 3% of patients survive.
- Differentials: hypertrophic obstructive cardiomyopathy.

Chapter 8

Abdominal Wall

Ⅰ **Abdominal Regions (Figure 8-1).** The abdomen can be topographically divided into nine regions, namely, the **right hypochondriac, epigastric, left hypochondriac, right lumbar, umbilical, left lumbar, right inguinal, hypogastric,** and **left inguinal.**

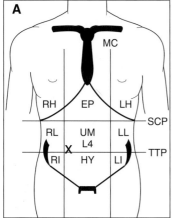 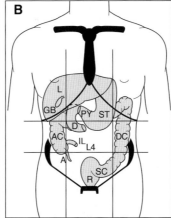 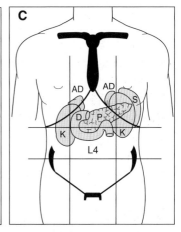

A = appendix	MC = midclavicular lines
AC = ascending colon	P = pancreas
AD = suprarenal gland	PY = pylorus
D = duodenum	R = rectum
DC = descending colon	RH = right hypochondriac
EP = epigastric	RI = right inguinal
GB = gallbladder	RL = right lumbar
HY = hypogastric	S = spleen
IL = ileum	SC = sigmoid colon
K = kidneys	ST = stomach
L = liver	SCP = subcostal plane
LH = left hypochondriac	TTP = transtubercular plane
LI = left inguinal	UM = umbilical
LL= left lumbar	X = McBurney point
L4 = vertebral level lumbar 4	

● **Figure 8-1 Abdominal regions. A:** Subdivisions of the abdomen. A commonly used clinical method for subdividing the abdomen into specific regions using the subcostal plane, transtubercular plane (joining the tubercles of the iliac crests), and midclavicular lines. **B:** Surface projections. Surface projection of the stomach, pylorus, duodenum, liver, gallbladder, ascending colon, appendix, ileum, descending colon, sigmoid colon, and rectum. **C:** Surface projections. Surface projection of the duodenum, pancreas, kidneys, suprarenal gland, and spleen. Many clinical vignette questions will describe pain associated with a particular region of the abdomen. Knowing what viscera are associated with each region will help in deciphering the clinical vignette (e.g., pain in the right lumbar region may be associated with appendicitis).

Ⅱ **Clinical Procedure.** **Paracentesis (Figure 8-2)** is a procedure whereby a needle is inserted through the layers of the abdominal wall to withdraw excess peritoneal fluid. Knife wounds to the abdomen will also penetrate the layers of the abdominal wall.

A. MIDLINE APPROACH. The needle or knife will pass through the following structures in succession: skin → superficial fascia (Camper and Scarpa) → linea alba → transversalis fascia → extraperitoneal fat → parietal peritoneum.

B. FLANK APPROACH. The needle or knife will pass through the following structures in succession: skin → superficial fascia (Camper and Scarpa) → external oblique muscle → internal oblique muscle → transverse abdominis muscle → transversalis fascia → extraperitoneal fat → parietal peritoneum.

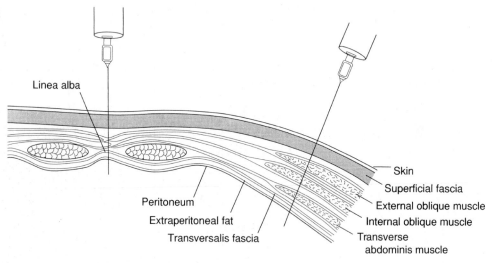

● **Figure 8-2 Anterior abdominal wall.** A transverse section through the anterior abdominal wall demonstrating the various layers that would be penetrated by a needle during paracentesis or a knife wound in a midline or flank approach.

Ⅲ **Inguinal Region (Figure 8-3).** The inguinal region is an area of weakness of the anterior abdominal wall due to the penetration of the testes and spermatic cord (in males) or the round ligament of the uterus (in females) during embryologic development.

A. INGUINAL LIGAMENT is the coiled lower border of the **external oblique muscle** and extends from the anterior-superior iliac spine to the pubic tubercle.

B. DEEP INGUINAL RING is an oval opening in the **transversalis fascia** located lateral to the inferior epigastric artery.

C. SUPERFICIAL INGUINAL RING is a triangular defect of the **external oblique muscle** located lateral to the pubic tubercle.

D. INGUINAL CANAL begins at the deep inguinal ring and ends at the superficial inguinal ring and transmits the **spermatic cord** (in males) or **round ligament of the uterus** (in females).

E. TYPES OF HERNIAS (FIGURE 8-3)
  1. **Direct inguinal hernia**
  2. **Indirect inguinal hernia**
  3. **Femoral hernia**
  4. **Surgical Repair.** Surgical hernia repair may damage the **iliohypogastric nerve,** causing anesthesia of the ipsilateral abdominal wall and inguinal region, and/or the **ilioinguinal nerve,** causing anesthesia of the ipsilateral penis, scrotum, and medial thigh.

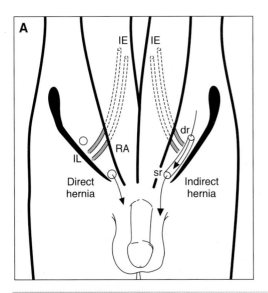 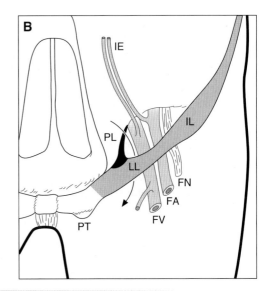

HERNIA CHARACTERISTICS

Type of Hernia	Characteristics
Direct inguinal	Protrudes directly through the anterior abdominal wall within the Hesselbach triangle[a] Protrudes *medial* to the inferior epigastric artery and vein[b] Common in *older* men; rare in women Clinical signs include mass in inguinal region that protrudes on straining and disappears at rest (i.e., easily reduced), constipation, prostate enlargement, and felt with pulp of finger
Indirect inguinal	Protrudes through the deep inguinal ring to enter the inguinal canal and may exit through the superficial inguinal ring into the scrotum Protrudes *lateral* to the inferior epigastric artery and vein[b] Protrudes *above* and *medial* to the pubic tubercle[c] Common in *young* men More common than a direct inguinal hernia Clinical signs include tender painful mass in the inguinal region that continues into the scrotum, and felt with the tip of the finger
Femoral	Protrudes through the femoral canal below the inguinal ligament Protrudes *below* and *lateral* to the pubic tubercle[c] Protrudes medial to the femoral vein More common in women on the right side Prone to early strangulation

[a]Hesselbach's (inguinal) triangle is bounded laterally by the inferior epigastric artery and vein, medially by the rectus abdominus muscle, and inferiorly by the inguinal ligament.
[b]Distinguishing feature of a direct hernia versus an indirect hernia.
[c]Distinguishing feature of an indirect hernia versus a femoral hernia.

FA = femoral artery
FN = femoral nerve
FV = femoral vein
IE = inferior epigastric artery and vein
IL = inguinal ligament
LL = lacunar ligament

PL = pectineal (Cooper) ligament
PT = pubic tubercle
RA = rectus abdominis muscle
dr = deep inguinal ring
sr = superficial inguinal ring

● **Figure 8-3 Inguinal hernias. A:** A schematic demonstrating the anatomy associated with a direct and indirect inguinal hernia. **B:** A schematic demonstrating the anatomy associated with a femoral hernia.

The Scrotum (Figure 8-4)

A. GENERAL FEATURES. The scrotum is an outpouching of the lower abdominal wall whereby layers of the abdominal wall continue into the scrotal area to cover the spermatic cord and testes.

B. CLINICAL CONSIDERATIONS
1. **Cancer of the scrotum** will metastasize to **superficial inguinal nodes**.
2. **Cancer of the testes** will metastasize to **deep lumbar nodes** due to the embryologic development of the testes within the abdominal cavity and subsequent descent into the scrotum.
3. **Extravasated urine** from a saddle injury will be found within the **superficial perineal space** located between the Colles fascia and dartos muscle (layer #2) and the external spermatic fascia (layer #3).
4. In a **vasectomy**, the scalpel will cut through the following layers in succession: skin → Colles fascia and dartos muscle → external spermatic fascia → cremasteric fascia and muscle → internal spermatic fascia → extraperitoneal fat. The tunica vaginalis is not cut since it is present only over the anterior aspect of the testes.
5. **Cremasteric Reflex.** Stroking the skin of the superior and medial thigh stimulates sensory fibers that run with the **ilioinguinal nerve** and serve as the afferent limb of the cremasteric reflex. Motor fibers that run with the **genital branch of the genitofemoral nerve** are distributed to the cremasteric muscle, where they cause contraction of the cremasteric muscle, thereby elevating the testis (i.e., the efferent limb of the cremasteric reflex).

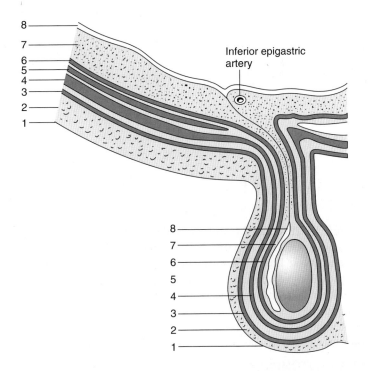

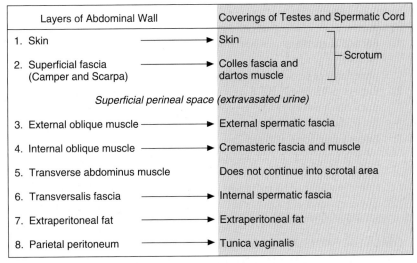

Layers of Abdominal Wall	Coverings of Testes and Spermatic Cord
1. Skin ⟶	Skin
2. Superficial fascia (Camper and Scarpa) ⟶	Colles fascia and dartos muscle
Superficial perineal space (extravasated urine)	
3. External oblique muscle ⟶	External spermatic fascia
4. Internal oblique muscle ⟶	Cremasteric fascia and muscle
5. Transverse abdominus muscle	Does not continue into scrotal area
6. Transversalis fascia ⟶	Internal spermatic fascia
7. Extraperitoneal fat ⟶	Extraperitoneal fat
8. Parietal peritoneum ⟶	Tunica vaginalis

● **Figure 8-4 The scrotum.** A schematic showing the layers of the abdominal wall continuing into the scrotal area as the coverings of the spermatic cord and testes. Note that the transverse abdominis muscle does not continue into the scrotal area, but instead joins with the tendon of the internal oblique muscle to form the conjoint tendon. Extravasated urine due to a straddle injury will leak between layers #2 and #3.

Chapter 9

Peritoneal Cavity

I **Peritoneal Cavity (Figure 9-1)** is a potential space between the visceral and parietal peritoneum. It is divided into the lesser peritoneal sac and greater peritoneal sac.

A. LESSER PERITONEAL SAC (OMENTAL BURSA) is an irregular-shaped sac that communicates with the greater peritoneal sac via the **omental (Winslow) foramen.** The lesser peritoneal sac forms due to the 90-degree clockwise rotation of the stomach during embryologic development. The boundaries of the lesser peritoneal sac are:
1. **Anterior:** the liver, stomach, and lesser omentum
2. **Posterior:** the diaphragm
3. **Right side:** liver
4. **Left side:** gastrosplenic and splenorenal ligaments

B. GREATER PERITONEAL SAC is the remainder of the peritoneal cavity and extends from the diaphragm to the pelvis. The greater peritoneal sac contains a number of pouches, recesses, and paracolic gutters through which peritoneal fluid circulates.
1. **Paracolic gutters** are channels that run along the ascending and descending colon. Normally, peritoneal fluid flows **upward** through the paracolic gutters to the **subphrenic recess**, where it enters the lymphatics associated with the diaphragm.
2. **Excess peritoneal fluid** due to peritonitis or ascites flows downward through the paracolic gutters to the **rectovesical pouch** (in males) or the **rectouterine pouch** (in females) when the patient is in a **sitting** or **standing position.**
3. **Excess peritoneal fluid** due to peritonitis or ascites flows upward through the paracolic gutters to the **subphrenic recess** and the **hepatorenal recess** when the patient is in the **supine position.** The patient may complain of shoulder pain (referred pain) due to irritation of the phrenic nerve (C3, C4, and C5 nerve roots). The hepatorenal recess is the **lowest** part of the peritoneal cavity when the patient is in the supine position.

C. OMENTAL (WINSLOW) FORAMEN is the opening (or connection) between the lesser peritoneal sac and greater peritoneal sac. If a surgeon places his or her finger in the omental foramen, the inferior vena cava (IVC) will lie posterior and the portal vein will lie anterior.

II **Omentum**

A. LESSER OMENTUM is a fold of peritoneum that extends from the porta hepatis of the liver to the lesser curvature of the stomach. It consists of the **hepatoduodenal ligament** and **hepatogastric ligament.** The **portal triad** lies in the free margin of the hepatoduodenal ligament and consists of the:
1. **Portal vein** lying posterior
2. **Common bile duct** lying anterior and to the right
3. **Hepatic artery** lying anterior and to the left

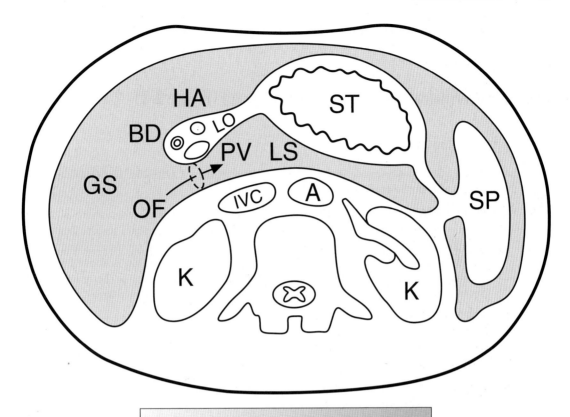

A = aorta	LO = lesser omentum
BD = common bile duct	LS = lesser peritoneal sac
GS = greater peritoneal sac	OF = omental foramen
HA = hepatic artery	PV = portal vein
IVC = inferior vena cava	SP = spleen
K = kidney	ST = stomach

INTRAPERITONEAL AND RETROPERITONEAL VISCERA

Intraperitoneal	Retroperitoneal
• Stomach	• Distal 3 cm of the superior part of the duodenum
• First 2 cm of the superior part of the duodenum (duodenal cap)	• Descending part of the duodenum
• Jejunum	• Horizontal part of the duodenum
• Ileum	• Ascending part of the duodenum
• Cecum	• Ascending colon
• Appendix	• Descending colon
• Transverse colon	• Rectum
• Sigmoid colon	• Head, neck, body of pancreas
• Liver	• Kidneys
• Gallbladder	• Ureters
• Tail of pancreas	• Suprarenal gland
• Spleen	• Abdominal aorta
	• Inferior vena cava

● **Figure 9-1 Cross section of the abdomen demonstrating the peritoneal cavity.** Note the greater peritoneal sac and lesser peritoneal sac connected by the omental foramen (*arrow*). The portal triad is shown at the free margin of the hepatoduodenal ligament of the lesser omentum.

B. GREATER OMENTUM is a fold of peritoneum that hangs down from the greater curvature of the stomach. It is known as the "abdominal policeman" because it adheres to areas of inflammation.

III Intraperitoneal and Extraperitoneal Viscera (Figure 9-1)

IV Clinical Considerations

A. ASCITES is an accumulation of fluid in the peritoneal cavity due to peritonitis from congestion of the venous drainage of the abdomen.

B. INFLAMMATION OF THE PARIETAL PERITONEUM occurs when there is an enlarged visceral organ or by escape of fluid from a visceral organ and results in a sharp, localized pain over the inflamed area. Patients exhibit rebound tenderness and guarding over the site of inflammation. **Rebound tenderness** is pain that is elicited after the pressure of palpation over the inflamed area is removed. **Guarding** is the reflex spasms of the abdominal muscles in response to palpation over the inflamed area.

C. PERITONITIS is inflammation and infection of the peritoneum and commonly occurs due to a burst appendix, a penetrating abdominal wound, a perforated ulcer, or poor sterile technique during surgery. Peritonitis is treated by rinsing the peritoneal cavity with large amounts of sterile saline and administering antibiotics.

D. PERITONEAL ADHESIONS occur after abdominal surgery, whereby scar tissue forms and limits the normal movement of the viscera. This tethering may cause chronic pain or emergency complications such as volvulus (i.e., twisting of the intestines).

Chapter 10

Abdominal Vasculature

❶ Abdominal Aorta (Figure 10-1)

A. MAJOR BRANCHES

1. **Celiac trunk** is located at the **T12** vertebral level and supplies viscera that derive embryologically from the **foregut** (i.e., intra-abdominal portion of esophagus, stomach, upper part of duodenum, liver, gallbladder, and pancreas). It further branches into the:
   a. **Left gastric artery**
   b. **Splenic artery**
   c. **Common hepatic artery**

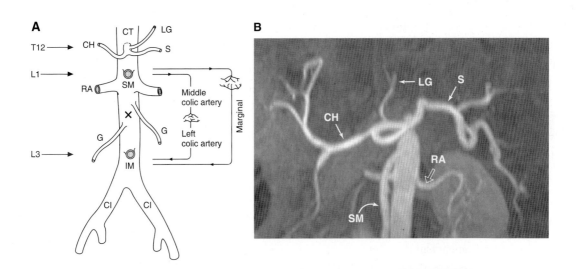

CH = common hepatic artery
CI = common iliac artery
CT = celiac trunk
G = gonadal artery
IM = inferior mesenteric artery
LG = left gastric artery

RA = renal artery
S = splenic artery
SM = superior mesenteric artery
T12, L1, and L3 indicate the vertebral level of the various branches

● **Figure 10-1 Abdominal aorta. A:** Diagram shows the major branches of the abdominal aorta. The abdominal vasculature has a fairly robust collateral circulation. Any blockage (see X) between the superior mesenteric artery (SM; at L1 vertebral level) and inferior mesenteric artery (IM; at L3 vertebral level) will cause blood to be diverted along two routes of collateral circulation. The first route uses the middle colic artery (a branch of the SM), which anastomoses with the left colic artery (a branch of the IM). The second route uses the marginal artery. **B:** An arteriogram showing the branches of the celiac trunk and other arteries in the vicinity.

2. **Superior mesenteric artery** is located at the **L1** vertebral level and supplies viscera that derive embryologically from the **midgut** (i.e., lower part of duodenum, jejunum, ileum, cecum, appendix, ascending colon, proximal two thirds of transverse colon).
3. **Renal arteries** supply the kidneys.
4. **Gonadal arteries** supply the testes or ovary.
5. **Inferior mesenteric artery** is located at the L3 vertebral level and supplies viscera that derive embryologically from the **hindgut** (i.e., distal one third of transverse colon, descending colon, sigmoid colon, upper portion of rectum).
6. **Common iliac arteries** are the terminal branches of the abdominal aorta.

B. **CLINICAL CONSIDERATIONS**
1. **Abdominal Aortic Aneurysm (AAA) (Figure 10-2).** AAA is most commonly seen in atherosclerotic elderly males below the L1 vertebral level (i.e., below the renal arteries and superior mesenteric artery). The most common site of a ruptured AAA is below the renal arteries in the **left posterolateral wall** (i.e., retroperitoneal). In a patient with a ruptured AAA, the first step is immediate compression of the aorta against the vertebral bodies **above the celiac trunk.** During a transabdominal surgical approach to correct a ruptured AAA, the **left renal vein** is put in jeopardy. The **inferior mesenteric artery** generally lies in the middle of an AAA. Clinical findings include sudden onset of severe, central abdominal pain, which may radiate to the back, and a pulsatile tender abdominal mass; **if rupture occurs, hypotension and delirium may occur.** Surgical complications include **ischemic colitis** due to ligation of the inferior mesenteric artery and **spinal cord ischemia** due to ligation of the great radicular artery (of Adamkiewicz). Diagram shows an abdominal aortic aneurysm.

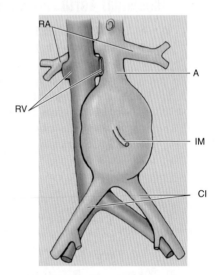

● **Figure 10-2 Abdominal aortic aneurysm.** A = aorta, CI = common iliac artery, IM = inferior mesenteric artery, RA = renal artery, RV = right ventricular artery.

2. **Acute mesenteric ischemia** is most commonly caused by an embolism within the **superior mesenteric artery.** Clinical signs include severe abdominal pain out of proportion to physical findings and no evidence of peritonitis; acute mesenteric ischemia usually occurs in elderly patients with a history of heart disease who are taking digoxin (a potent splanchnic vasoconstrictor).
3. **Gradual occlusion** is most commonly seen in atherosclerotic patients at the bifurcation of the abdominal aorta. It may result in **claudication** (i.e., pain in the legs when walking) and **impotence** due to the lack of blood to the internal iliac arteries.

II Venous Drainage of the Abdomen (Figure 10-3)

A. AZYGOS VENOUS SYSTEM

1. **Azygos Vein**

   a. The unpaired **azygos vein** is formed by the union of the **right ascending lumbar vein** and the **right subcostal vein.**

   b. The lower end of the azygos vein communicates with the **inferior vena cava (IVC).**

   c. The azygos vein ascends on the right side of the vertebral column and forms a collateral pathway from the IVC to the superior vena cava (SVC).

   d. The azygos vein communicates with the **posterior intercostal veins** and the **external and internal vertebral venous plexuses.**

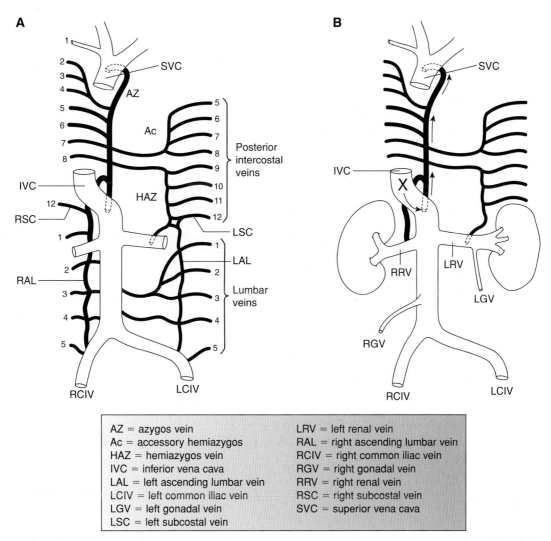

AZ = azygos vein	LRV = left renal vein
Ac = accessory hemiazygos	RAL = right ascending lumbar vein
HAZ = hemiazygos vein	RCIV = right common iliac vein
IVC = inferior vena cava	RGV = right gonadal vein
LAL = left ascending lumbar vein	RRV = right renal vein
LCIV = left common iliac vein	RSC = right subcostal vein
LGV = left gonadal vein	SVC = superior vena cava
LSC = left subcostal vein	

● **Figure 10-3 Azygos venous system and inferior vena cava (IVC). A:** Diagram shows the diffuse origin of the azygos, hemiazygos, and accessory hemiazygos veins along with the anatomic relationship to the IVC. The lower end of the azygos vein communicates with the IVC. The lower end of the hemiazygos vein communicates with the renal vein. **B:** Diagram shows how the azygos vein provides a route of collateral venous return (*arrows*) to the superior vena cava (SVC) in cases where the IVC is blocked (see X). Note the differences in drainage of the right and left gonadal veins.

2. **Hemiazygos Vein**
   a. The unpaired **hemiazygos vein** is formed by the union of the **left ascending lumbar vein** and the **left subcostal vein.**
   b. The lower end of the hemiazygos vein communicates with the **renal vein.**
   c. The hemiazygos vein ascends on the left side of the vertebral column and crosses to the left side at vertebral level T9 to join the azygos vein.

B. INFERIOR VENA CAVA
   1. The IVC is formed by the union of the **right and left common iliac veins** at vertebral level L5.
   2. The IVC drains all the blood from below the diaphragm (even portal blood from the gastrointestinal (GI) tract after it percolates through the liver) to the right atrium.
   3. The IVC is in jeopardy during surgical repair of a herniated intervertebral disc.
   4. The IVC above the kidneys (suprarenal) should never be ligated (there is a 100% mortality rate).
   5. The IVC below the kidneys (infrarenal) may be ligated (there is a 50% mortality rate).
   6. The **right gonadal vein** drains directly into the IVC, whereas the **left gonadal vein** drains into the left renal vein.
      a. This is important in females, where the appearance of a **right-side hydronephrosis** may indicate thrombosis of the right ovarian vein that constricts the ureter, since the right ovarian vein crosses the ureter to drain into the IVC.
      b. This is also important males, where the appearance of a **left-side testicular varicocele** may indicate occlusion of the **left testicular vein** and/or **left renal vein** due to a malignant tumor of the kidney.
   7. **Routes of collateral venous return** exist in case the IVC is blocked by either a malignant retroperitoneal tumor or a large blood clot (thrombus). These include:
      a. Azygos vein → superior vena cava → right atrium
      b. Lumbar veins → external and internal vertebral venous plexuses → cranial dural sinuses → internal jugular vein → right atrium

Ⅲ Hepatic Portal System (Figure 10-4)

A. In general, the term "portal" refers to a vein interposed between two capillary beds (i.e., capillary bed → vein → capillary bed).

B. The hepatic portal system consists specifically of the following vascular structures: capillaries of GI tract → portal vein → hepatic sinusoids.

C. The **portal vein** is formed posterior to the neck of the pancreas by the union of the **splenic vein** and **superior mesenteric vein.** The **inferior mesenteric vein** usually ends by joining the splenic vein. The blood within the portal vein carries high levels of nutrients from the GI tract and products of red blood cell destruction from the spleen.

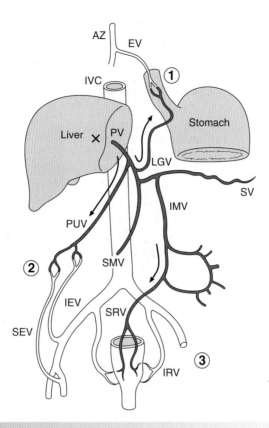

CLINICAL SIGNS OF PORTAL HYPERTENSION

Site of Anastomosis	Clinical Sign	Veins Involved in Portal ↔ Inferior Venal Caval Anastomosis
Esophagus (site 1)	Esophageal varices	Left gastric vein ↔ esophageal vein
Umbilicus (site 2)	Caput medusa	Paraumbilical vein ↔ superficial and inferior epigastric veins
Rectum (site 3)	Anorectal varices	Superior rectal vein ↔ middle and inferior rectal veins

AZ = azygos vein	PUV = paraumbilical vein
EV = esophageal vein	PV = portal vein
IEV = inferior epigastric vein	SEV = superficial epigastric vein
IMV = inferior mesenteric vein	SMV = superior mesenteric vein
IRV = inferior rectal vein	SRV = superior rectal vein
IVC = inferior vena cava	SV = splenic vein
LGV = left gastric vein	

● **Figure 10-4 Hepatic portal system.** Diagram shows the three main sites (1, 2, 3) of portal inferior venal cava (IVC; caval) anastomosis. In case of portal hypertension where blood flow through the liver is severely reduced (see X), these anastomoses provide collateral circulation (*arrows*) through the IVC back to the heart. Table shows the various clinical signs of portal hypertension.

D. CLINICAL CONSIDERATION: PORTAL HY-PERTENSION (FIGURE 10-5). Portal IVC (caval) anastomosis becomes clinically relevant when **portal hypertension** occurs. Portal hypertension will cause blood within the portal vein to reverse its flow and enter the IVC in order to return to the heart. There are three main sites of portal IVC anastomosis: **esophagus**, **umbilicus**, and **rectum**. Clinical signs of portal hypertension include vomiting copious amounts of blood, enlarged abdomen due to ascites fluid, and splenomegaly. Portal hypertension may be caused by alcoholism, liver cirrhosis, and schistosomiasis. The photograph shows an elderly man with portal hypertension demonstrating caput medusae.

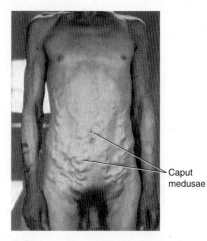

Caput medusae

● **Figure 10-5 Portal hypertension demonstrating caput medusae.**

Chapter 11

Abdominal Viscera

① Esophagus (Figure 11-1)

A. GENERAL FEATURES

1. The esophagus begins at the **cricoid cartilage** (at vertebral level C6) and ends at the **gastroesophageal junction**. The esophagus pierces the diaphragm through the **esophageal hiatus** (at vertebral level T10).

2. The **upper 5%** of the esophagus consists of *skeletal muscle* only. The **middle 45%** of the esophagus consists of both **skeletal muscle and smooth muscle** interwoven together. The **distal 50%** of the esophagus consists of *smooth muscle* only.

3. In clinical practice, endoscopic distances are measured from the incisor teeth, and in the average male the gastroesophageal junction is 38 to 43 cm away from the incisor teeth.

4. For purposes of classification, staging, and reporting of esophageal malignancies, the esophagus is divided into four segments based on the distance from the incisor teeth: **cervical segment**, **upper thoracic segment**, **midthoracic segment**, and **lower thoracic segment**.

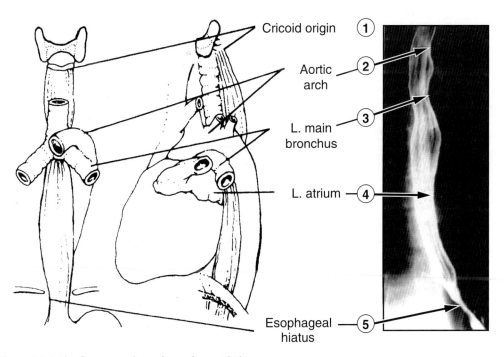

Cricoid origin ①

Aortic arch ②

L. main bronchus ③

L. atrium ④

Esophageal hiatus ⑤

● Figure 11-1 The five normal esophageal constrictions.

B. CONSTRICTIONS. There are five main sites where the esophagus is constricted:
1. The junction of the pharynx and esophagus (cricoid origin)
2. At the aortic arch
3. At the tracheal bifurcation (vertebral level T4) where the left main bronchus crosses the esophagus
4. At the left atrium
5. At the esophageal hiatus

C. SPHINCTERS
1. **Upper Esophageal Sphincter (UES)**
    a. The UES is *skeletal muscle* that separates the pharynx from the esophagus.
    b. The UES is composed of **opening muscles** (i.e., thyrohyoid and geniohyoid muscles) and **closing muscles** (i.e., inferior pharyngeal constrictor and **cricopharyngeus** [main player]).
2. **Lower Esophageal Sphincter (LES)**
    a. The LES is *smooth muscle* that separates the esophagus from the stomach.
    b. The LES **prevents gastroesophageal reflux.**

D. ARTERIAL SUPPLY
1. The arterial supply of the **cervical esophagus** is from the **inferior thyroid arteries** (subclavian artery → thyrocervical trunk → inferior thyroid artery), which give off ascending and descending branches that anastomose with each other across the midline.
2. The arterial supply of the **thoracic esophagus** is from **four to five branches from the descending thoracic aorta.**
3. The arterial supply of the **abdominal esophagus** is from the **left gastric artery** (abdominal aorta → celiac trunk → left gastric artery).

E. VENOUS DRAINAGE
1. The venous drainage of the **cervical esophagus** is to the **inferior thyroid veins** (inferior thyroid veins → brachiocephalic veins → superior vena cava).
2. The venous drainage of the **thoracic esophagus** is to an **esophageal plexus of veins** (esophageal plexus of veins → azygous veins → superior vena cava).
3. The venous drainage of the **abdominal esophagus** is to the **left gastric vein** (left gastric vein → portal vein → hepatic sinusoids → central veins → hepatic veins → inferior vena cava).

F. INNERVATION. The innervation of the esophagus is by the **somatic nervous system** (upper portion only) and by the **enteric nervous system,** which in the esophagus consists of the myenteric plexus of Auerbach only. The enteric nervous system is modulated by the parasympathetic and sympathetic nervous systems.
1. **Somatic Innervation**
    a. Somatic neuronal cell bodies are located in the ventral horn of the spinal cord at **cervical level 1 (C1)** and travel with the **hypoglossal nerve (cranial nerve [CN] XII)** to innervate the **opening muscles of the UES (thyrohyoid and geniohyoid muscles).**
2. **Parasympathetic**
    a. Preganglionic neuronal cell bodies are located in the **dorsal nucleus of the vagus.** Preganglionic axons run in **CN X** and enter the **esophageal plexus.**
    b. Postganglionic neuronal cell bodies are located in the **enteric nervous system;** some of these are the "traditional" postganglionic parasympathetic neurons that release acetylcholine (ACh) as a neurotransmitter.
    c. The postganglionic axons terminate on mucosal glands, submucosal glands, and smooth muscle.

    d. Neuronal cell bodies located in the **nucleus ambiguus** send axons that run in CN X (**recurrent laryngeal nerves**) and enter the **esophageal plexus**. These axons terminate on the **closing muscles of the UES** (**inferior pharyngeal constrictor and cricopharyngeus muscle**) and the **esophageal skeletal muscle**.

3. Sympathetic

    a. Preganglionic neuronal cell bodies are located in the **intermediolateral cell column** of the spinal cord (T5–9). Preganglionic axons form the **greater splanchnic nerve**.

    b. Postganglionic neuronal cell bodies are located in **diffuse ganglia** along the esophagus and **celiac ganglion**.

    c. Postganglionic axons synapse in the complex circuitry of the enteric nervous system.

G. CLINICAL CONSIDERATIONS

1. Enlarged left atrium may constrict the esophagus due to their close anatomic relationship.

2. Bronchogenic carcinoma may indent the esophagus due to the enlargement of mediastinal lymph nodes. This indentation can be observed radiologically during a barium swallow.

3. Malignant tumors of the esophagus most commonly occur in the lower one third of the esophagus and metastasize below the diaphragm to the **celiac lymph nodes**.

4. Forceful vomiting is commonly seen in alcoholism, bulimia, and pregnancy and may tear the posterior wall of the esophagus. Clinical findings include severe retrosternal pain after vomiting and extravasated contrast medium. **Mallory-Weiss tears** involve only the mucosal and submucosal layers. **Boerhaave syndrome** involves tears through all layers of the esophagus.

5. Sliding hiatal hernia occurs when the stomach, along with the gastroesophageal junction, herniates through the diaphragm into the thorax. Clinical findings include deep burning retrosternal pain and reflux of gastric contents into the mouth (i.e., heartburn), which are accentuated in the supine position.

6. Paraesophageal hiatal hernia occurs when only the stomach herniates through the diaphragm into the thorax. Clinical findings include no reflux of gastric contents, but strangulation or obstruction may occur.

7. Achalasia is failure of LES to relax during swallowing, probably due to absence of the myenteric plexus. Clinical findings include progressive dysphagia (difficulty swallowing), and the barium swallow shows a dilated esophagus above the LES and distal stenosis at the LES ("**bird beak**"). **Chagas disease** (caused by *Trypanosoma cruzi*) may lead to achalasia.

8. Esophageal reflux is caused by LES dysfunction that allows gastric acid reflux into the lower esophagus. Clinical findings include substernal pain and heartburn, which may worsen with bending or lying down. **Scleroderma** may be a systemic cause of esophageal reflux.

9. Esophageal Strictures (Narrowing). Caustic strictures are caused by ingestion of caustic agents (e.g., drain openers, oven cleaners, etc.). **Other strictures** are caused by recurrent mucosal destruction due to gastric acid reflux. These strictures most often occur at the gastroesophageal junction.

10. Esophageal varices refer to the dilated subepithelial and submucosal venous plexuses of the esophagus that drain into the **left gastric (coronary) vein**. The left gastric vein empties into the portal vein from the distal esophagus and proximal stomach. Esophageal varices are caused by **portal hypertension** due to cirrhosis of the liver.

11. Barrett Esophagus (Figure 11-2). The gastroesophageal (GE) junction in gross anatomy is fairly easy to demarcate. However, the histologic GE junction does *not* correspond to the gross anatomic GE junction. The mucosal lining of the cardiac portion of the stomach **extends about 2 cm into the esophagus** such that the distal 2 cm of the esophagus is lined by a simple columnar epithelium instead of stratified squamous epithelium. The junction where stratified squamous epithelium changes to simple columnar epithelium (or the mucosal GE junction) can be seen macroscopically as a **zig-zag line** (called the **Z-line**). This distinction is very important clinically, especially when dealing with Barrett esophagus. Barrett esophagus can be defined as the replacement of esophageal stratified squamous epithelium with metaplastic "intestinalized" simple columnar epithelium (with Goblet cells) extending **at least 3 cm** into the esophagus. The clinical importance of this metaplastic invasion is that virtually all lower esophageal adenocarcinomas occur as a sequelae. The photograph of the esophagus shows the Z-line where the stratified squamous epithelium (*white portion*) changes to a simple columnar epithelium (*dark portion*). The light micrograph shows the mucosal GE junction where the stratified squamous epithelium (SE) of the esophagus abruptly changes (*arrow*) to simple columnar epithelium (CE), similar to the stomach.

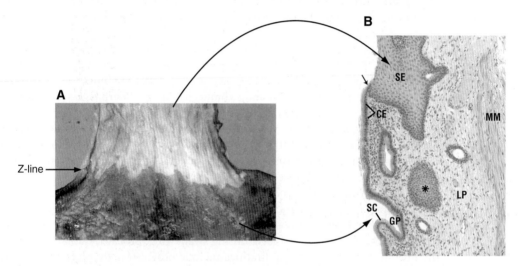

● **Figure 11-2 The mucosal gastroesophageal (GE) junction.** Asterisk indicates the tangential cut of esophageal epithelium. CE = columnar epithelium, GP = gastric pit, LP = lamina propria, MM = muscularis mucosa, SC = surface-lining cells of the stomach, SE = stratified squamous epithelium.

Ⅱ Stomach (Figure 11-3)

A. GENERAL FEATURES. The stomach is divided into four parts:
1. **Cardia** is near the gastroesophageal junction.
2. **Fundus** is above the gastroesophageal junction.
3. **Body** is between the fundus and antrum.
4. **Pylorus**
   a. The pylorus is the distal part of the stomach and is divided into the **pyloric antrum** (wide part) and the **pyloric canal** (narrow part).
   b. The **pyloric orifice** is surrounded by the **pyloric sphincter**, which is a well-defined muscular sphincter that controls movement of food out of the stomach and prevents reflux of duodenal contents into the stomach.

A

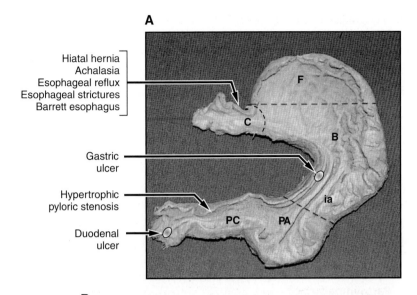

Hiatal hernia
Achalasia
Esophageal reflux
Esophageal strictures
Barrett esophagus

Gastric
ulcer

Hypertrophic
pyloric stenosis

Duodenal
ulcer

B

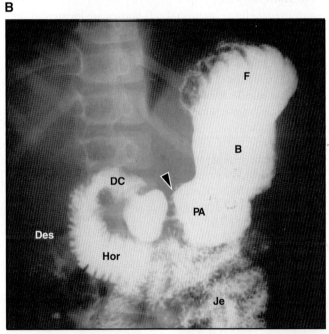

B = body	Hor = horizontal part of duodenum
C = cardia	ia = incisura angularis
DC = duodenal cap or superior part of duodenum	Je = jejunum
Des = descending part of duodenum	PA = pyloric antrum
F = fundus	PC = pyloric canal

● **Figure 11 2 Stomach A:** Photograph of the stomach. Note the various parts of the stomach. High-yield clinical considerations associated with the esophagus, stomach, and duodenum are indicated. **B:** Radiograph after barium swallow. Note the parts of the stomach and duodenum. *Arrowhead* indicates the peristaltic wave.

B. ARTERIAL SUPPLY. The arterial supply of the stomach is from the:
1. **Right and left gastric arteries,** which supply the lesser curvature (abdominal aorta → celiac trunk → common hepatic artery → right gastric artery; abdominal aorta → celiac trunk → left gastric artery)
2. **Right and left gastroepiploic arteries,** which supply the greater curvature (abdominal aorta → celiac trunk → common hepatic artery → gastroduodenal artery

→ right gastroepiploic artery; abdominal aorta → celiac trunk → splenic artery → left gastroepiploic artery)

3. **Short gastric arteries,** which supply the fundus (abdominal aorta → celiac trunk → splenic artery → short gastric arteries)

C. VENOUS DRAINAGE. The venous drainage of the stomach is to the:

1. **Right and left gastric veins** (right and left gastric veins → portal vein → hepatic sinusoids → central veins → hepatic veins → inferior vena cava)

2. **Left gastroepiploic vein** and **short gastric veins** (left gastroepiploic vein and short gastric veins → splenic vein → portal vein → hepatic sinusoids → central veins → hepatic veins → inferior vena cava)

3. **Right gastroepiploic vein** (right gastroepiploic vein → superior mesenteric vein → portal vein → hepatic sinusoids → central veins → hepatic veins → inferior vena cava)

D. INNERVATION. The innervation of the stomach is by the **enteric nervous system,** which in the stomach consists of the myenteric plexus of Auerbach only. The enteric nervous system is modulated by the parasympathetic and sympathetic nervous systems.

1. **Parasympathetic**
   a. Preganglionic neuronal cell bodies are located in the **dorsal nucleus of the vagus.** Preganglionic axons run in CN X and enter the **anterior and posterior vagal trunks.**
   b. Postganglionic neuronal cell bodies are located in the enteric nervous system; some of these are the "traditional" postganglionic parasympathetic neurons that release ACh as a neurotransmitter.
   c. The postganglionic axons terminate on mucosal glands and smooth muscle.

2. **Sympathetic**
   a. Preganglionic neuronal cell bodies are located in the **intermediolateral cell column** of the spinal cord (T5–9). Preganglionic axons form the **greater splanchnic nerve.**
   b. Postganglionic neuronal cell bodies are located in the **celiac ganglion.**
   c. Postganglionic axons synapse in the complex circuitry of the enteric nervous system.

E. CLINICAL CONSIDERATIONS

1. **Gastric ulcers (Figure 11-4A)** most often occur within the **body of the stomach** along the **lesser curvature** above the **incisura angularis.**

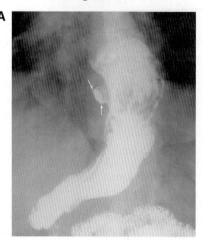
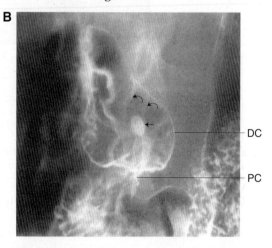

● **Figure 11-4 Gastric and duodenal ulcers. A:** Radiograph shows a gastric ulcer (*arrows*) along the lesser curvature of the stomach. **B:** Radiograph shows a duodenal ulcer (*straight arrow*) located in the duodenal cap (DC). The duodenal mucosal folds (*curved arrows*) radiate toward the ulcer crater. The table shows the comparison of gastric and duodenal ulcers. PC = pyloric canal. (*continued*)

COMPARISON OF GASTRIC AND DUODENAL ULCERS

	Gastric Ulcer	Duodenal Ulcer
% of ulcer cases	25%	75%

Epidemiology	Male-to-female ratio = 1:1
	Increased risk with blood type A
	No association with MEN I or II
	COPD
	Renal failure
		Male-to-female ratio = 2:1
		Increased risk with blood type O
		Associated with Zollinger-Ellison
		  syndrome (MEN)
		Liver cirrhosis or alcoholism
		COPD
		Renal failure
		Hyperparathyroidism
		Family history with an autosomal
		  dominant pattern

Pathogenesis	*Helicobacter pylori* infection in 80% of cases
	Damage to mucosal barrier caused by smoking,
	  salicylate or NSAID ingestion, type B chronic
	  atrophic gastritis, mucosal ischemia
	  because of reduced PGE production, or
	  bile reflux
		H. pylori infection in 95% of cases
		Damage to mucosal barrier
		Gastric acid hypersecretion caused by
		  increased parietal cell mass, increased
		  secretion to stimuli, increased
		  nocturnal secretion, or rapid gastric
		  emptying

Location	Single ulcer within the body of the stomach
	  along the lesser curvature above the
	  incisura angularis
		Single ulcer on the anterior wall of the
		  first part of the duodenum (i.e., at
		  the **duodenal cap**) most common
		Single ulcer on the posterior wall
		  (danger of perforation into the
		  pancreas)

Malignant Potential	No malignant potential Cancer may be associated with a benign ulcer in 1%–3% of cases (biopsy necessary)	No malignant potential

Complications	Bleeding from left gastric artery
	Perforation
	Both are less common than seen in duodenal
	  ulcers
		Bleeding from gastroduodenal artery
		Perforation (air under diaphragm,
		  pain radiates to left shoulder)
		Gastric outlet obstruction
		Pancreatitis

Clinical Findings	Burning epigastric pain **soon after eating** Pain increases with food intake Pain is relieved by antacids Patient is afraid to eat and loses weight	Burning epigastric pain **1–3 hours after eating** Pain decreases with food intake Pain is relieved by antacids Patient does not lose weight Patient wakes at night because of pain

COPD = chronic obstructive pulmonary disease, MEN = multiple endocrine neoplasia, NSAID = nonsteroidal anti-inflammatory drug, PGE = prostaglandin E.

● **Figure 11-4** *Continued.*

2. **Carcinomas of the stomach** are most commonly found in the **pylorus** of the stomach and may metastasize to **supraclavicular lymph nodes (Virchow nodes)** on the left side, which can be palpated within the posterior triangle of the neck. Carcinomas of the stomach may also metastasize to the ovaries, which is called a **Krukenberg tumor.**

 **Duodenum**

A. **GENERAL FEATURES.** The duodenum pursues a C-shaped course around the head of the pancreas. The duodenum is divided into four parts:

1. **Superior Part (First Part)**
   a. The first 2 cm of the superior part is intraperitoneal and therefore has a mesentery and is mobile; the remaining distal 3 cm is retroperitoneal.
   b. Radiologists refer to the first 2 cm of the superior part of the duodenum as the **duodenal cap** or **bulb.**
   c. The superior part begins at the pylorus of the stomach (**gastroduodenal junction**), which is marked by the **prepyloric vein.**
   d. Posterior relationships include the **common bile duct** and **gastroduodenal artery.** The **hepatoduodenal ligament** attaches superiorly and the **greater omentum** attaches inferiorly.

2. **Descending Part (Second Part).** The descending part is retroperitoneal and receives the **common bile duct** and **main pancreatic duct** on its posterior/medial wall at the **hepatopancreatic ampulla (ampulla of Vater).**

3. **Horizontal Part (Third Part)**
   a. The horizontal part is retroperitoneal and runs horizontally across the L3 vertebra between the superior mesenteric artery anteriorly and the aorta and inferior vena cava (IVC) posteriorly.
   b. In severe abdominal injuries, this part of the duodenum may be crushed against the L3 vertebra.

4. **Ascending Part (Fourth Part)**
   a. The ascending part is intraperitoneal and ascends to meet the jejunum at the **duodenojejunal junction,** which occurs approximately at the L2 vertebral level about 2 to 3 cm to the left of the midline.
   b. This junction usually forms an acute angle called the **duodenojejunal flexure,** which is supported by the **ligament of Treitz** (represents the **cranial end of the dorsal mesentery**).
   c. The ligament of Treitz serves as the anatomic landmark for the distinction between **upper and lower gastrointestinal (GI) tract bleeds.**

B. **ARTERIAL SUPPLY.** The arterial supply of the duodenum is from the:

1. **Supraduodenal artery,** which supplies the upper portion of the duodenum (abdominal aorta → celiac trunk → common hepatic artery → gastroduodenal artery → supraduodenal artery)
2. **Anterior and posterior superior pancreaticoduodenal arteries** (abdominal aorta → celiac trunk → common hepatic artery → gastroduodenal artery → anterior and posterior superior pancreaticoduodenal arteries)
3. **Anterior and posterior inferior pancreaticoduodenal arteries** (abdominal aorta → superior mesenteric artery → anterior and posterior inferior pancreaticoduodenal arteries)

C. **VENOUS DRAINAGE.** The venous drainage of the duodenum is to the:

1. **Anterior and posterior superior pancreaticoduodenal veins** (anterior and posterior superior pancreaticoduodenal veins → portal vein → hepatic sinusoids → central veins → hepatic veins → inferior vena cava)
2. **Anterior and posterior inferior pancreaticoduodenal veins** (anterior and posterior inferior pancreaticoduodenal veins → superior mesenteric vein → portal vein → hepatic sinusoids → central veins → hepatic veins → inferior vena cava)

D. **INNERVATION.** See Chapter 11 VI.

E. CLINICAL CONSIDERATIONS

1. **Duodenal Ulcers (Figure 11-4B)** most often occur on the anterior wall of the first part of the duodenum (i.e., at the **duodenal cap**) followed by the posterior wall (danger of perforation into the pancreas).

2. **Perforations of the Duodenum** occur most often with ulcers on the **anterior** wall of the duodenum. Perforations occur less often with ulcers on the **posterior** wall; however, these may erode the **gastroduodenal artery,** causing severe hemorrhage, and perforate into the pancreas. Clinical findings include air under the diaphragm and pain radiating to the left shoulder.

Ⅳ Jejunum

A. GENERAL FEATURES (Table 11-1)

B. ARTERIAL SUPPLY.
The arterial supply of the jejunum is from the **vasa rectae** (abdominal aorta → superior mesenteric artery → jejunal arteries → one to two arterial arcades → vasa rectae).

C. VENOUS DRAINAGE.
The venous drainage of the jejunum is to the **veins associated with the arcades** (veins associated with the arcades → jejunal veins → superior mesenteric vein → portal vein → hepatic sinusoids → central veins → hepatic veins → inferior vena cava).

D. INNERVATION.
See Chapter 11 VI.

TABLE 11-1	GENERAL FEATURES OF THE SMALL AND LARGE INTESTINE	
Jejunum	**Ileum**	**Large Intestine**
Villi present (long, finger shaped)	Villi present (short, club shaped)	Villi absent
Intestinal glands (crypts) present	Intestinal glands (crypts) present	Intestinal glands (crypts) present
>3 cm in diameter	<3 cm in diameter	~6–9 cm in diameter
Large, numerous, and palpable circular folds[1]	Small and few circular folds that disappear distally	No circular folds; inner luminal surface is smooth
Initial two fifths of small intestine	Terminal three fifths of small intestine	
Located in the umbilical region on left side of abdomen	Located in the hypogastric and inguinal regions on the right side of abdomen	
Long vasa recta with one to two arterial arcades	Short vasa recta with three to four arterial arcades	
Main site of nutrient absorption	Site of vitamin B$_{12}$ and H$_2$O/electrolyte absorption	Site of H$_2$O/electrolyte absorption
Often empty (no fecal contents)	Site of bile recirculation Peyer patches prominent	Site where sedatives, anesthetics, and steroids may also be absorbed when medications cannot be delivered orally
Thicker wall, more vascular, and redder in the living person than ileum	Terminal ileum ends several centimeters above the cecal tip	**Taeniae coli** (three longitudinal bands of smooth muscle) are present **Appendices epiploicae** (fatty tags) are present **Haustra** (sacculations of the wall) are present

[1]Are folds of the mucosa and submucosa (also called **plicae circularis**).

Ileum

A. GENERAL FEATURES (see Table 11-1)

B. ARTERIAL SUPPLY. The arterial supply of the jejunum is from the **vasa rectae** (abdominal aorta → superior mesenteric artery → jejunal arteries → one to two arterial arcades → vasa rectae).

C. VENOUS DRAINAGE. The venous drainage of the jejunum is to the **veins associated with the arcades** (veins associated with the arcades → jejunal veins → superior mesenteric vein → portal vein → hepatic sinusoids → central veins → hepatic veins → inferior vena cava).

D. INNERVATION. See Chapter 11 VI.

Innervation of the Small Intestine. The innervation of the small intestine is by the enteric nervous system, which in the small intestine consists of the **submucosal plexus of Meissner** and the **myenteric plexus of Auerbach**. The motor component of the submucosal plexus controls primarily **mucosal and submucosal gland secretion and blood flow**, whereas the sensory component consists of **mucosal mechanosensitive neurons**. The motor component of the myenteric plexus controls primarily **GI motility (contraction/relaxation of GI smooth muscle)**, whereas the sensory component consists of **tension-sensitive neurons** and **chemosensitive neurons**. The enteric nervous system is modulated by the parasympathetic and sympathetic nervous systems.

A. PARASYMPATHETIC
1. Preganglionic neuronal cell bodies are located in the **dorsal nucleus of the vagus**. Preganglionic axons run in CN X and reach the small intestine via the **posterior vagal trunk**.
2. Postganglionic neuronal cell bodies are located in the enteric nervous system; some of these are the "traditional" postganglionic parasympathetic neurons that release ACh as a neurotransmitter.
3. The postganglionic axons terminate on mucosal glands and smooth muscle.

B. SYMPATHETIC
1. Preganglionic neuronal cell bodies are located in the **intermediolateral cell column of the spinal cord (T8–12)**. Preganglionic axons form the **greater splanchnic nerve, lesser splanchnic nerve, and least splanchnic nerve**.
2. Postganglionic neuronal cell bodies are located in the **celiac ganglion** and **superior mesenteric ganglion**.
3. Postganglionic axons synapse in the complex circuitry of the enteric nervous system.

Large Intestine

A. GENERAL FEATURES (see Table 11-1)

B. ARTERIAL SUPPLY
1. The arterial supply of the **ascending colon** is from the **ileocolic artery** and **right colic artery** (abdominal aorta → superior mesenteric artery → ileocolic artery and right colic artery). The **marginal artery** is formed by an anastomotic connection between the superior mesenteric artery and inferior mesenteric artery.
2. The arterial supply of the **proximal two thirds of the transverse colon** is from the **middle colic artery** (abdominal aorta → superior mesenteric artery → middle colic artery).
3. The arterial supply of the **distal one third of the transverse colon** is from the **left colic artery** (abdominal aorta → inferior mesenteric artery → left colic artery).
4. The arterial supply of the **upper portion of the descending colon** is from the **left colic artery** (abdominal aorta → inferior mesenteric artery → left colic artery).

5. The arterial supply of the **lower descending colon** is from the **sigmoid arteries** (abdominal aorta → inferior mesenteric artery → sigmoid arteries).

C. VENOUS DRAINAGE

1. The venous drainage of the **ascending colon** is to the **ileocolic vein** and **right colic vein** (ileocolic vein and right colic vein → superior mesenteric vein → portal vein → hepatic sinusoids → central veins → hepatic veins → inferior vena cava).

2. The venous drainage of the **proximal two thirds of the transverse colon** is to the **middle colic vein** (middle colic vein → superior mesenteric vein → portal vein → hepatic sinusoids → central veins → hepatic veins → inferior vena cava).

3. The venous drainage of the **distal one third of the transverse colon** is to the **left colic vein and other unnamed veins** (left colic vein and other unnamed veins → inferior mesenteric vein → portal vein → hepatic sinusoids → central veins → hepatic veins → inferior vena cava).

4. The venous drainage of the **upper portion of the descending colon** is to the **left colic vein** (left colic vein → inferior mesenteric vein → portal vein → hepatic sinusoids → central veins → hepatic veins → inferior vena cava).

5. The venous drainage of the **lower portion of the descending colon** is to the **sigmoid veins** (sigmoid veins → inferior mesenteric vein → portal vein → hepatic sinusoids → central veins → hepatic veins → inferior vena cava).

D. INNERVATION. See Chapter 11 VIII.

E. CLINICAL CONSIDERATIONS

1. **Crohn disease (CD; Figure 11-5A,B)** is a chronic inflammatory bowel disease that most commonly affects the **ileum** and involves an abundant accumulation

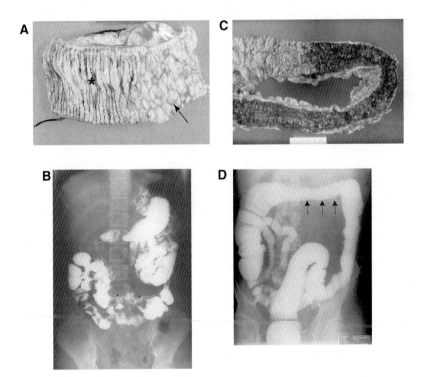

● **Figure 11-5 Crohn disease and ulcerative colitis. A,B:** Crohn disease. **A:** Photograph of a gross specimen of the ileum shows prominent cobblestoning (*arrow*) due to multiple transverse and linear ulcers. The other portion of the ileum is normal (*). **B:** Radiograph shows the luminal narrowing ("string sign") and cobblestone pattern of the affected small intestines. **C,D:** Ulcerative colitis. **C:** Photograph of a gross specimen of the colon shows inflammatory pseudopolyps and ulceration. **D:** Radiograph shows the "lead pipe" appearance of the affected transverse and descending colon. Note the small ulcerations extending from the colon lumen (*arrows*). (*continued*)

COMPARISON OF CROHN DISEASE AND ULCERATIVE COLITIS

	Crohn Disease	Ulcerative Colitis
Epidemiology	More common in whites vs. blacks More common in Jews vs. non-Jews More common in women Affects young adults	More common in whites vs. blacks No sex predilection Affects young adults
Extent	Transmural	Mucosal and submucosal
Location	Terminal ileum alone (30%) Ileum and colon (50%) Colon alone (20%) Involves other areas of GI tract (mouth to anus)	Mainly the rectum May extend into descending colon May involve entire colon Does not involve other areas of GI tract

Gross Features	Thick bowel wall and narrow lumen (leads to
	    obstruction)
	Aphthous ulcers (early sign)
	Skip lesions, strictures, and fistulas
	Deep linear ulcers with cobblestone pattern
	Fat creeping around the serosa
		Inflammatory pseudopolyps
		Areas of friable, bloody residual
		    mucosa
		Ulceration and hemorrhage
		Collar-button ulcers

Microscopic Findings	Noncaseating granulomas Lymphoid aggregates Dysplasia or cancer less likely	Ulcers and intestinal gland abscesses with neutrophlis Dysplasia or cancer may be present

Clinical Findings	Recurrent right lower quadrant colicky pain
	    with diarrhea
	Bleeding occurs with colon or anal involvement
		Recurrent left sided abdominal
		    cramping with bloody diarrhea and
		    mucus

Radiography	**"String" sign** in terminal ileum because of
	    luminal narrowing
		"Lead pipe" appearance in chronic
		    state

Complications	Fistulas, obstruction Calcium oxalate renal calculi Malabsorption because of bile deficiency Macrocytic anemia because of vitamin B_{12} deficiency	Toxic megacolon Primary sclerosing cholangitis Adenocarcinoma

GI = gastrointestinal.

● **Figure 11-5** *Continued.*

of lymphocytes forming a **granuloma** (a typical feature of CD) within the submucosa that may further extend into the muscularis externa. Neutrophils infiltrate the intestinal glands and ultimately destroy them, leading to ulcers. With progression of CD, the ulcers coalesce into long, **serpentine ulcers** ("**linear ulcers**") oriented along the long axis of the bowel. A classic feature of CD is the clear demarcation between diseased bowel segments located directly next to uninvolved normal bowel and a cobblestone appearance that can be seen grossly and radiographically. The etiology of CD is unknown. Clinical findings include intermittent bouts of diarrhea, weight loss, and weakness. Complications include strictures of the intestinal lumen, formation of fistulas, and perforation.

2. **Ulcerative colitis (Figure 11-5C,D)** is a type of idiopathic inflammatory bowel disease. It always involves the rectum and extends proximally for varying distances. The inflammation is continuous (i.e., there are no "skip areas," as in Crohn disease). The etiology of ulcerative colitis is unknown. Clinical signs include bloody diarrhea with mucus and pus, malaise, fever, weight loss, and anemia; ulcerative colitis may lead to toxic megacolon.

Innervation of the Large Intestine. The innervation of the large intestine is by the **enteric nervous system,** which in the large intestine consists of the submucosal plexus of Meissner and the myenteric plexus of Auerbach. The motor component of the submucosal plexus controls primarily **mucosal and submucosal gland secretion and blood flow,** whereas the sensory component consists of **mucosal mechanosensitive neurons.** The motor component of the myenteric plexus controls primarily **GI motility (contraction/relaxation of GI smooth muscle),** whereas the sensory component consists of **tension-sensitive neurons** and **chemosensitive neurons.** The enteric nervous system is modulated by the parasympathetic and sympathetic nervous systems.

A. PARASYMPATHETIC

    1. Proximal to the Splenic Flexure

        a. Preganglionic neuronal cell bodies are located in the **dorsal nucleus of the vagus.** Preganglionic axons run in **CN X** and reach the large intestine proximal to the splenic flexure via the **superior mesenteric nerve plexus.**

        b. Postganglionic neuronal cell bodies are located in the enteric nervous system; some of these are the "traditional" postganglionic parasympathetic neurons that release ACh as a neurotransmitter.

        c. Postganglionic axons terminate on mucosal glands and smooth muscle.

    2. Distal to the Splenic Flexure, Rectum, and Upper Anal Canal

        a. Preganglionic neuronal cell bodies are also located in the **gray matter of the S2–4 spinal cord.** Preganglionic axons form the **pelvic splanchnic nerves** and reach the large intestine distal to the splenic flexure via the **inferior hypogastric plexus.**

        b. Postganglionic neuronal cell bodies are located in the enteric nervous system; some of these are the "traditional" postganglionic parasympathetic neurons that release ACh as a neurotransmitter.

        c. Postganglionic axons terminate on mucosal glands and smooth muscle.

B. SYMPATHETIC

    1. Proximal to the Splenic Flexure

        a. Preganglionic neuronal cell bodies are located in the **intermediolateral cell column** of the spinal cord (T8–12). Preganglionic axons form the **greater splanchnic nerve, lesser splanchnic nerve,** and **least splanchnic nerve.**

        b. Postganglionic neuronal cell bodies are located in the **celiac ganglion** and **superior mesenteric ganglion.**

        c. Postganglionic axons reach the large intestine proximal to the splenic flexure via the **superior mesenteric plexus** and synapse in the complex circuitry of the enteric nervous system.

    2. Distal to the Splenic Flexure

        a. Preganglionic neuronal cell bodies are located in the **intermediolateral cell column** of the spinal cord (L1–2). Preganglionic axons form the **lumbar (L1, L2) splanchnic nerves.**

        b. Postganglionic neuronal cell bodies are located in the **inferior mesenteric ganglion.**

        c. Postganglionic axons reach the large intestine distal to the splenic flexure via the **inferior mesenteric plexus** and synapse in the complex circuitry of the enteric nervous system.

3. **Rectum and Upper Anal Canal**
   a. Preganglionic neuronal cell bodies are located in the **intermediolateral cell column** of the spinal cord (L2–3). Preganglionic axons form the **lumbar (L3, L4) splanchnic nerves.**
   b. Postganglionic neuronal cell bodies are located in the **superior hypogastric plexus.**
   c. Postganglionic axons reach the rectum and upper anal canal via the **rectal plexus** and synapse in the complex circuitry of the enteric nervous system.

IX Appendix

A. GENERAL FEATURES
   1. The appendix is an intraperitoneal (**mesoappendix**), narrow, muscular tube attached to the posteromedial surface of the cecum.
   2. The appendix is located ~2.5 cm below the ileocecal valve.
   3. The appendix may lie in the following positions: **retrocecal (65%), pelvis (32%), subcecal (2%), anterior juxta-ileal (1%), and posterior juxt-aileal (0.5%).**

B. ARTERIAL SUPPLY. The arterial supply of the appendix is from the **appendicular artery** (abdominal aorta → superior mesenteric artery → ileocolic artery → posterior cecal artery → appendicular artery).

C. VENOUS DRAINAGE. The venous drainage of the appendix is to the **posterior cecal vein** (posterior cecal vein → superior mesenteric vein → portal vein → hepatic sinusoids → central veins → hepatic veins → inferior vena cava).

D. CLINICAL CONSIDERATION. **Appendicitis** begins with the obstruction of the appendix lumen with a fecal concretion (fecalith) and lymphoid hyperplasia followed by distention of the appendix. Clinical findings include initial pain in the umbilical or epigastric region (later pain localizes to the right lumbar region), nausea, vomiting, anorexia, and tenderness to palpation and percussion in the right lumbar region. Complications may include peritonitis due to rupture of the appendix. **McBurney point** is located by drawing a line from the right anterior superior iliac spine to the umbilicus. The midpoint of this line locates the root of the appendix. The appendix is suspended by the **mesoappendix** (i.e., intraperitoneal) and is generally found in the **retrocecal fossa** (although its position is variable).

X Gallbladder

A. GENERAL FEATURES
   1. The gallbladder is divided into the **fundus** (anterior portion), **body**, and the **neck** (posterior portion).
   2. A small pouch (**Hartmann pouch**) may extend from the neck as a sequela to pathologic changes and is a common site for gallstones to lodge.
   3. **Rokitansky-Aschoff sinuses** occur when the mucosa of the gallbladder penetrates deep into the muscularis externa. They are an early indicator of pathologic changes (e.g., acute cholecystitis or gangrene).

B. ARTERIAL SUPPLY. The arterial supply of the gallbladder is from the **cystic artery** (abdominal aorta → celiac trunk → common hepatic artery → proper hepatic artery → right hepatic artery → cystic artery).

C. VENOUS DRAINAGE. The venous drainage of the gallbladder is to the **cystic vein** (cystic vein → portal vein → hepatic sinusoids → central veins → hepatic veins → inferior vena cava).

D. INNERVATION

1. Parasympathetic

   a. Preganglionic neuronal cell bodies are located in the **dorsal nucleus of the vagus.** Preganglionic axons run in **CN X.**

   b. Postganglionic neuronal cell bodies are located within the wall of the gallbladder.

   c. Postganglionic parasympathetic axons terminate on smooth muscle and **stimulate gallbladder contraction.**

2. Sympathetic

   a. Preganglionic neuronal cell bodies are located in the **intermediolateral cell column** of the spinal cord (T5–9). Preganglionic axons form the **greater splanchnic nerve.**

   b. Postganglionic neuronal cell bodies are located in the **celiac ganglion.**

   c. Postganglionic axons terminate on smooth muscle and **inhibit gallbladder contraction.**

   d. Sensory nerve fibers for pain from the gallbladder travel with the **greater thoracic splanchnic nerve** to T7–10 spinal levels.

3. Somatic. Sensory neuronal cell bodies located in dorsal root ganglion (C3–5) of the **right phrenic nerve** send peripheral processes to the gallbladder. These sensory nerve fibers are probably responsible for the **somatic referred pain** associated with gallbladder disease.

XI Extrahepatic Biliary Ducts (Figure 11-6)

A. GENERAL FEATURES

1. The **right and left hepatic ducts** join together after leaving the liver to form the **common hepatic duct.**

2. The common hepatic duct is joined at an acute angle by the **cystic duct** to form the **common bile duct.**

3. The cystic duct drains bile from the gallbladder. The mucosa of the cystic duct is arranged in a spiral fold with a core of smooth muscle known as the **spiral valve (valve of Heister).** The spiral valve keeps the cystic duct constantly open so that bile can flow freely in either direction.

4. The common bile duct passes posterior to the pancreas and ends at the **hepatopancreatic ampulla (ampulla of Vater),** where it joins the **pancreatic duct.**

5. The **sphincter of Oddi** is an area of thickened smooth muscle that surrounds the bile duct as it traverses the ampulla. The sphincter of Oddi **controls bile flow** (sympathetic innervation causes contraction of the sphincter).

B. CLINICAL CONSIDERATIONS.
The term **cholelithiasis** refers to the presence or formation of gallstones either in the gallbladder (called **cholecystolithiasis**) or common bile duct (called **choledocholithiasis**).

1. Gallstones form when bile salts and lecithin are overwhelmed by cholesterol. Most stones consist of **cholesterol (major component), bilirubin, and calcium.** There are three main types of gallstones:

A

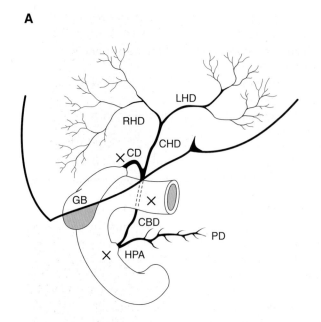

B

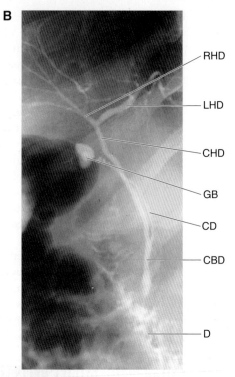

RHD
LHD
CHD
GB
CD
CBD
D

C

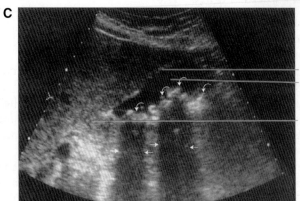

Gallbladder wall
Gallbladder lumen
Liver

D

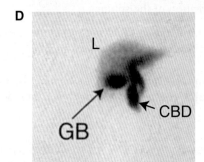

L

GB

CBD

E

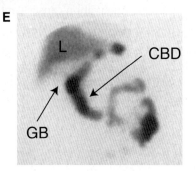

L

CBD

GB

a. **Cholesterol stones (Figure 11-7)** are yellow, large, smooth, and composed mainly of cholesterol. These stones are associated with obesity, Crohn disease, cystic fibrosis, clofibrate, estrogens, rapid weight loss, and the U.S. or Native American population (4 Fs: female, fat, fertile, over forty). The photograph shows a solitary cholesterol gallstone.

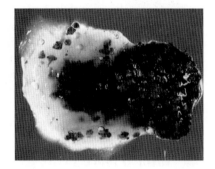

● **Figure 11-7 Solitary cholesterol gallstone.**

b. **Pigment (bilirubin) stones (Figure 11-8)** are brown or black, smooth, and composed mainly of bilirubin salts. These stones are associated with chronic red blood cell hemolysis (e.g., sickle cell anemia or spherocytosis), alcoholic cirrhosis, biliary infection, and the Asian population. The photograph shows a pigment gallstone embedded in a mucous gel.

c. **Calcium bilirubinate stones** are associated with infection and/or inflammation of the biliary tree.

● **Figure 11-8 Pigment gallstone embedded in a mucous gel.**

● **Figure 11-6 Gallbladder and extrahepatic biliary ducts. A:** The diagram shows the gallbladder and biliary tree. Note the termination of the common bile duct (CBD) at the hepatopancreatic ampulla (HPA) along with the pancreatic duct (PD). Note the three main sites (X) of gallstone obstruction. **B:** Endoscopic retrograde cholangiograph shows the normal gallbladder and biliary tree. Note that the cystic duct normally lies on the right side of the common hepatic duct and joins it superior to the duodenal cap. **C:** Longitudinal decubitus sonogram shows gallstones within the gallbladder (cholelithiasis; *curved arrows*), which cast acoustic shadows (between the *straight arrows*) because the sound waves cannot penetrate the dense gallstones. Ultrasonography generally elicits the Murphy sign, where a patient reports pain as the operator presses on the gallbladder. **D:** Hepatobiliary iminodiacetic acid (HIDA) scan of a normal patient. HIDA makes use of the radionuclide 99mTc attached to bilirubin analogs bound to iminodiacetic acid. This compound is injected intravenously, processed by hepatocytes, and excreted into the bile. In a normal person, filling of the liver, gallbladder, and biliary tract occurs within 60 minutes after injection. Note the filling of the liver (L), gallbladder (GB), and common bile duct (CBD) within 60 minutes after injection. **E:** HIDA scan of a patient with acute cholecystitis. Note the absence of filling of the gallbladder (GB) minutes after morphine injection, which contracts the sphincter of Oddi and leads to a rise of biliary system pressure. Even after morphine injection, the gallbladder does not fill, which is diagnostic of a blockage. However, the liver (L) and common bile duct (CBD) are filled. Because most gallstones are composed of cholesterol (and therefore radiolucent), plain abdominal radiographic films are often of little value. Therefore, ultrasonography and HIDA are the methods of choice for diagnosis. CBD = common bile duct, CD = cystic duct, CHD = common hepatic duct, D = duodenum, GB = gallbladder, LHD = left hepatic duct, PD = pancreatic duct, RHD = right hepatic duct.

2. **Gallstone Obstruction.** There are three clinically important sites of gallstone obstruction:

a. **Within the cystic duct.** A stone may transiently lodge within the cystic duct and cause pain (**biliary colic**) within the epigastric region due to the distention of the duct. If a stone becomes entrapped within the cystic duct, bile flow from the gallbladder will be obstructed, resulting in inflammation of the gallbladder (**acute cholecystitis**) and pain will shift to the right hypochondriac region. Bile becomes concentrated and precipitates in the gallbladder, forming a layer of high-density material called "**milk of calcium**" **bile** due to the large amount of calcium carbonate. Bile flow from the liver remains open (i.e., **no jaundice**). This may lead to **Mirizzi syndrome**, where impaction of a large gallstone in the cystic duct extrinsically obstructs the nearby common hepatic duct.

b. **Within the common bile duct.** If a stone becomes entrapped within the common bile duct, bile flow from both the gallbladder and liver will be obstructed, resulting in inflammation of the gallbladder and liver. **Jaundice** is frequently observed and is first observed clinically **under the tongue.** The jaundice is moderate and fluctuates since a stone rarely causes complete blockage of the lumen.

c. **At the hepatopancreatic ampulla.** If a stone becomes entrapped at the ampulla, bile flow from both the gallbladder and liver will be obstructed. In addition, the pancreatic duct may be blocked. In this case, **jaundice** and **pancreatitis** are frequently observed.

XII Liver (Figure 11-9)

A. GENERAL FEATURES

1. The liver stroma begins as a thin connective tissue capsule called the **Glisson capsule** that extends into the liver, around the portal triads, and around the periphery of a hepatic lobule; extends into the perisinusoidal space of Disse to surround hepatocytes; and then terminates around the central vein.

2. The components of the **porta hepatis** are the:
   a. **Common bile duct**
   b. **Portal vein**
   c. **Hepatic artery**
   d. **Lymphatics**

B. LOBES OF THE LIVER

1. The liver is classically divided into the **right lobe** and **left lobe** by the **interlobar fissure** (an invisible line running from the gallbladder to the inferior vena cava), **quadrate lobe**, and **caudate lobe**.

2. The left lobe contains the **falciform ligament** (a derivative of the ventral mesentery) with the **ligamentum teres** (a remnant of the left umbilical vein) along its inferior border.

3. The **bare area** of the liver is located on the diaphragmatic surface and is devoid of peritoneum.

C. SEGMENTS AND SUBSEGMENTS OF THE LIVER

1. There are five liver segments: **anterior segment of the right lobe, posterior segment of the right lobe, medial segment of the left lobe, lateral segment of the left lobe,** and **caudate lobe.** The hepatic veins define the boundaries of the liver segments.

2. There are nine liver subsegments: **posterior superior; posterior inferior; anterior superior; anterior inferior; medial superior; medial inferior,** which corresponds to the classic quadrate lobe; **lateral superior; lateral inferior;** and the classic **caudate lobe.**

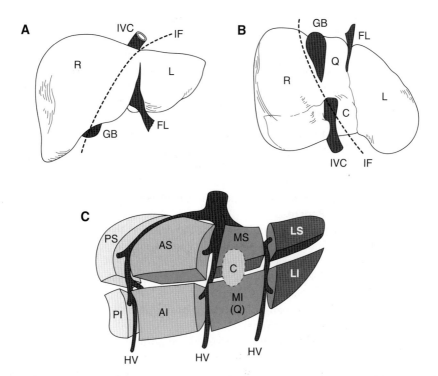

● **Figure 11-9 Liver. A:** Diagram of the anterior surface of the liver. Note the right lobe (R) and left lobe (L) divided by the interlobar fissure (IF). FL = falciform ligament, GB = gallbladder, IVC = inferior vena cava. **B:** Diagram of the inferior surface of the liver. Note the quadrate lobe (Q) and caudate lobe (C). **C:** Diagram of the five liver segments and nine liver subsegments used in liver resectioning. The five liver segments are the posterior segment and anterior segment of the right lobe, the medial segment and lateral segment of the left lobe, and the caudate lobe. Note the hepatic veins (HV) at the periphery of the liver segments. The nine liver subsegments are the posterior superior (PS), posterior inferior (PI), anterior superior (AS), anterior inferior (AI), medial superior (MS), medial inferior (MI), which corresponds to the classic quadrate lobe (Q), lateral superior (LS), lateral inferior (LI), and the classic caudate lobe (C).

D. ARTERIAL SUPPLY. The arterial supply of the liver is from the **right hepatic artery** and **left hepatic artery** (abdominal aorta → celiac trunk → common hepatic artery → proper hepatic artery → right hepatic artery and left hepatic artery → hepatic sinusoids).

E. PORTAL SUPPLY
   1. The portal supply of the liver is from the **portal vein** (superior mesenteric vein, inferior mesenteric vein, and splenic vein → portal vein → hepatic sinusoids).
   2. The portal vein is formed by the union of the splenic vein and superior mesenteric vein.
   3. The inferior mesenteric vein joins the splenic vein.
   4. The arterial blood and portal blood mix in the hepatic sinusoids.

F. VENOUS DRAINAGE. The venous drainage of the liver is to the **central veins** located at the center of a classic liver lobule (central veins → hepatic veins → inferior vena cava).

G. INNERVATION. The exact function of both the parasympathetic and sympathetic innervation is unclear, except that sympathetics play a role in vasoconstriction.

H. CLINICAL CONSIDERATIONS
   1. Liver biopsies are frequently performed by needle puncture through the right intercostal space 8, 9, or 10 when the patient has exhaled. The needle will pass through the following structures: skin → superficial fascia → external oblique

muscle → intercostal muscles → costal parietal pleura → costodiaphragmatic recess → diaphragmatic parietal pleura → diaphragm → peritoneum.

2. **Surgical resection of the liver** may be performed by removing one of the **liver segments** (five total segments) or one of the **liver subsegments** (nine total subsegments). **Hepatic veins** form the surgical landmarks that mark the periphery (or border) of a liver segment during segmental resection of the liver.

XIII Pancreas (Figure 11-10)

A. **GENERAL FEATURES** In the adult, the pancreas is a retroperitoneal organ that measures 15 to 20 cm in length and weighs about 85 to 120 g. The pancreas is both an exocrine gland and an endocrine gland. The pancreas consists of four parts:

1. **Head of the Pancreas**
   a. The head is the expanded part of the pancreas that lies in the concavity of the C-shaped curve of the duodenum and is firmly attached to the descending and horizontal parts of the duodenum.
   b. The **uncinate process** is a projection from the inferior portion of the pancreatic head.
   c. The head of the pancreas is related posterior to the IVC, right renal artery, right renal vein, and left renal vein.

2. **Neck of the Pancreas.** The neck is related posteriorly to the confluence of the superior mesenteric vein and splenic vein to form the portal vein.

3. **Body of the Pancreas.** The body is related posteriorly to the aorta, superior mesenteric artery, left suprarenal gland, left kidney, renal artery, and renal vein.

4. **Tail of the Pancreas.** The tail is related to the splenic hilum and the left colic flexure.

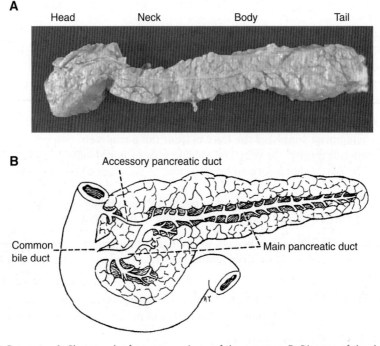

● **Figure 11-10 Pancreas. A:** Photograph of a gross specimen of the pancreas. **B:** Diagram of the duct system of the pancreas.

B. **ARTERIAL SUPPLY.** The arterial supply of the pancreas is from the:
1. **Anterior and posterior superior pancreaticoduodenal arteries,** which supply the head and neck of the pancreas (abdominal aorta → celiac trunk → common hepatic artery → gastroduodenal artery → anterior and posterior superior pancreaticoduodenal arteries)
2. **Anterior and posterior inferior pancreaticoduodenal arteries,** which supply the head and neck of the pancreas (abdominal aorta → superior mesenteric artery → anterior and posterior inferior pancreaticoduodenal arteries)
3. **Dorsal pancreatic artery,** which supplies the body and tail of the pancreas (abdominal aorta → celiac trunk → splenic artery → dorsal pancreatic artery)
4. **Great pancreatic artery,** which supplies the body and tail of the pancreas (abdominal aorta → celiac trunk → splenic artery → great pancreatic artery)
5. **Caudal pancreatic arteries,** which supply the body and tail of the pancreas (abdominal aorta → celiac trunk → splenic artery → caudal pancreatic arteries)

C. **VENOUS DRAINAGE.** The venous drainage of the pancreas is to the:
1. **Splenic vein** (splenic vein → portal vein → hepatic sinusoids → central veins → hepatic veins → inferior vena cava)
2. **Superior mesenteric vein** (superior mesenteric vein → portal vein → hepatic sinusoids → central veins → hepatic veins → inferior vena cava)

D. **INNERVATION**
1. **Parasympathetic**
   a. Preganglionic neuronal cell bodies are located in the **dorsal nucleus of the vagus.** Preganglionic axons run in **CN X.**
   b. Postganglionic neuronal cell bodies are located within and/or nearby the pancreas.
   c. The postganglionic axons terminate on pancreatic acinar cells and ductal epithelium. Parasympathetics **stimulate pancreatic secretion** of pancreatic enzymes and pancreatic juice.
2. **Sympathetic**
   a. Preganglionic neuronal cell bodies are located in the **intermediolateral cell column** of the spinal cord (T5–9). Preganglionic axons form the **greater splanchnic nerve.**
   b. Postganglionic neuronal cell bodies are located in the **celiac ganglion.**
   c. Postganglionic axons terminate on pancreatic acinar cells, ductal epithelium, and smooth muscle of blood vessels. Sympathetics **inhibit pancreatic secretion,** probably by reducing blood flow.

E. **CLINICAL CONSIDERATION: ANNULAR PANCREAS (FIGURE 11-11).** An annular pancreas occurs when the ventral pancreatic bud fuses with the dorsal bud both dorsally and ventrally, thereby forming a **ring of pancreatic tissue** around the duodenum, causing severe **duodenal obstruction** noticed shortly after birth. Newborns and infants are intolerant of oral feeding and often have bilious vomiting. Radiographic evidence of an annular pancreas is indicated by a duodenal obstruction, where a "double bubble" sign is often seen due to dilation of the stomach and distal duodenum. The barium contrast radiograph shows partial duodenal obstruction consistent with an annular pancreas.

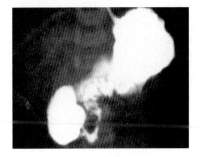

● **Figure 11-11** Annular pancreas.

XIV Cross-sectional Anatomy

A. AT ABOUT T12, WHERE THE PORTAL TRIAD IS LOCATED (FIGURE 11-12)

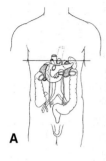

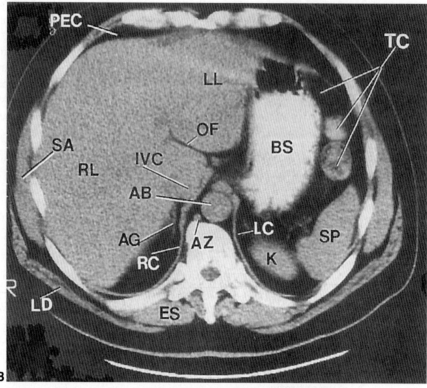

AB = abdominal aorta	LL = left lobe of the liver
AG = adrenal gland	OF = oblique fissure
AZ = azygos vein	PEC = peritoneal cavity
BS = body of the stomach	RC = right crus of the diaphragm
ES = erector spinae muscle	RL = right lobe of the liver
IVC = inferior vena cava	SA = serratus anterior
K = kidney	SP = spleen
LC = left crus of the diaphragm	TC = transverse colon
LD = latissimus dorsi	

● **Figure 11-12 A computed tomography (CT) scan at about T12, where the portal triad is located. A:** A schematic to show where the cross section was taken. **B:** A CT scan. Note the various structures, as indicated by the key. In addition, note the psoas major and quadratus lumborum muscles along the sides of the vertebral body. The right and left lobes of the liver are shown in their relationship to the portal vein, common hepatic artery, and inferior vena cava. The right adrenal gland lies posterolateral to the IVC. The left adrenal gland lies between the body of the stomach and the abdominal aorta.

B. AT THE LEVEL OF THE GALLBLADDER (FIGURE 11-13)

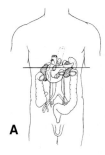

A

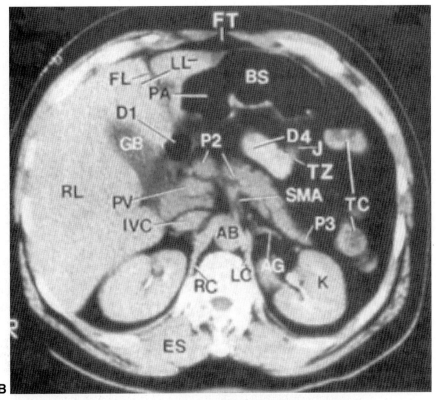

B

AB = abdominal aorta	LC = left crus of the diaphragm
AG = adrenal gland	LL = left lobe of the liver
BS = body of the stomach	PA = pyloric antrum of the stomach
D1 = first part of the duodenum	PV = portal vein
D4 = fourth part of the duodenum	P2 = body of the pancreas
ES = erector spinae muscle	P3 = tail of the pancreas
FL = falciform ligament	RC = right crus of the diaphragm
FT = fat	RL = right lobe of the liver
GB = gallbladder	SMA = superior mesenteric artery
IVC = inferior vena cava	TC = transverse colon
J = jejunum	TZ = ligament of Treitz
K = kidney	

● **Figure 11-13 A computed tomography (CT) scan at the level of the gallbladder. A:** A schematic to show where the cross section was taken. **B:** A CT scan. Note the various structures, as indicated by the key. The second part of the duodenum is adjacent to the head of the pancreas. The body of the pancreas extends to the left, posterior to the stomach. The tail of the pancreas reaches the spleen. The uncinate process of the pancreas lies posterior to the superior mesenteric artery. The gallbladder lies between the right and left lobes of the liver, just to the right of the antrum of the stomach. Note the location of the adrenal gland. A large mass in this area is indicative of a pheochromocytoma or neuroblastoma, both of which are associated with the adrenal medulla.

C. AT THE LEVEL OF THE HILUM OF THE KIDNEYS (FIGURE 11-14)

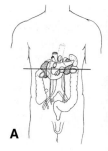

A

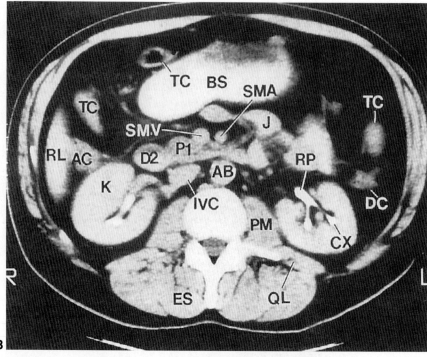

B

AB = abdominal aorta	K = kidney
AC = ascending colon	P1 = head of the pancreas
BS = body of the stomach	PM = psoas major muscle
CX = minor renal calyx	QL = quadratus lumborum muscle
DC = descending colon	RL = right lobe of the liver
D2 = second part of the duodenum	RP = renal pelvis
ES = erector spinae muscle	SMA = superior mesenteric artery
IVC = inferior vena cava	SMV = superior mesenteric vein
J = jejunum	TC = transverse colon

● **Figure 11-14 A computed tomography (CT) scan at the level of the hilum of the kidney. A:** A schematic to show where the cross section was taken. **B:** A CT scan. Note the various structures as indicated by the key. The inferior vena cava and the abdominal aorta lie side by side as both vessels pass posterior to the pancreas. The second part of the duodenum contacts the right kidney and the right lobe of the liver. The left renal vein lies anterior to the renal artery.

XV Radiology

A. RADIOGRAPH OF THE STOMACH AND SMALL INTESTINES AFTER A BARIUM MEAL (FIGURE 11-15A)
B. ANTEROPOSTERIOR RADIOGRAPH OF THE LARGE INTESTINE AFTER A BARIUM ENEMA (FIGURE 11-15B)

A

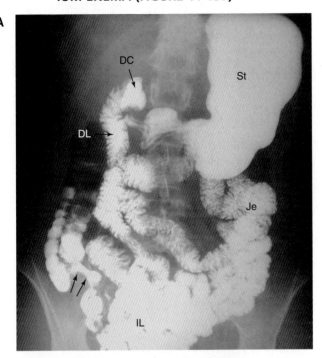

B

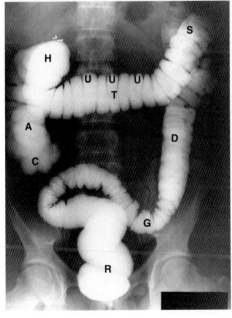

A = ascending colon	IL = ileum
C = cecum	Je = jejunum
D = descending colon	R = rectum
DC = duodenal cap	S = splenic or left colic flexure
DL = duodenal C-loop	St = stomach
G = sigmoid colon	T = transverse colon
H = hepatic or right colic flexure	U = haustra

● **Figure 11-15 Stomach, small intestine, and large intestine. A:** Radiograph of the stomach and small intestines after a barium meal. Note the structures indicated. Note the barium-filled stomach, duodenal cap, duodenal C-loop, feathered jejunum in the upper portion and left side of the abdomen, and relatively formless mucosa of the ileum in the lower portion and right side of the abdomen. The terminal ileum (*arrows*) entering the cecum is shown. **B:** Radiograph of large intestine after a barium enema.

Case Study 11-1

A 25-year-old woman comes to your office complaining that "I have pain on the right side of my abdomen down low and sometimes the pain moves to my bellybutton. It just started last night." She also tells you that "the pain feels like cramps, I am nauseated and feel like I am going to vomit, I have a fever, and I get chills." After some discussion, she tells you that she is sexually active, had her last menstrual period 3 weeks ago, and has not noticed any vaginal discharge or bleeding. She has a past medical history of endometriosis and irritable bowel syndrome. She also has had surgery to repair a direct inguinal hernia in the recent past.

Relevant Physical Exam Findings

- Heat rate = 92 bpm
- Blood pressure = 130/90 mm Hg
- Temperature = 102.5°F
- Tenderness in right and left lower abdominal quadrants
- Guarding and rebound tenderness in right lower abdominal quadrant at the McBurney point
- Positive psoas and obturator signs
- No tenderness in right upper abdominal quadrant
- Pelvic exam unremarkable

Relevant Lab Findings

- White blood cells = 22.5×10^9/L (moderately high) with a left shift (85% neutrophils)
- Amylase and lipase levels normal
- β-Human chorionic gonadotropin (β-hCG) = negative
- Cervical culture = negative

Diagnosis

Acute Appendicitis

- Acute appendicitis occurs when the appendix is obstructed by a fecalith, a foreign body, inflammation, or a tumor, which results in increased intraluminal pressure, engorgement, infection, and thrombosis of the blood vessels in the walls of the appendix.
- Acute appendicitis presents with colicky or crampy periumbilical pain, which migrates to the lower right abdominal quadrant (i.e., McBurney point) within 12 hours and which is reproduced when the patient coughs; anorexia; nausea; vomiting; constipation; and moderate leukocytosis. Positive psoas and obturator signs are indicative of peritoneal irritation.
- When a woman of child-bearing age presents with abdominal complaints, it is imperative to rule out an ectopic pregnancy. Ectopic pregnancy presents with abdominal pain and cramping, uterine bleeding, and shock; the β-hCG is positive, and the pelvic exam reveals cervical motion tenderness.
- Endometriosis presents with abdominal pain that coincides with menstruation; as the disease progresses, the pain may occur at any time during the menstrual cycle.
- Crohn disease and ulcerative colitis present with symptoms similar to appendicitis but usually associated with diarrhea.
- Pelvic inflammatory disease (PID) is most commonly caused by *Neisseria gonorrhoeae* and *Chlamydia trachomatis*. Risk factors include a previous PID, multiple sex partners, use of an intrauterine contraceptive device (IUD), history of sexually transmitted diseases, and nulliparity. PID presents with lower abdominal pain, fever, abnormal cervical or vaginal discharge, uterine bleeding, cervical motion tenderness, and adnexal tenderness.

Case Study 11-2

A 50-year-old man is brought into the emergency room in the early morning by his son. The man says that "I woke up from my sleep with a lot of pain in the abdomen. The pain is like a broad band across my stomach and goes all the way to my back." He also tells you that "the pain was real bad when I was laying down in the bed but when I sat up in bed the pain decreased a lot." He has no fever, chills, or diarrhea. He also complains that during the past week he had intermittent pain in the abdomen ("it felt like I had a lot of gas and I felt like I had to belch"). After some discussion, he informs you that he is an alcoholic and drinks six beers per day and a half pint of whiskey per day. He takes a nonsteroidal anti-inflammatory drug (NSAID) every morning for his hangover.

Relevant Physical Exam Findings

- Heart rate = 87 bpm
- Blood pressure = 110/65 mm Hg
- Temperature = 95.7°F
- The man is sweating (diaphoretic) and writhing around in the gurney
- No jaundice is apparent (sclera of the eyes are white)
- Heart exam unremarkable
- Lungs are clear to auscultation
- Tenderness in the epigastric and bilateral upper quadrants of the abdomen
- Guarding and rebound tenderness apparent

Relevant Lab Findings

- Anteroposterior chest radiograph is normal with no air visualized under the diaphragm
- Abdominal ultrasound shows no signs of gallstones
- Electrocardiogram (ECG) = normal
- Complete blood count (CBC) = normal
- Hepatic panel, lipase, and troponin = normal
- Computed tomography (CT) scan shows air underneath the diaphragm and fluid in the region of the distal antrum of the stomach

Diagnosis

Perforated Peptic Ulcer

- Uncomplicated peptic ulcer disease (PUD) is highly prevalent in the United States. There are ~500,000 new cases per year and ~4 million recurrences per year.
- After *Helicobacter pylori* infection, the abuse of NSAIDs is the next most common cause of PUD.
- Perforation of a peptic or duodenal ulcer into the peritoneal cavity is associated with significant morbidity and mortality.
- The majority of perforated duodenal ulcers involve the anterior wall of the duodenal cap.
- With the presentation of epigastric pain, there is a broad range of differential diagnoses that need to be considered, which include acute coronary syndrome, aortic dissection, gallbladder disease, acute hepatitis, acute cholangitis, acute pancreatitis, acute appendicitis (may present initially with upper abdominal pain), and pneumonia.

Chapter 12

Sigmoid Colon, Rectum, and Anal Canal

❶ Sigmoid Colon (Figure 12-1)

A. GENERAL FEATURES

1. The sigmoid colon is a segment of the large intestine between the descending colon and rectum whose primary function is **storage of feces**.
2. The sigmoid colon begins at vertebral level S1 (sacral promontory; pelvic inlet) and ends at S3 (**rectosigmoid junction**), where **teniae coli** (longitudinal bands of smooth muscle) are replaced by a complete circular layer of smooth muscle of the rectum.
3. The sigmoid colon is suspended by the **sigmoid mesocolon** (i.e., intraperitoneal).
4. The **left ureter** and **left common iliac artery** lie at the apex of the sigmoid mesocolon.

B. ARTERIAL SUPPLY

1. The arterial supply of the sigmoid colon is from the **sigmoid arteries** (abdominal aorta → inferior mesenteric artery → sigmoid arteries).

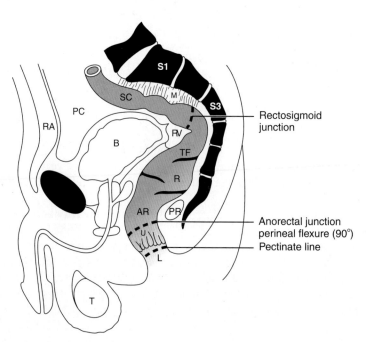

● **Figure 12-1 Sagittal view of the male pelvis.** The sigmoid colon (SC) extends from vertebral level S1 to S3 suspended by the sigmoid mesocolon (M) and ends at the rectosigmoid junction (*dotted line*). The rectum (R) and ampulla of the rectum (AR) are shown along with the transverse rectal folds (TF; Houston valve). The rectum ends at the anorectal junction (*dotted line*) at the tip of the coccyx, where the puborectalis muscle (PR) maintains a perineal flexure of 90 degrees. The anal canal is divided into the upper anal canal (U) and lower anal canal (L) by the pectinate line. B = urinary bladder, PC = peritoneal cavity, RA = rectus abdominus muscle, RV = rectovesical pouch, T = testes.

C. VENOUS DRAINAGE

1. The venous drainage of the sigmoid colon is to the **sigmoid veins** (sigmoid veins → inferior mesenteric vein → portal vein → hepatic sinusoids → central veins → hepatic veins → inferior vena cava).

D. INNERVATION. See Chapter 11 VIII.

E. CLINICAL CONSIDERATIONS

1. **Colonic aganglionosis (Hirschsprung disease; Figure 12-2)** is caused by the arrest of the caudal migration of neural crest cells. The hallmark is the absence of ganglionic cells in the myenteric and submucosal plexuses, most commonly in the sigmoid colon and rectum, resulting in a narrow segment of colon (i.e., the colon fails to relax). Although the ganglionic cells are absent, there is a proliferation of hypertrophied nerve fiber bundles. The most characteristic functional finding is the failure of the internal anal sphincter to relax following rectal distention (i.e., abnormal rectoanal reflex). Mutations of the **RET proto-oncogene** (chromosome 10q.11.2) have been associated with Hirschsprung disease. Clinical findings include a distended abdomen, inability to pass meconium, gushing of fecal material upon a rectal digital exam, fecal retention, and a loss of peristalsis in the colon segment distal to the normal innervated colon. The barium radiograph shows a narrowed rectum and a classic transition zone (*arrows*). The upper segment (*) of the normal colon is distended with fecal material. The distal segment (**) of the colon is narrow and is the portion of colon where the myenteric plexus of ganglion cells is absent.

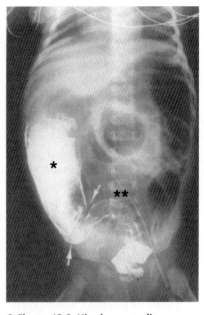

● Figure 12-2 Hirschsprung disease.

2. **Diverticulosis (Figure 12-3)** is the presence of diverticula (abnormal pouches or sacs) most commonly found in the sigmoid colon in patients older than 60 years of age. It is associated with a low-fiber, modern Western-world diet. Perforation and/or inflammation of the diverticula result in **diverticulitis.** Clinical findings include pain in the left lumbar region, palpable inflammatory mass in the left lumbar region, fever, leukocytosis, **ileus,** and **peritonitis.** The postevacuation barium radiograph shows numerous small outpouchings or diverticula (*arrows*) from the colonic lumen. These diverticula are filled with barium and fecal material. Note the hernia (H) on the right.

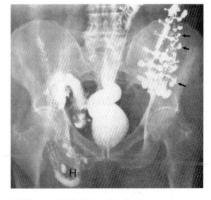

● Figure 12-3 Diverticulosis.

3. **Flexible sigmoidoscopy** permits examination of the sigmoid colon and rectum. During sigmoidoscopy, the large intestine may be punctured if the angle at the rectosigmoid junction is not negotiated properly. At the rectosigmoid junction, the sigmoid colon bends in an **anterior direction and to the left.** During sigmoidoscopy, the transverse rectal folds (Houston valves) must be negotiated also.

4. **Colostomy.** The sigmoid colon is often used in a **colostomy** due to the mobility rendered by the sigmoid mesocolon (mesentery). An ostomy is an intestinal diversion that brings out a portion of the gastrointestinal (GI) tract through the **rectus abdominis muscle.** A colostomy may ablate the pelvic nerve plexus, which results in loss of ejaculation, loss of erection, urinary bladder retention, and decreased peristalsis in the remaining colon.

Rectum

A. GENERAL FEATURES

1. The rectum is a segment of the large intestine between the sigmoid colon and the anal canal.
2. The rectum begins at vertebral level S3 and ends at the tip of the coccyx (**anorectal junction**), where the **puborectalis muscle** forms a U-shaped sling causing a 90-degree **perineal flexure.**
3. The **ampulla of the rectum** lies just above the pelvic diaphragm and generates the **urge to defecate** when feces moves into it.
4. The rectum contains three **transverse rectal folds (Houston valves)** formed by the mucosa, submucosa, and inner circular layer of smooth muscle that permanently extend into the lumen of the rectum to support the fecal mass.

B. ARTERIAL SUPPLY. The arterial supply of the rectum is from the:

1. **Superior rectal artery** (abdominal aorta → inferior mesenteric artery → superior rectal artery)
2. **Middle rectal artery** (abdominal aorta → common iliac artery → internal iliac artery → middle rectal artery)
3. **Inferior rectal artery** (abdominal aorta → common iliac artery → internal iliac artery → internal pudendal artery → inferior rectal artery)

C. VENOUS DRAINAGE. The venous drainage of the rectum is to the:

1. **Superior rectal vein** (superior rectal vein → inferior mesenteric vein → portal vein → hepatic sinusoids → central veins → hepatic veins → inferior vena cava)
2. **Middle rectal vein** (middle rectal vein → internal iliac vein → common iliac vein → inferior vena cava)
3. **Inferior rectal vein** (inferior rectal vein → internal pudendal vein → internal iliac vein → common iliac vein → inferior vena cava)

D. INNERVATION. See Chapter 11 VIII.

E. CLINICAL CONSIDERATION. Rectal prolapse is the protrusion of the **full thickness of the rectum** through the anus (this should be distinguished from **mucosal prolapse,** which is the protrusion of only the rectal mucosa through the anus). Clinical findings include bowel protruding through the anus, bleeding, anal pain, mucous discharge, and anal incontinence caused by stretching of the **internal and external anal sphincters** or stretch injury to the **pudendal nerve.**

Anal Canal (Figure 12-4)

A. GENERAL FEATURES. The entire anal canal is~4 cm long, which extends from the rectum at the anorectal junction to the surface of the body at the anus. The anal canal is divided into the **upper anal canal** and **lower anal canal** by the **pectinate line.**

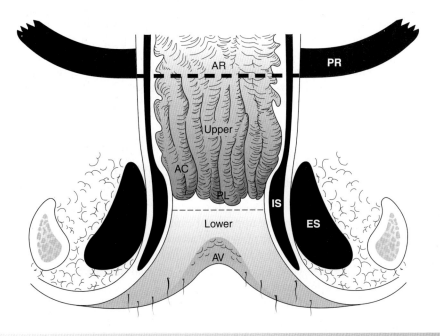

Feature	Upper Anal Canal	Lower Anal Canal
Arterial supply	Superior rectal artery (branch of inferior mesenteric artery)	Inferior rectal artery (branch of internal pudendal artery)
Venous drainage	Superior rectal vein → inferior mesenteric vein → hepatic portal system	Inferior rectal vein → internal pudendal vein → internal iliac vein → IVC
Lymphatic drainage	Deep nodes	Superficial inguinal nodes
Innervation	Motor: autonomic innervation of internal anal sphincter (smooth muscle) Sensory: stretch sensation; no pain sensation	Motor somatic innervation (pudendal nerve) of external anal sphincter (striated muscle) Sensory pain, temperature, touch sensation
Embryologic derivation	Endoderm (hindgut)	Ectoderm (proctodeum)
Epithelium	Simple columnar	Stratified squamous
Tumors	Palpable enlarged superficial nodes will not be found Patients do not complain of pain	Palpable enlarged superficial nodes will be found Patients do complain of pain
Hemorrhoids	Internal hemorrhoids (varicosities of superior rectal veins) Covered by rectal mucosa Patients do not complain of pain	External hemorrhoids (varicosities of inferior rectal veins) Covered by skin Patients do complain of pain

IVC = inferior vena cava.

● **Figure 12-4 Diagram of the anal canal.** Note the following structures: anal columns (AC), anal verge (AV), pectinate line (PL), internal anal sphincter (IS), and external anal sphincter (ES). AR = ampulla of the rectum, PR = puborectalis muscle.

1.   **Upper Anal Canal**
     a.   The upper anal canal extends from the anorectal junction (perineal flexure) to the pectinate line.
     b.   The mucosa of the upper anal canal is thrown into longitudinal folds called the **anal columns (of Morgagni)**. The base of the anal columns defines the **pectinate line**.

c. At the base of the anal columns are folds of tissue called the **anal valves.** Behind the anal valves are small, blind pouches called the **anal sinuses,** into which the **anal glands** open.

d. The upper anal canal is predominately surrounded by the **internal anal sphincter,** which is a continuation of smooth muscle from the rectum with involuntary control via autonomic innervation.

2. Lower Anal Canal

a. The lower anal canal extends from the pectinate line to the **anal verge** (the point at which perianal skin begins).

b. The lower anal canal is predominately surrounded by the **external anal sphincter,** which is striated muscle under voluntary control via the pudendal nerve.

B. ARTERIAL SUPPLY

1. The arterial supply of the **upper anal canal** is from the **superior rectal artery** (abdominal aorta → inferior mesenteric artery → superior rectal artery).

2. The arterial supply of the **lower anal canal** is from the **inferior rectal artery** (abdominal aorta → common iliac artery → internal iliac artery → internal pudendal artery → inferior rectal artery).

3. The **middle rectal artery** (abdominal aorta → common iliac artery → internal iliac artery → middle rectal artery) forms an anastomosis with the superior and inferior rectal arteries.

C. VENOUS DRAINAGE

1. The venous drainage of the **upper anal canal** is to the **superior rectal vein** (superior rectal vein → inferior mesenteric vein → portal vein → hepatic sinusoids → central veins → hepatic veins → inferior vena cava).

2. The venous drainage of the **lower anal canal** is to the **inferior rectal vein** (inferior rectal vein → internal pudendal vein → internal iliac vein → common iliac vein → inferior vena cava).

D. INNERVATION

1. The innervation of the **upper anal canal** is via the **autonomic nervous system** (parasympathetic and sympathetic nervous systems) such that the internal anal sphincter is under autonomic, nonvoluntary control and sensation is limited to stretch sensation. See Chapter 11 VIII.

2. The innervation of the **lower anal canal** is via the **somatic nervous system** by the **pudendal nerve** such that the external anal sphincter is under voluntary control and sensation is expanded to pain, temperature, and touch.

E. CLINICAL CONSIDERATIONS

1. **Internal hemorrhoids** are varicosities of the **superior rectal veins.** They are located above the pectinate line and are covered by rectal mucosa. Clinical findings include bleeding, mucous discharge, prolapse, pruritus, and are painless.

2. **External hemorrhoids** are varicosities of the **inferior rectal veins.** They are located below the pectinate line near the anal verge and are covered by skin. Clinical findings include bleeding, swelling, and are painful.

IV   Defecation Reflex. Sensory impulses from **pressure-sensitive receptors** within the ampulla of the rectum travel to sacral spinal cord levels when feces is present. Motor impulses travel with the **pelvic splanchnic nerves (parasympathetics; S2–4),** which increase peristalsis and relax the internal anal sphincter. If the external anal sphincter and puborectalis muscle are also relaxed, defecation takes place with the help of contraction of the anterior abdominal wall muscles and closure of the glottis. If the external anal sphincter and puborectalis muscle are voluntarily contracted via the **pudendal nerve,** defecation is delayed and the feces moves back into the sigmoid colon for storage. The **hypogastric**

plexus and **lumbar splanchnic nerves (sympathetics)** decrease peristalsis and maintain tone of the internal anal sphincter.

Ⓥ Radiology

A. ANTEROPOSTERIOR BARIUM RADIOGRAPH (FIGURE 12-5A)

B. LATERAL BARIUM RADIOGRAPH (FIGURE 12-5B)

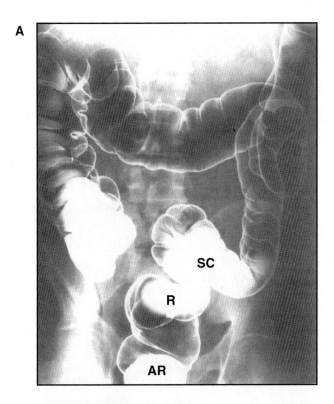

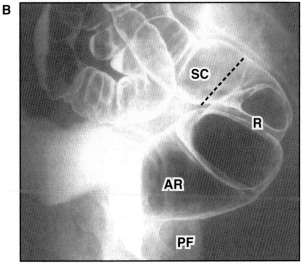

● **Figure 12-5 Radiology. A:** Anteroposterior barium radiograph shows the sigmoid colon (SC), rectum (R), and ampulla of the rectum (AR). **B:** Lateral barium radiograph shows the sigmoid colon (SC), rectosigmoid junction (*dotted line*), rectum (R), ampulla of the rectum (AR), and perirenal flexure (PF).

Chapter 13

Spleen

I ⬤ General Features (Figure 13-1)

A. The spleen is located in the left hypochondriac region posterior to the 9th, 10th, and 11th ribs, which puts the spleen in jeopardy in the case of rib fractures.

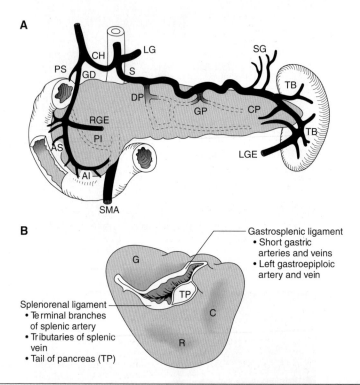

AI = anterior-inferior pancreaticoduodenal artery
AS = anterior-superior pancreaticoduodenal artery
C = colon depression
CH = common hepatic artery
CP = caudal pancreatic arteries
DP = dorsal pancreatic artery
G = gastric depression
GD = gastroduodenal artery
GP = great pancreatic artery
LG = left gastric artery

LGE = left gastroepiploic artery
PI = posterior-inferior pancreaticoduodenal artery
PS = posterior-superior pancreaticoduodenal artery
R = renal depression
RGE = right gastroepiploic artery
SG = short gastric arteries
SMA = superior mesenteric artery
TB = terminal branches of splenic artery
TP = tail of pancreas

⬤ **Figure 13-1 Spleen. A:** Diagram of arterial supply of the spleen. The splenic artery (S) is the largest branch of the celiac trunk. **B:** Diagram of visceral surface of the spleen. The gastrosplenic ligament and splenorenal ligament are shown along with the structures they contain.

B. The spleen does not extend below the costal margin and therefore is not palpable unless splenomegaly is present.

C. The spleen is attached to the stomach by the **gastrosplenic ligament**, which contains the **short gastric arteries and veins** and the **left gastroepiploic artery and vein.**

D. The spleen is attached to the kidney by the **splenorenal ligament**, which contains the **five terminal branches of the splenic artery, tributaries of the splenic vein,** and the **tail of the pancreas.**

E. **ACCESSORY SPLEENS** occur in 20% of the population and are commonly located near the hilum, near the tail of the pancreas, or within the gastrosplenic ligament.

F. The functions of the spleen include removal of old or abnormal red blood cells (RBCs), removal of inclusion bodies from RBCs (e.g., **Howell-Jolly bodies** [nuclear remnants], **Pappenheimer bodies** [iron granules], **Heinz bodies** [denatured hemoglobin]), removal of poorly opsonized pathogens, IgM production by plasma cells, storage of platelets, and protection from infection.

Ⅱ Arterial Supply

A. The arterial supply is from the **splenic artery** (the largest branch of the celiac trunk), which gives off the following branches: **dorsal pancreatic artery, great pancreatic artery, caudal pancreatic arteries, short gastric arteries,** and **left gastroepiploic artery;** it ends with about **five terminal branches.**

B. The five terminal branches of the splenic artery supply individual segments of the spleen with no anastomosis between them (i.e., **end arteries**) so that obstruction or ligation of any terminal branch will result in **splenic infarction** (i.e., the spleen is very prone to infarction).

C. **SPLENIC ARTERY ANEURYSMS** show a particularly high incidence of rupture in **pregnant women** such that these aneurysms should be resected in women of childbearing age.

Ⅲ Venous Drainage

A. The venous drainage is to the **splenic vein** via tributaries.

B. The splenic vein joins the superior mesenteric vein to form the portal vein.

C. The inferior mesenteric vein usually joins the splenic vein.

D. **SPLENIC VEIN THROMBOSIS** is most commonly associated with **pancreatitis** and shows the following clinical signs: gastric varices and upper gastrointestinal bleeding.

Ⅳ Clinical Considerations

A. **SPLENECTOMY** is the surgical removal of the spleen. Nearby anatomic structures may be injured during a splenectomy, including the **gastric wall (stomach)** if the short gastric arteries are compromised, **tail of pancreas** if the caudal pancreatic arteries are compromised or during manipulation of the splenorenal ligament, and **left kidney** during manipulation of the splenorenal ligament. The most common complication of a splenectomy is **atelectasis of the left lower lobe of the lung. Thrombocytosis** (i.e., increased number of platelets within the blood) is common postoperatively such that anticoagulation therapy may be necessary to prevent spontaneous thrombosis. Abnormal RBCs with bizarre shapes, some of which contain **Howell-Jolly bodies** (nuclear remnants), are found in the blood postoperatively.

B. SPLENIC VEIN THROMBOSIS most commonly is associated with pancreatitis and is one of the causes of splenomegaly. Clinical signs include gastric varices and upper gastrointestinal bleeding.

C. SPLENOMEGALY. The causes of splenomegaly include autoimmune disease (e.g., systemic lupus erythematous, rheumatoid arthritis), infectious disease (e.g., mononucleosis, visceral leishmaniasis), infiltrative disease (e.g., lysosomal storage disease, leukemias), extramedullary hematopoiesis (e.g., myeloproliferative diseases such as myelofibrosis and myeloid metaplasia), and vascular congestion (portal hypertension in cirrhosis). In the United States, myeloproliferative disease and lymphoid malignancies (e.g., chronic lymphocytic leukemia) are the most common causes of massive splenomegaly.

D. SPLENIC INFARCT (FIGURE 13-2). An infarction is a process by which coagulating necrosis develops in an area distal to the occlusion of an end artery. The necrotic tissue or zone is called an infarct. The computed tomography (CT) scan shows multiple wedge-shaped areas of diminished contrast enhancement in the spleen representing multiple areas of embolic infarction (*arrows*).

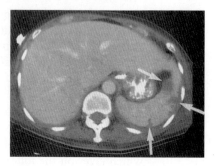

● Figure 13-2 Splenic infarction.

Ⓥ Radiology

A. CT SCAN AT THE LEVEL OF THE LIVER AND SPLEEN (FIGURE 13-3)

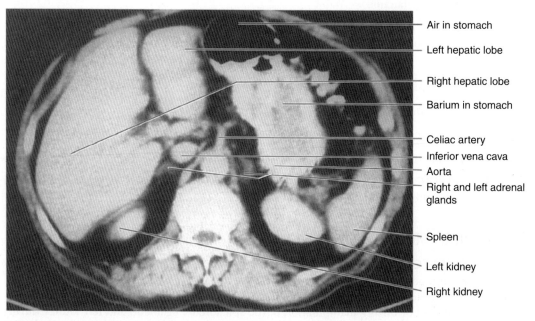

Air in stomach
Left hepatic lobe
Right hepatic lobe
Barium in stomach
Celiac artery
Inferior vena cava
Aorta
Right and left adrenal glands
Spleen
Left kidney
Right kidney

● Figure 13-3 Computed tomography scan at the level of the liver and spleen.

Chapter 14

Kidney, Ureter, Bladder, and Urethra

I. General Features (Figure 14-1A)

    A. The kidneys are retroperitoneal organs that lie on the ventral surface of the quadratus lumborum muscle and lateral to the psoas muscle and vertebral column.

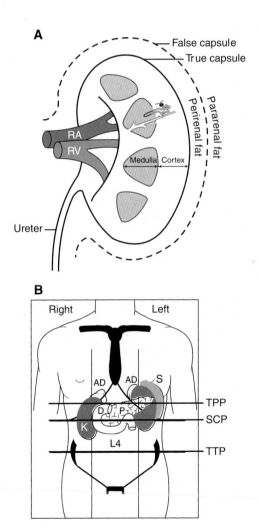

● **Figure 14-1 A:** Diagram of general features of the kidney. Note the true capsule and false capsule. Also note the location of the perirenal fat and pararenal fat. **B:** Anterior kidney surface projections. A commonly used clinical method for subdividing the abdomen into specific regions using the subcostal plane (SCP), transtubercular plane (TTP; joining the tubercles of the iliac crest), and midclavicular lines is shown. Note also the transpyloric plane (TPP). RA = renal artery, RV = renal vein, AD = adrenal glands, D = duodenum, K = kidney, P = pancreas, S = spleen.

B. The kidneys are directly covered by a fibrous capsule called the **renal capsule (or true capsule)**, which can be readily stripped from the surface of the kidney except in some pathologic conditions where it is strongly adherent due to scarring.

C. The kidneys are further surrounded by the **perirenal fascia of Gerota (or false capsule)**, which is important in staging renal cell carcinoma. The perirenal fascia of Gerota defines the **perirenal space** that contains the **kidney, adrenal gland, ureter, gonadal artery and vein,** and **perirenal fat.**

D. Any fat located outside the perirenal space is called **pararenal fat,** which is most abundant posterolaterally.

E. At the concave medial margin of each kidney is a vertical cleft called the **renal hilum,** where the following anatomic structures are arranged in an anterior-to-posterior direction: **renal vein (most anterior)** → renal artery → **renal pelvis (most posterior).**

F. The renal hilum is continuous with a space called the **renal sinus** that contains the renal pelvis, major and minor calyces, renal blood vessels, nerves, lymphatics, and a variable amount of fat.

Ⅱ Kidney Surface Projections (Figure 14-1B)

A. LEFT KIDNEY
1. The upper pole of the left kidney is located at about **vertebral level T11. The left kidney is higher than the right kidney.**
2. The left kidney is related to rib 11 and rib 12.
3. The renal hilum of the left kidney lies 5 cm from the median plane along the transpyloric plane (which passes through vertebral level L1).

B. RIGHT KIDNEY
1. The upper pole of the right kidney is located at about **vertebral level T12. The right kidney is lower than the left kidney** due to the presence of the liver on the right side.
2. The right kidney is related to rib 12.
3. The renal hilum of the right kidney lies 5 cm from the median plane just below the transpyloric plane (which passes through vertebral level L1).

Ⅲ Internal Macroscopic Anatomy of the Kidney. A coronal section through the kidney reveals the following macroscopic structures:

A. RENAL CORTEX
1. The renal cortex lies under the renal capsule and also extends between the renal pyramids as the **renal columns of Bertin.**
2. The renal cortex may be divided into the **cortical labyrinth** and the **medullary rays.**

B. RENAL MEDULLA
1. The renal medulla is composed of **5 to 11 renal pyramids of Malpighi** whose tips terminate as **5 to 11 renal papillae.** The base of a renal pyramid abuts the renal cortex, whereas the tip of a renal pyramid (i.e., the renal papillae) abuts a minor calyx.
2. The renal medulla may be divided into the **outer medulla** and **inner medulla.**
3. The **papillary ducts of Bellini** open onto the surface of the renal papillae at the **area cribrosa.**

C. 5 TO 11 MINOR CALYCES
1. The minor calyces are cup-shaped structures that abut the renal papillae.
2. Each minor calyx may receive one to three renal papillae.

D. 2 TO 3 MAJOR CALYCES
1. The major calyces are continuous with the minor calyces.

E. RENAL PELVIS
1. The renal pelvis is continuous with the major calyces.
2. The renal pelvis tapers inferomedially as it traverses the renal hilum to become continuous with the ureter at the **ureteropelvic junction**.

IV Arterial Supply (Figure 14-2A)

A. RENAL ARTERY
1. The abdominal aorta branches between vertebral level L1–2 into the **right renal artery** and **left renal artery**.
2. The longer right renal artery passes posterior to the inferior vena cava (IVC) on its path to the right kidney.
3. Each renal artery gives rise to the **inferior suprarenal arteries**.

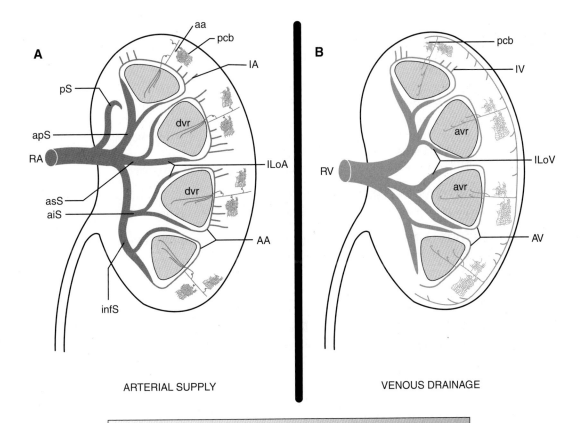

ARTERIAL SUPPLY   VENOUS DRAINAGE

AA = arcuate artery	IA = interlobular artery
Aa = afferent arteriole	ILoA = interlobar artery
aa = afferent arteriole	ILoV = interlobar vein
aiS = anteroinferior segmental artery	infS = inferior segmental artery
apS = apical segmental artery	IV = interlobular vein
asS = anterosuperior segmental artery	pcb = peritubular capillary bed
AV = arcuate vein	pS = posterior segmental artery
avr = ascending vasa recta	RA = renal artery
dvr = descending vasa recta	RV = renal vein

● **Figure 14-2 Vasculature of the kidney. A:** Arterial supply of the kidney. **B:** Venous drainage of the kidney.

4. Near the renal hilum, each renal artery divides into an **anterior division** and a **posterior division**.

B. ANTERIOR AND POSTERIOR DIVISIONS

1. The anterior division branches into **four anterior segmental arteries**, which supply anterior segments of the kidney called the **apical segmental artery, anterosuperior segmental artery, anteroinferior segmental artery,** and **inferior segmental artery.**

2. The posterior division continues as the **posterior segmental artery**, which supplies the posterior segment of the kidney.

C. SEGMENTAL ARTERIES

1. The segmental arteries are **end arteries** (i.e., they do not anastomose) and are distributed to various segments of the kidney. Segmental arteries have the following clinical importance:

   a. Since there is very little collateral circulation between segmental arteries (i.e., end arteries), an **avascular line (the Brodel white line)** is created between anterior and posterior segments such that a longitudinal incision through the kidney will produce minimal bleeding. This approach is useful for surgical removal of renal (staghorn) calculi.

   b. Ligation of a segmental artery results in necrosis of the entire segment of the kidney.

   c. **Supernumerary (or aberrant) segmental arteries** are arteries that form during fetal development and persist in the adult. They may arise from either the renal artery **(hilar)** or directly from the aorta **(polar)**. Ligation of a supernumerary segmental artery results in necrosis of the entire segment of the kidney.

2. The segmental arteries branch into **5 to 11 interlobar arteries.**

D. INTERLOBAR ARTERIES.
The interlobar arteries branch into the **arcuate arteries**, which travel along the base of the renal pyramids at the corticomedullary junction.

E. ARCUATE ARTERIES.
The arcuate arteries branch into the **interlobular arteries**, which travel through the cortex toward the capsule.

F. INTERLOBULAR ARTERIES.
The interlobular arteries branch into numerous **afferent arterioles.**

G. AFFERENT ARTERIOLES.
Each afferent arteriole forms a capillary bed (or tuft) called the **renal glomerulus**, which is drained by an **efferent arteriole.**

H. EFFERENT ARTERIOLES

1. The efferent arteriole of renal glomeruli from cortical and midcortical nephrons branches into a **cortical peritubular capillary bed.**

2. The efferent arteriole of renal glomeruli from juxtamedullary nephrons branches into **12 to 25 descending vasa recta**, which are long, straight capillaries that run to varying depths of the medulla.

3. The ends of the descending vasa recta give rise to a **medullary peritubular capillary bed.**

Ⓥ Venous Drainage (Figure 14-2B)

A. The venous drainage of the kidney is to the interlobular veins → arcuate veins → interlobar veins → renal vein → IVC.

B. The arcuate veins drain into **interlobar veins**, which anastomose and converge to form several renal veins that unite in a variable fashion to form the **renal vein.**

C. The veins draining the kidney have no segmental organization like the arterial supply.

D. The renal veins lie anterior to the renal arteries at the renal hilum.

E. The longer left renal vein passes anterior to the aorta on its path to the IVC. The renal veins ultimately drain into the IVC.

VI   Innervation. The kidney is innervated by the **renal plexus**, which is intimately associated with the renal artery. The lower part of the celiac ganglion is more or less detached as the **aorticorenal ganglion**, which is located at the origin of the renal artery from the abdominal aorta. The aorticorenal ganglion receives predominately the **lesser thoracic splanchnic nerve** and **least thoracic splanchnic nerve** and forms most of the **renal plexus**. The renal plexus contains only **sympathetic components**. There is no (or at least very minimal) parasympathetic innervation of the kidney.

A. PARASYMPATHETIC. None (or at least very minimal)

B. SYMPATHETIC
1. Preganglionic neuronal cell bodies are located in the **intermediolateral cell column** of the spinal cord. Preganglionic axons pass through the paravertebral ganglia (do not synapse) to become the **lesser thoracic splanchnic nerve, least thoracic splanchnic nerve, first lumbar splanchnic nerve,** and **second lumbar splanchnic nerve** and travel to the **aorticorenal ganglion.**
2. Postganglionic neuronal cell bodies are located in the aorticorenal ganglion.
3. Postganglionic axons enter the **renal plexus** and are distributed to **renal vasculature** including the **juxtaglomerular cells**, where they play an important role in the regulation of blood pressure by effecting **renin release.**

C. SENSORY INNERVATION
1. Afferent (sensory) neurons whose cell bodies are located in the **dorsal root ganglion** run with the **least thoracic splanchnic nerve, first lumbar splanchnic nerve,** and **second lumbar splanchnic nerve** and relay **pain** sensation from the kidney to T12–L2 spinal cord segments within the central nervous system (CNS).
2. The pain associated with kidney pathology may be referred over the **T12–L2 dermatomes** (i.e., lumbar region, inguinal region, and anterosuperior thigh).
3. Note that the sensory innervation runs with the sympathetic component.

VII   Clinical Considerations of the Kidney

A. ROTATION OF THE KIDNEY. During the relative ascent of the kidneys in fetal development, the kidneys **rotate 90 degrees medially** so that the renal hilus is normally orientated in a medial direction.

B. ASCENT OF THE KIDNEY. The fetal metanephros is located in the sacral region, whereas the adult kidneys are normally located at vertebral levels T12–L3. The change in location (i.e., ascent) results from a disproportionate growth of the fetus caudal to the metanephros.

C. HORSESHOE KIDNEY occurs when the inferior poles of both kidneys fuse during fetal development. The horseshoe kidney gets trapped behind the **inferior mesenteric artery** as the kidney attempts to ascend toward the normal adult location.

D. KIDNEY TRAUMA. Kidney trauma should be suspected in the following situations: fracture of the lower ribs, fracture of the transverse processes of lumbar vertebrae, gunshot or knife wound over the lower rib cage, and after a car accident when seat belt marks are present. Right kidney trauma is associated with liver trauma, whereas left kidney trauma is associated with spleen trauma. Clinical findings include flank mass and/or tenderness, flank ecchymosis, hypotension, and hematuria. One of the absolute indications for renal exploration is the presence of a **pulsatile or expanding retroperitoneal hematoma** found at laparotomy.

E. SURGICAL APPROACH TO THE KIDNEY. An incision is made below and parallel to the 12th rib in order to prevent inadvertent entry into the pleural space. The incision may be extended to the front of the abdomen by traveling parallel to the inguinal ligament.

VIII Ureter

A. GENERAL FEATURES
1. The ureters begin at the **ureteropelvic junction**, where the renal pelvis joins the ureter.
2. Within the abdomen, the ureters descend **retroperitoneal** and anterior to the **psoas major** muscle, where they cross the pelvic inlet to enter the minor (or true) pelvis.
3. Within the minor (or true) pelvis, the ureters descend **retroperitoneal** and anterior to the **common iliac artery and vein**, where they may be compromised by an aneurysm of the common iliac artery.
4. The ureters end at the **ureterovesical junction** surrounded by the **vesical venous plexus**.
5. The ureters end by traveling obliquely through the wall of the urinary bladder (i.e., the **intramural portion of the ureter**) and define the upper limit of the **urinary bladder trigone**.
6. The intramural portion of the ureter functions as a check valve (**ureterovesical valve of Sampson**) to prevent urine reflux.

B. URETER RELATIONSHIPS TO NEIGHBORING STRUCTURES
1. **In the male,** the ureters pass posterior to the **ductus deferens**.
2. **In the female,** the ureters pass posterior and inferior to the **uterine artery**, which lies in the **transverse cervical ligament** (or **cardinal ligament of Mackenrodt**), and lie 1 to 2 cm lateral to the **cervix of the uterus**. During gynecologic operations (e.g., hysterectomy), the ureters may be inadvertently injured. The most common sites of injury are at the pelvic brim, where the ureter is close to the ovarian blood vessels, and where the uterine artery crosses the ureter along the side of the cervix.

C. NORMAL CONSTRICTIONS OF THE URETER. The ureters are normally constricted at three sites, where kidney stones most commonly cause obstruction:
1. **At the ureteropelvic junction**
2. **Where the ureters cross the pelvic inlet**
3. **At the ureterovesical junction (along the intramural portion of the ureter)**

D. ARTERIAL SUPPLY
1. The arteries supplying the ureter are derived from the **abdominal aorta, renal artery, testicular artery, ovarian artery, common iliac artery, internal iliac artery, inferior vesical artery,** and **uterine artery**. Branches from these arteries supply different parts of the ureter along its course and are subject to much variation.
2. The most constant arterial supply of the lower part of the ureter is the **uterine artery** in the female and the **inferior vesical artery** in the male.

3. The longitudinal anastomosis between these branches may be weak so that inadvertent damage of one of these branches may lead to necrosis of a ureteral segment about 1 week postoperatively.

E. VENOUS DRAINAGE. The veins draining the ureter follow the arterial supply, although there is a conspicuous **vesical venous plexus** surrounding the end of the ureter.

F. INNERVATION. The ureter is innervated by the **ureteric plexus**. In the upper part of the ureter, the ureteric plexus receives input from the **renal plexus** and **abdominal aortic plexus**. In the intermediate part of the ureter, the ureteric plexus receives input from the **superior hypogastric plexus**. In the lower part of the ureter, the ureteric plexus receives input from the **inferior hypogastric plexus**. The ureteric plexus contains both **parasympathetic** and **sympathetic components**, although they do not play a major role in ureteral peristalsis but only a modulatory role.

G. CLINICAL CONSIDERATION (FIGURE 14-3). Renal calculi ("kidney stones") obstruction occurs most often at the three sites where the ureter normally constricts (see earlier), causing a **unilateral hydronephrosis**. Clinical findings include intermittent, excruciating pain in the flank area, abdomen, or testicular or vulvar region radiating onto the inner thigh depending on the obstruction site; fever, hematuria, and decreased urine output may be present, and the patient assumes a posture with a severe ipsilateral costovertebral angle. There are four types of kidney stones:

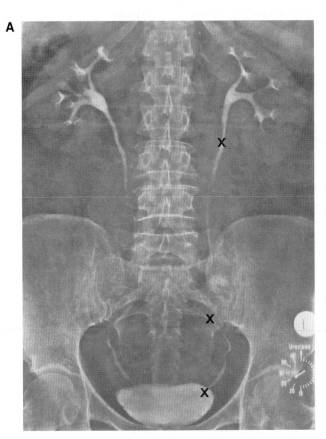

● **Figure 14-3 Intravenous urogram (IVU) of normal kidney and computed tomography (CT) scan of renal calculi. A:** The IVU shows the normal collecting system of the kidney and the ureter. The ureters are normally constricted at three sites (X), where kidney stones most commonly cause obstruction. (*continued*)

B

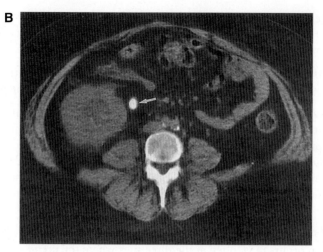

● **Figure 14-3** *Continued* **B:** The CT scan shows a large, obstructing calculus ("kidney stone") in the ureter (*arrow*).

1. **Calcium oxalate calculi (Figure 14-4)** are radiopaque. By urinalysis, they are colorless, octahedral-shaped crystals that look like small squares crossed by diagonal lines; rarely, they are dumbbell shaped. They are the most common (80%) type of calculi and form when urine pH is <6.0 (acid pH) or with neutral pH. Calcium oxalate calculi are associated with absorptive hypercalcemia, vitamin D intoxication, hyperparathyroidism, milk-alkali syndrome, and renal tubular acidosis, all of which result in **hypercalcemia**; diabetes; liver disease; and ethylene glycol poisoning. The photograph of calcium oxalate calculi shows that these kidney stones are colorless, octahedral-shaped crystals that look like small squares crossed by intersecting diagonal lines.

● **Figure 14-4** Calcium oxalate calculi.

2. **Magnesium ammonium sulfate (struvite; triple phosphate) calculi (Figure 14-5)** are radiopaque. By urinalysis, they are colorless, rectangular, prism-shaped crystals. They are the second most common (15%) type of calculi, generally progress to form staghorn calculi, and form when urine pH is <7.4 (alkaline pH). Magnesium ammonium sulfate calculi are associated with urinary tract infections by urea-splitting bacteria (e.g., *Proteus mirabilis*, *Proteus vulgaris*, *Providencia*, *Pseudomonas*, *Klebsiella*, and *Staphylococcus*). The photograph of magnesium ammonium sulfate (struvite or triple phosphate) calculi shows that these kidney stones are colorless, rectangular, prism-shaped crystals.

● **Figure 14-5** Magnesium ammonium sulfate calculi.

3. **Uric acid calculi (Figure 14-6)** are radiolucent. By urinalysis, they are yellow or red-brown diamond-shaped crystals. They are the third most common (5%) type of calculi and form when urine pH is <6.0 (acid pH). Uric acid calculi are associated with gout, leukemia, Lesch-Nyhan syndrome, and myeloproliferative disorders. The photograph of uric acid calculi shows that these kidney stones are yellow or red-brown in color and diamond-shaped crystals.

● **Figure 14-6 Uric acid calculi.**

4. **Cystine calculi (Figure 14-7)** are faintly radiopaque. By urinalysis, they are colorless, refractile, hexagonal-shaped crystals that may have a layered appearance. They are the least common (1%) type of calculi and form when urine pH is <6.0 (acid pH). Cystine calculi are caused by **cystinuria**, which is an autosomal recessive disorder that results in defective renal tubular reabsorption of the amino acids cystine, ornithine, arginine, and lysine. The photograph of cystine calculi shows that these kidney stones are colorless, refractile, hexagonal-shaped crystals that may have a layered appearance.

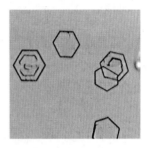

● **Figure 14-7 Cystine calculi.**

IX Urinary Bladder (Figure 14-8)

A. GENERAL FEATURES

1. The urinary bladder is a hollow structure with prominent smooth muscle walls that is a temporary reservoir for urine and has a capacity of 120 to 320 mL.
2. The empty bladder is tetrahedral shaped and consists of a **posterior surface (fundus or base)**, **anterior surface**, **superior surface**, **apex**, and **neck**.

B. URINARY BLADDER RELATIONSHIPS TO NEIGHBORING STRUCTURES

1. **Posterior Surface (Fundus or Base)**
    a. In the male, the posterior surface is related to the **rectovesical pouch, rectum, seminal vesicles**, and **ampulla of the ductus deferens.**
    b. In the female, the posterior surface is related to the **anterior wall of the vagina.**
2. **Anterior Surface**
    a. In the male and female, the anterior surface is related to the **pubic symphysis** and **retropubic space (of Retzius).**
3. **Superior Surface**
    a. In the male, the superior surface is related to the **peritoneal cavity** (completely covered by peritoneum), sigmoid colon, and terminal coils of the ileum.
    b. In the female, the superior surface is related to the **peritoneal cavity** (largely covered by peritoneum but reflected posteriorly to the uterus, forming the **vesicouterine pouch**) and **uterus.**
4. **Apex**
    a. The apex is located posterior to the upper part of the pubic symphysis.
    b. In the male and female, the apex is related to the one **median umbilical ligament** or **urachus** (a remnant of the allantois in the fetus), the two **medial umbilical**

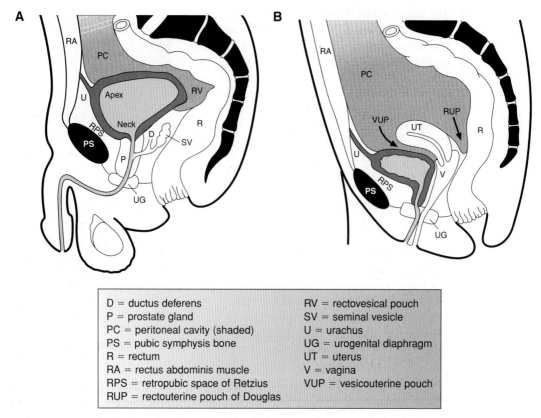

A

RA
PC
U
Apex
RV
RPS
Neck
PS
P
D
SV
R
UG

B

RA
PC
VUP
RUP
UT
R
U
V
RPS
PS
UG

D = ductus deferens	RV = rectovesical pouch
P = prostate gland	SV = seminal vesicle
PC = peritoneal cavity (shaded)	U = urachus
PS = pubic symphysis bone	UG = urogenital diaphragm
R = rectum	UT = uterus
RA = rectus abdominis muscle	V = vagina
RPS = retropubic space of Retzius	VUP = vesicouterine pouch
RUP = rectouterine pouch of Douglas	

● **Figure 14-8 Sagittal view of male and female pelvis. A:** Male pelvis. A sagittal section through the male pelvis demonstrating the various anatomic relationships of the urinary bladder. **B:** Female pelvis. A sagittal section through the female pelvis demonstrating the various anatomic relationships of the urinary bladder.

ligaments (remnants of the right and left umbilical arteries in the fetus), and the two **lateral umbilical ligaments** (elevations formed by the right and left inferior epigastric arteries and veins).

5. **Neck**
   a. The neck is the lowest region of the bladder and is located posterior to the lower part of the pubic symphysis. The neck is pierced by the **internal urethral orifice**.
   b. In the male, the neck is related to the **prostate gland** and **prostatic urethra**.
   c. In the female, the neck is related to the **urogenital diaphragm**.

C. **SUPPORT OF THE BLADDER.** The support of the urinary bladder involves the:
   1. **Urogenital Diaphragm**
   2. **Pubovesical Ligaments**
      a. The pubovesical ligaments are extensions of the **puboprostatic ligaments (in the male)** and **pubourethral ligaments (in the female)**.
      b. The pubovesical ligaments extend from the lower portion of the pubic bone to the neck of the bladder.
   3. **Median Umbilical Ligament or Urachus**
      a. The median umbilical ligament or urachus (a remnant of the allantois in the fetus) extends from the umbilicus to the apex of the bladder.
   4. **False Ligaments**
      a. The false ligaments are reflections or folds of peritoneum.

D. INTERNAL ANATOMY OF THE BLADDER

1. The **trigone of the bladder** is always **smooth surfaced** because the mucosa is tightly attached to the detrusor muscle.
2. The **trigone of the bladder** is located on the posterior surface of the bladder (fundus or base) and its limits are defined superiorly by the **openings of the ureters** and inferiorly by the **internal urethral orifice.**

E. ARTERIAL SUPPLY.
The arterial supply of the bladder is from the **superior vesical artery** (a branch of the internal iliac artery), **inferior vesical artery** (a branch of the internal iliac artery), **obturator artery,** and **inferior gluteal artery.** In the female, branches of the **uterine artery** and **vaginal artery** also supply the bladder.

F. VENOUS DRAINAGE.
The venous drainage of the bladder is to a complicated **venous plexus** along the inferolateral portion of the bladder → **internal iliac vein** → prostatic **venous plexus.**

G. INNERVATION.
The bladder is innervated by the **vesical plexus,** which receives input from the **inferior hypogastric plexus.** The vesical plexus contains both **parasympathetic** and **sympathetic** components.

1. **Parasympathetic**
   a. Preganglionic neuronal cell bodies are located in the intermediolateral cell column of the S2–4 spinal cord segments. Preganglionic axons travel to the vesical plexus as the **pelvic splanchnic nerves.**
   b. Postganglionic neuronal cell bodies are located in the vesical plexus and the bladder wall.
   c. Postganglionic axons are distributed to the detrusor muscle of the bladder, where they cause **contraction of the detrusor muscle** and **relaxation of the internal urethral sphincter** (i.e., efferent limb of the micturition reflex).

2. **Sympathetic**
   a. Preganglionic neuronal cell bodies are located in the **intermediolateral cell column** of the spinal cord. Preganglionic axons pass through the paravertebral ganglia (do not synapse) to become the **lesser thoracic splanchnic nerve, least thoracic splanchnic nerve, first lumbar splanchnic nerve,** and **second lumbar splanchnic nerve** and travel to the **inferior hypogastric plexus** by way of the superior hypogastric plexus.
   b. Postganglionic neuronal cell bodies are located in the inferior hypogastric plexus.
   c. Postganglionic axons enter the vesical plexus and are distributed to the detrusor muscle of the bladder, where they cause **relaxation of the detrusor muscle** and **contraction of the internal urethral sphincter** (although some investigators claim their action is strictly on the smooth muscle of blood vessels).

3. **Sensory Innervation.** Sensory information from the bladder is carried by both parasympathetics (mainly) and sympathetics.
   a. Parasympathetic
      i. Afferent (sensory) neurons whose cell bodies are located in the **dorsal root ganglion** run with the **pelvic splanchnic nerves** and relay **pain** and **stretch** information from the bladder to the S2–4 spinal segments within the CNS.
      ii. The **pain** associated with bladder pathology may be referred over the **S2–4 dermatomes** (i.e., perineum and posterior thigh).
      iii. The **stretch** information associated with bladder fullness from stretch receptors in the bladder wall runs with the pelvic splanchnic nerves and serves as the **afferent limb in the micturition reflex.**

b. **Sympathetic**
    i. Afferent (sensory) neurons whose cell bodies are located in the **dorsal root ganglion** run with the lesser thoracic splanchnic nerve, least thoracic splanchnic nerve, first lumbar splanchnic nerve, and second lumbar splanchnic nerve and relay pain information from the bladder to the T11–L2 spinal cord segments with the CNS.
    ii. The pain associated with bladder pathology may be referred over the **T11–L2** dermatomes (i.e., lumbar region, inguinal region, and anterosuperior thigh).

4. **Micturition Reflex.** As the bladder fills with urine, **stretch** information associated with bladder fullness from stretch receptors in the bladder wall runs with the pelvic splanchnic nerves and serves as the afferent limb in the micturition reflex. Pelvic splanchnic nerves are distributed to the detrusor muscle of the bladder, where they cause **contraction of the detrusor muscle** and **relaxation of the internal urethral sphincter** (i.e., efferent limb of the micturition reflex). The **external urethral sphincter** is innervated by the **pudendal nerve** and is voluntarily relaxed.

H. CLINICAL CONSIDERATIONS

1. **Location of the Urinary Bladder.** In the adult, the empty bladder lies within the **minor (true) pelvis.** In the infant, the empty bladder lies within the **abdominal cavity.** As the bladder fills in the adult, it rises out of the minor pelvis above the pelvic inlet and may extend as high as the umbilicus. In acute retention of urine, a needle may be passed through the anterior abdominal wall (skin → superficial fascia [Camper and Scarpa] → linea alba → transversalis fascia → extraperitoneal fat → parietal peritoneum) without entering the peritoneal cavity in order to remove the urine (**suprapubic cystostomy**).

2. **Urine Leakage Due to Trauma (Figure 14-9)**
    a. **Rupture of the superior wall (dome)** results in an intraperitoneal extravasation of urine within the **peritoneal cavity.** It is caused by a compressive force on a full bladder.
    b. **Rupture of the anterior wall** results in an extraperitoneal extravasation of urine within the **retropubic space (of Retzius).** It is caused by a fractured pelvis (e.g., car accident) that punctures the bladder.
    c. **Type I urethral injury** occurs when the posterior urethra is stretched but intact due to the rupture of the puboprostatic ligaments. Type I urethral injuries are rare.
    d. **Type II urethral injury** occurs when the posterior urethra is torn **above the urogenital diaphragm.** This results in an extraperitoneal extravasation of urine within the **retropubic space of Retzius.** It may be caused by a fractured pelvis (e.g., car accident) or improper insertion of a catheter.
    e. **Type III urethral injury** occurs when the anterior urethra (i.e., bulbous urethra) is torn **below the urogenital diaphragm** along with a disruption of the urogenital diaphragm so that the membranous urethra is torn also. Radiologists consider a type III urethral injury as a combined anterior/posterior urethral injury. This results in an extraperitoneal extravasation of urine within the **superficial perineal space** extending into the scrotal, penile, and anterior abdominal wall areas (urine will *not* extend into the thigh region or anal triangle). The superficial perineal space is located between Colles fascia and dartos muscle and the external spermatic fascia. It is caused by a **straddle injury** (e.g., a boy slips off a bicycle seat and falls against the crossbar) and is the most common type of urine leakage injury. Clinical findings include blood at the urethral meatus, ecchymosis, painful swelling of the scrotal and perineal areas, and tender enlargement in the suprapubic region due to a full bladder.

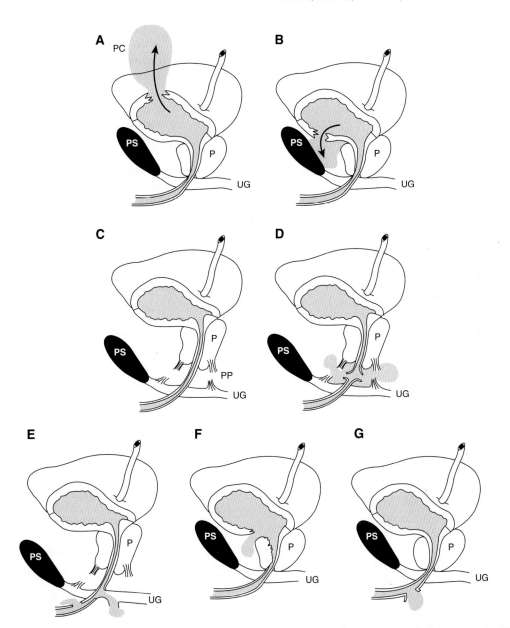

● **Figure 14-9 Urine leakage due to trauma. A:** Diagram shows a rupture of the superior wall of the urinary bladder that results in extravasation of urine into the peritoneal cavity (PC). **B:** Diagram shows a rupture of the anterior wall of the urinary bladder that results in extravasation of urine into the retropubic space of Retzius. **C:** Type I urethral injury. Diagram shows a stretched but intact posterior urethra. Note the rupture of the puboprostatic ligaments (PP). **D:** Type II urethral injury. Diagram shows a torn posterior urethra above the urogenital diaphragm that results in the extravasation of urine into the retropubic space of Retzius. **E:** Type III urethral injury. Diagram shows a torn bulbous urethra below the urogenital diaphragm along with a disruption of the urogenital diaphragm so that the membranous urethra is torn also. This results in extravasation of urine within the superficial perineal space. This is the most common type of urine leakage injury and is sometimes called a "straddle injury." **F:** Type IV urethral injury. Diagram shows injury to the neck of the bladder and the proximal prostatic urethra that may result in extravasation of urine into the retropubic space of Retzius. **G:** Type V urethral injury. Diagram shows a torn penile urethra that results in extravasation of urine beneath the deep fascia of Buck. P = prostate gland, PS = pubic symphysis, UG = urogenital diaphragm.

f. **Type IV urethral injury** occurs when the neck of the bladder and proximal prostatic urethra are injured. This may result in an extraperitoneal extravasation of urine within the **retropubic space of Retzius**. Type IV urethral injuries may be serious if the internal urethral sphincter is injured, which leads to incontinence.

g. **Type V urethral injury** occurs when the penile urethra is torn. This is a pure anterior urethral injury. This results in an extraperitoneal extravasation of urine **beneath the deep fascia (of Buck)** and will be confined to the penis if the deep fascia of Buck is not torn. However, if the trauma also tears the deep fascia of Buck, then extravasation of urine will occur within the **superficial perineal space**. It is caused by a crushing injury to the penis.

X Urethra

A. FEMALE URETHRA

1. The female urethra is about 3 to 5 cm long and begins at the **internal urethral orifice** of the bladder, where the detrusor muscle extends longitudinally into the urethra but does not form a significant internal urethral sphincter.

2. The female urethra courses through the **urogenital diaphragm**, where it becomes related to the **deep transverse perineal muscle** and **sphincter urethrae muscle** (also called **external urethral sphincter**), both of which are skeletal muscles innervated by the **pudendal nerve**.

3. The posterior surface of the female urethra fuses with the anterior wall of the vagina such that the **external urethral sphincter** does not completely surround the female urethra. This may explain the high incidence of stress incontinence in women, especially after childbirth.

4. The female urethra terminates as the **navicular fossa** at the **external urethral orifice**, which opens into the **vestibule of the vagina** between the labia minora just below the clitoris.

B. MALE URETHRA

1. The male urethra is about 18 to 20 cm long and begins at the **internal urethral orifice** of the bladder, where the detrusor muscle extends longitudinally into the prostatic urethra and forms a complete collar around the neck of the bladder called the **internal urethral sphincter**. The male urethra is divided into five parts:

    a. **Prostatic urethra**

    i. The prostatic urethra courses through and is surrounded by the **prostate gland**.

    ii. The posterior wall has an elevation called the **urethral crest**.

    iii. The **prostatic sinus** is a groove on either side of the urethral crest that receives most of the prostatic ducts from the prostate gland.

    iv. At a specific site along the urethral crest there is an ovoid enlargement called the **seminal colliculus** (also called the **verumontanum**), where the ejaculatory ducts open and the **prostatic utricle** (a vestigial remnant of the paramesonephric duct in males that is involved in the embryologic development of the vagina and uterus) is found.

    b. **Membranous urethra**

    i. The membranous urethra courses through the **urogenital diaphragm**, where it becomes related to the **deep transverse perineal muscle** and **sphincter urethrae muscle** (also called **external urethral sphincter**), both of which are skeletal muscles innervated by the **pudendal nerve**.

    ii. The external urethral sphincter completely surrounds the male urethra.

    iii. The prostatic urethra plus the membranous urethra are called the **posterior urethra** by radiologists.

c. **Bulbous urethra**
   i. The bulbous urethra courses through the **bulb of the penis** and develops endodermal outgrowths into the surrounding mesoderm to form the **bulbourethral glands of Cowper.**
   ii. The bulbous urethra contains the openings of the bulbourethral glands of Cowper.
d. **Proximal part of the penile (spongy or cavernous) urethra**
   i. The proximal part of the penile urethra courses through and is surrounded by the **corpus spongiosum.**
e. **Distal part of the penile urethra**
   i. The distal part of the penile urethra courses through the **glans penis** and terminates as the **navicular fossa** at the **external urethral orifice,** which opens onto the surface of the glans penis.
   ii. The bulbous urethra plus the proximal and distal parts of the penile urethra are called the **anterior urethra** by radiologists.

XI Radiology

A. INTRAVENOUS UROGRAMS (FIGURE 14-10)

A

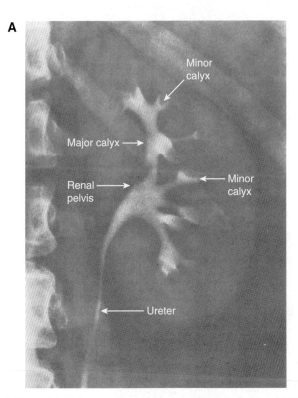

B

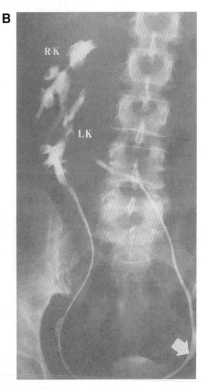

● **Figure 14-10 Intravenous urograms (IVUs). A:** A normal IVU shows the details of the collecting system and upper ureter of the left kidney. **B:** IVU shows a congenital malformation called crossed renal ectopia with kidney fusion. The left kidney (LK) is ectopic on the right side and is fused with the right kidney (RK). Note that the left ureter (*arrow*) inserts normally into the bladder. (*continued*)

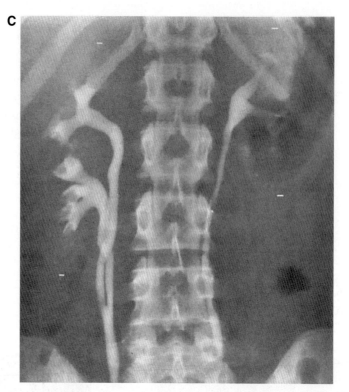

● **Figure 14-10** *Continued* **C:** IVU shows a congenital malformation called a ureteropelvic duplication of the right side.

B. VOIDING CYSTOURETHROGRAM AND RETROGRADE URETHROGRAM (FIGURE 14-11)

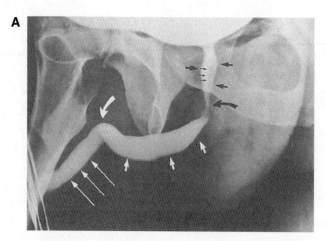

● **Figure 14-11 Voiding cystourethrogram and retrograde urethrogram. A:** A normal voiding cystourethrogram in a male (right posterior projection) shows the prostatic urethra (*large black arrows*); seminal colliculus (or verumontanum), which appears as a filling defect (*small black arrows*); and short membranous urethra (*curved black arrow*). This makes up the posterior urethra. The bulbous urethra (*short white arrows*) and penile urethra (*long white arrows*) are also shown. The penoscrotal junction (*curved white arrow*) is shown. (*continued*)

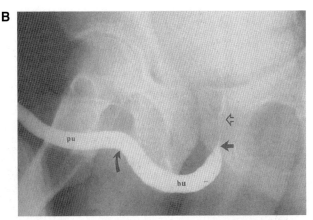

B

● **Figure 14-11** *Continued* **B:** A normal retrograde urethrogram in a male shows the penile urethra (pu) and bulbous urethra (bu) demarcated by the suspensory ligament of the penis at the penoscrotal junction (*curved arrow*). Note that the urethra tapers to a point at the urogenital diaphragm, marking the location of the membranous urethra. The seminal colliculus (or verumontanum; *open arrow*) indicates the location of the prostatic urethra.

C. COMPUTED TOMOGRAPHY SCAN (FIGURE 14-12A–C)

A

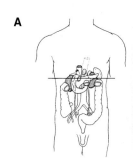

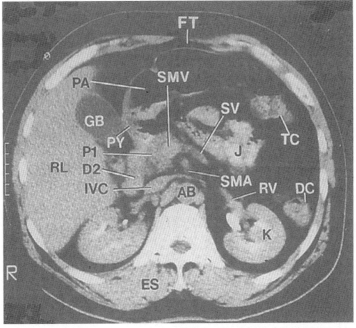

AB = abdominal aorta	PA = pyloric antrum of stomach
DC = descending colon	PV = portal vein
D2 = second part of duodenum	P1 = head of pancreas
ES = erector spinae muscle	RL = right lobe of liver
FT = fat	RV = renal vein
GB = gall bladder	SMA = superior mesenteric artery
IVC = inferior vena cava	SMV = superior mesenteric vein
J = jejunum	SV = splenic vein
K = kidney	TC = transverse colon

● **Figure 14-12 Computed tomography (CT) images. A:** A normal CT image with contrast material at the upper border of vertebral level L2. (*continued*)

B

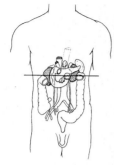

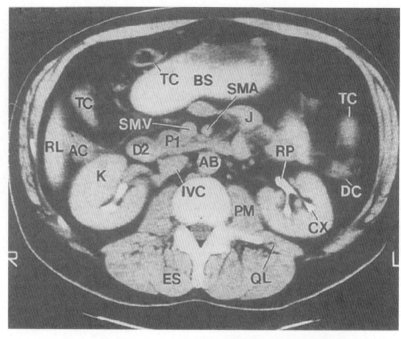

AB = abdominal aorta	K = kidney
AC = ascending colon	PM = psoas major muscle
BS = body of stomach	P1 = head of pancreas
CX = renal calyx	QL = quadratus lumborum
DC = descending colon	RL = right lobe of liver
D2 = second part of duodenum	RP = renal pelvis
ES = erector spinae muscle	SMA = superior mesenteric artery
IVC = inferior vena cava	SMV = superior mesenteric vein
J = jejunum	TC = transverse colon

● **Figure 14-12** *Continued* **B:** A normal CT image with contrast material at the lower border of vertebral level L2. (*continued*)

C

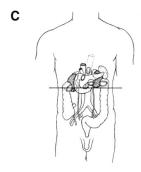

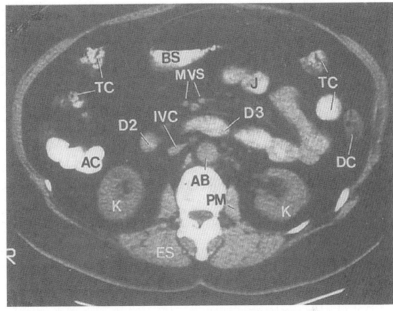

AB = abdominal aorta	IVC = inferior vena cava
AC = ascending colon	J = jejunum
BS = body of stomach	K = kidney
DC = descending colon	MVS = superior mesenteric vessels
D2 = second part of duodenum	PM = psoas major muscle
D3 = third part of duodenum	TC = transverse colon
ES = erector spinae muscle	

● **Figure 14-12** *Continued* **C:** A normal CT image with contrast material at about vertebral level L3.

Chapter 15

Suprarenal (Adrenal) Glands

Ⅰ General Features (Figure 15-1)

A. The right suprarenal gland is shaped like a **pyramid**, with its apex projecting superior and its base embracing the kidney.

B. The left suprarenal gland is shaped like a **half-moon** covering the superior aspect of the kidney and extending inferiorly along the medial aspect of the kidney.

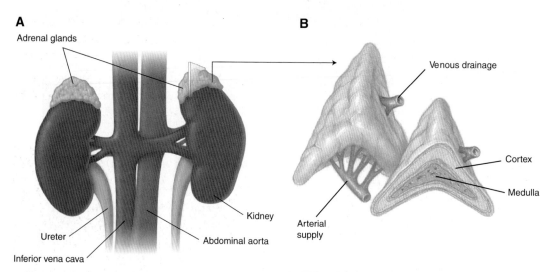

A

Adrenal glands

Kidney

Ureter

Abdominal aorta

Inferior vena cava

B

Venous drainage

Cortex

Medulla

Arterial supply

● **Figure 15-1 Gross anatomy and cut section of the suprarenal glands.**

Ⅱ Arterial Supply

A. The arterial supply of the adrenal gland is from the **superior suprarenal artery**, which arises from the inferior phrenic artery; the **middle suprarenal artery**, which arises from the aorta; and the **inferior suprarenal artery**, which arises from the renal artery.

Ⅲ Venous Drainage

A. The venous drainage of the adrenal gland is to the **right suprarenal vein** (which empties into the inferior vena cava) and the **left suprarenal vein** (which empties into the left renal vein).

B. The venous drainage is particularly important during an adrenalectomy, since the suprarenal vein must be ligated as soon as possible to prevent catecholamine (epinephrine and norepinephrine) release into the circulation.

C. In addition, the adrenal medulla receives venous blood draining the cortex that has a high concentration of cortisol. The synthesis of phenylethanolamine–N-methyltransferase (a key enzyme in the synthesis of epinephrine) is dependent on high levels of cortisol received via venous blood from the cortex.

Ⓘ Innervation

A. SYMPATHETIC

1. Preganglionic neuronal cell bodies are located in the intermediolateral cell column of the spinal cord (T10–L1). Preganglionic axons run with the splanchnic nerves.
2. Modified postganglionic neuronal cell bodies called **chromaffin cells** are located in the adrenal medulla.

Ⓥ Adrenal Cortex is derived embryologically from mesoderm and is divided into three zones.

A. ZONA GLOMERULOSA (ZG) constitutes 15% of the cortical volume. The ZG secretes **aldosterone**, which is controlled by the **renin-angiotensin system**.

B. ZONA FASCICULATA (ZF) constitutes 78% of the cortical volume. The ZF secretes **cortisol**, which is controlled by **corticotropin-releasing factor (CRF)** and **adrenocorticotropic hormone (ACTH)** from the hypothalamus and adenohypophysis, respectively.

C. ZONA RETICULARIS (ZR) constitutes 7% of the cortical volume. The ZR secretes **dehydroepiandrosterone (DHEA)** and **androstenedione**, which are controlled by **CRF** and ACTH from the hypothalamus and adenohypophysis, respectively.

D. CLINICAL CONSIDERATIONS

1. **Primary hyperaldosteronism (Figure 15-2)** is caused by elevated levels of aldosterone, which are commonly caused either by an aldosterone-secreting adenoma (**Conn syndrome**) within the ZG or adrenal hyperplasia. Clinical findings include hypertension, hypernatremia due to increased sodium ion reabsorption, weight gain due to water retention, hypokalemia due to increased K^+ secretion, and decreased plasma renin levels. The magnetic resonance image (MRI) shows a right adrenal mass (*arrows*) that proved to be a benign hyperfunctioning adenoma causing Conn syndrome.

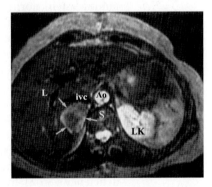

● **Figure 15-2 Conn syndrome.** Ao = aorta, ivc = inferior vena cava, L = liver, LK = left kidney, s = spine.

2. **Cushing syndrome (Figure 15-3)** is caused by elevated levels of **cortisol** (i.e., hypercortisolism), which are commonly due to either an ACTH-secreting adenoma within the adenohypophysis (70% of the cases; strictly termed **Cushing** *disease*), an adrenal adenoma (25% of the cases), or adrenal hyperplasia. An oat cell carcinoma of the lung may also ectopically produce ACTH. However, Cushing syndrome is most commonly caused by iatrogenic corticosteroid drug therapy. Clinical features include mild hypertension with cardiac hypertrophy, buffalo hump, osteoporosis with back pain, central obesity, moon facies, purple skin striae, skin ulcers (poor wound healing), thin wrinkled skin, amenorrhea, purpura, impaired glucose tolerance, and emotional disturbances. The photograph shows a woman with an ACTH-secreting pituitary adenoma with a moon face, buffalo hump, and increased facial hair. The computed tomography (CT) scan shows Cushing syndrome due to adrenal hyperplasia. Both adrenal glands are enlarged (*arrows*) while maintaining their normal anatomic shapes. Note that except for the increased size, the adrenal glands appear normal, which may confound the diagnosis. In some cases of adrenal hyperplasia, the adrenal glands may demonstrate bilateral nodularity.

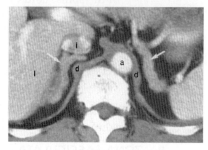

● **Figure 15-3 Cushing syndrome.** a = aorta, d = crura of diaphragm, i = inferior vena cava, l = right lobe of the liver.

3. **Congenital adrenal hyperplasia (Figure 15-4)** is most commonly caused by mutations in genes for enzymes involved in adrenocortical steroid biosynthesis (e.g., **21-hydroxylase deficiency, 11β-hydroxylase deficiency**). In 21-hydroxylase deficiency (90% of all cases), there is virtually no synthesis of aldosterone or cortisol so that intermediates are funneled into androgen biosynthesis, thereby elevating androgen levels. Clinical findings include increased urine 17-ketosteroids and **virilization of a female fetus** ranging from mild clitoral enlargement to complete labioscrotal fusion with a phalloid organ due to elevated levels of androgens; **adrenal hyperplasia** occurs because cortisol cannot be synthesized and therefore the negative feedback to the adenohypophysis does not occur, so

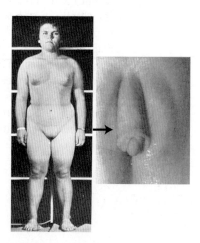

● **Figure 15-4 Congenital adrenal hyperplasia.**

ACTH continues to stimulate the adrenal cortex. The photograph shows a patient (XX genotype) with female pseudo-intersexuality due to congenital adrenal hyperplasia. Masculinization of female external genitalia is apparent with fusion of the labia majora and enlarged clitoris.

4. **Primary Adrenal Insufficiency (Addison Disease).** Addison disease is commonly caused by autoimmune destruction of the adrenal cortex. Other causes include adrenal tuberculosis, fungal infections, and adrenal hemorrhage. Clinical findings include fatigue, anorexia, nausea, weight loss, hypoglycemia, hypotension, and hyperpigmentation of the skin due to increased secretion of melanocyte-stimulating hormone (MSH).

VI Adrenal Medulla

A. **GENERAL FEATURES.** The adrenal medulla contains **chromaffin cells** that are **modified postganglionic sympathetic neurons** derived embryologically from neural crest cells. Preganglionic sympathetic axons (via splanchnic nerves) synapse on chromaffin cells and cause chromaffin cells to secrete catecholamines. The secretion product is **90% epinephrine** and **10% norepinephrine**.

B. **CLINICAL CONSIDERATIONS**

1. **Pheochromocytoma (Figure 15-5)** is a relatively rare neoplasm (usually not malignant) of **neural crest origin** that contains both epinephrine and norepinephrine. It occurs within families (mainly in adults) as part of the **multiple endocrine neoplasia (MEN) type IIa syndrome** (pheochromocytoma, hyperparathyroidism, and medullary carcinoma of the thyroid) or associated with **von Recklinghausen neurofibromatosis**. It is generally found in the region of the adrenal gland but is also found in extra-adrenal sites (e.g., near the aortic bifurcation called the **organ of Zuckerkandl**). Clinical features include persistent or paroxysmal hypertension, anxiety, tremor, profuse sweating, pallor, chest pain, and abdominal pain. The photograph shows a pheochromocytoma. Pheochromocytomas vary in size from 3 to 5 cm in diameter. They are gray-white to pink-tan in color. Exposure of the cut surface often results in darkening of the surface due to formation of yellow-brown adeno-chrome pigment.

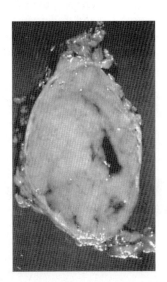

● **Figure 15-5 Pheochromocytoma.**

2. **Neuroblastoma (Figure 15-6)** is an extracranial neoplasm containing primitive neuroblasts of **neural crest origin** and is associated with the amplification of the N-*myc* **oncogene**. It is the most common solid tumor in children and may metastasize to the bone marrow, liver, and orbit. This tumor may be found in extra-adrenal sites, usually along the sympathetic chain ganglia (60%) or within the adrenal medulla (40%). Clinical features include **opsoclonus** (rapid, irregular movements of the eye in the horizontal and vertical directions: "dancing eyes"). The photograph shows a neuroblastoma. Neuroblastomas vary in size from 1 cm to filling the entire abdomen. They are generally soft and white to gray-pink in color. As the size increases, the tumors become hemorrhagic and undergo calcification and cyst formation. Note the nodular appearance of this tumor with the kidney apparent on the left border (*arrow*).

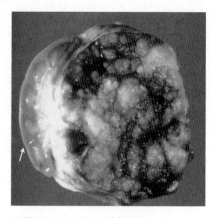

● **Figure 15-6 Neuroblastoma.**

VII   Laboratory Findings Used to Diagnose Suprarenal Gland Disorders (Table 15-1)

TABLE 15-1	LABORATORY FINDINGS USED TO DIAGNOSE SUPRARENAL GLAND DISORDERS[a]				
		Plasma Levels			
Clinical Condition	**Suppression with Dexamethasone Test[b]**	**Aldosterone**	**Cortisol**	**Androgens**	**ACTH**
Primary hyperaldosteronism (Conn syndrome)		High			
Cushing syndrome ACTH adenoma Adrenal adenoma Normal patient	Positive Negative Positive		High High Normal		High Low Normal
Congenital adrenal hyperplasia 21-Hydroxylase deficiency 11β-Hydroxylase deficiency		Low	Low	High	High
Addison disease (Primary adrenal insufficiency)		Low	Low	Low	High

ACTH = adrenocorticotropic hormone.
[a]Many clinical vignette questions will include the plasma levels of various hormones and/or the results of a dexamethasone test. Knowing which hormones are increased, decreased, normal, or not applicable in certain clinical conditions will be of great assistance in your diagnosis and answering the question.
[b]The **dexamethasone suppression test** is based on the ability of dexamethasone (a synthetic glucocorticoid) to inhibit ACTH and cortisol secretion. If the adenohypophysis–adrenal cortex axis is normal, dexamethasone will inhibit ACTH and cortisol secretion by negative feedback.

Chapter 16

Female Reproductive System

I Ovaries (Figure 16-1)

A. GENERAL FEATURES

1. The ovaries are almond-shaped structures that are located **posterior** to the broad ligament.
2. The ovaries are attached to the lateral pelvic wall by the suspensory ligament of the ovary (a region of the broad ligament), which contains the ovarian artery, vein, and nerve.
3. The surface of the ovaries is not covered by peritoneum, but instead is covered by a simple cuboidal epithelium called the **germinal epithelium**.

B. ARTERIAL SUPPLY.
The arterial supply of the ovaries is from the **ovarian arteries**, which arise from the abdominal aorta, and **ascending branches of the uterine arteries**, which arise from the internal iliac artery.

C. VENOUS DRAINAGE.
The venous drainage of the ovaries is to the **right ovarian vein** (which empties into the inferior vena cava) and the **left ovarian vein** (which empties into the left renal vein).

D. CLINICAL CONSIDERATIONS

1. **Right side hydronephrosis** may indicate thrombosis of the right ovarian vein that constricts the ureter, since the right ovarian vein crosses the ureter to enter the inferior vena cava (IVC).
2. **Ovarian pain** is often referred down the inner thigh via the obturator nerve.

II Uterine Tubes (Figure 16-1)

A. GENERAL FEATURES

1. The function of the uterine tubes is to convey fertilized and unfertilized oocytes to the uterine cavity by ciliary action and muscular contractions and to transport sperm in the opposite direction for fertilization to take place.
2. The uterine tubes are supported by the **mesosalpinx**, which is a region of the broad ligament.
3. The uterine tube has four divisions:
   a. The **infundibulum** is funnel shaped, is fimbriated, and opens into the peritoneal cavity.
   b. The **ampulla** is the longest and widest part of the uterine tube. It is the site of fertilization.
   c. **Isthmus**
   d. The **intramural** division opens into the uterine cavity.

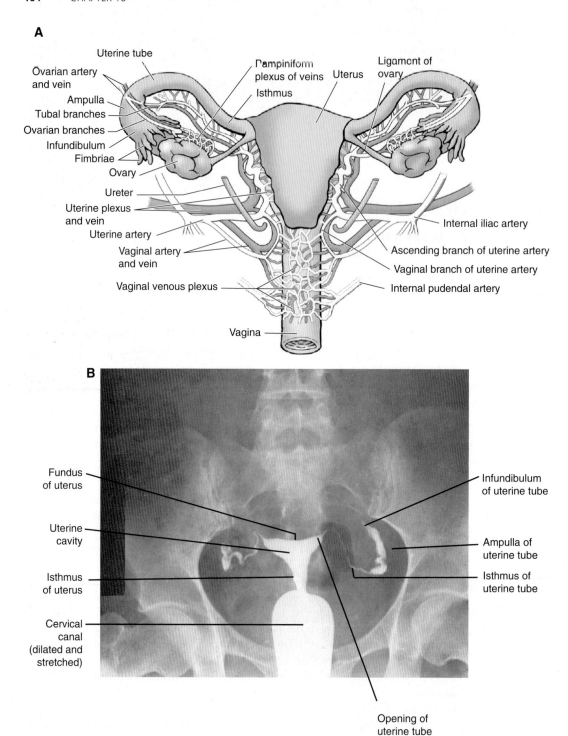

● **Figure 16-1 Female reproductive system. A:** Diagram of the arterial supply and venous drainage of the ovaries, uterine tubes, and vagina. **B:** Anteroposterior radiograph of the female pelvis after injection of a radiopaque compound into the uterine cavity (hysterosalpingography).

B. **ARTERIAL SUPPLY.** The arterial supply of the uterine tubes is from the **ovarian arteries,** which arise from the abdominal aorta, and the **ascending branches of the uterine arteries,** which arise from the internal iliac artery.

C. VENOUS DRAINAGE. The venous drainage of the uterine tubes is to the **right ovarian vein** (which empties into the inferior vena cava), the **left ovarian vein** (which empties into the left renal vein), and the **uterine veins**.

D. CLINICAL CONSIDERATIONS

1. **Acute and chronic salpingitis (Figure 16-2)** is a bacterial infection (most commonly *Neisseria gonorrhea* or *Chlamydia trachomatis*) of the uterine tube with acute inflammation (neutrophil infiltration) or chronic inflammation, which may lead to scarring of the uterine tube, predisposing to **ectopic tubal pregnancy**. Salpingitis is probably the most common cause of female sterility. The photograph shows that the uterine tube is markedly distended, the fimbriated end is closed, and there is hemorrhage on the serosal surface.

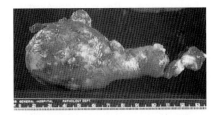

● **Figure 16-2** Salpingitis.

2. **Ectopic tubal pregnancy (Figure 16-3)** most often occurs in the **ampulla** of the uterine tube. Risk factors include salpingitis, pelvic inflammatory disease, pelvic surgery, or exposure to diethylstilbestrol (DES). Clinical signs include sudden onset of abdominal pain, which may be confused with appendicitis in a young woman; last menses 60 days ago; positive human chorionic gonadotropin (hCG) test; and culdocentesis showing intraperitoneal blood. The photograph shows an enlarged uterine due to the growing embryo.

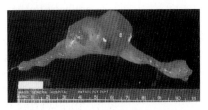

● **Figure 16-3** Ectopic tubal pregnancy.

III Uterus (Figure 16-1)

A. GENERAL FEATURES. The uterus is divided into four regions.

1. The **fundus** is located superior to the cornua and contributes largely to the upper segment of the uterus during pregnancy. At term, the fundus may extend as high as the xiphoid process (vertebral level T9).
2. The **cornu** is located near the entry of the uterine tubes.
3. The **body** is located between the cornu and cervix. The **isthmus** is part of the body and is the dividing line between the body of the uterus and the cervix. The isthmus is the preferred site for a surgical incision during a delivery by cesarean section.
4. The **cervix** is located inferior to the body of the uterus and protrudes into the vagina. The cervix contains the **internal os**, **cervical canal**, and **external os**. The external os in a nulliparous woman is round. The external os in a parous woman is transverse.

B. ARTERIAL SUPPLY. The arterial supply of the uterus is from the **uterine arteries**, which arise from the internal iliac artery. There is a potential collateral supply from the ovarian arteries.

C. VENOUS DRAINAGE. The venous drainage of the uterus is to the **internal iliac veins** (which empties into the inferior vena cava).

D. SUPPORT OF THE UTERUS. The uterus is supported by the following structures:

1. **Pelvic diaphragm (levator ani muscles)**
2. **Urogenital diaphragm**
3. **Urinary bladder**
4. **Round ligament of the uterus,** which is a remnant of the gubernaculum in the embryo.
5. **Transverse cervical ligament (cardinal ligament of Mackenrodt),** which extends laterally from the cervix to the side wall of the pelvis. It is located at the base of the broad ligament and contains the **uterine artery** (a branch of the internal iliac artery).
6. **Uterosacral ligament,** which extends posteriorly from the cervix to the sacrum and is responsible for bracing the uterus in its normal **anteverted** position.
7. **Pubocervical ligament,** which extends anteriorly from the cervix to the pubic symphysis and helps to prevent a **cystocele** (a herniation of the urinary bladder into the anterior wall of the vagina).
8. **Broad ligament**
   a. The broad ligament is a double fold of parietal peritoneum, which extends laterally from the uterus to the side wall of the pelvis.
   b. The broad ligament is divided into four regions: **mesosalpinx** (which supports the uterine tubes), **mesovarium** (which supports the ovary), **mesometrium** (which supports the uterus), and the **suspensory ligament of the ovary.**
   c. The broad ligament contains the following structures:
      i. **Ovarian artery, vein, and nerves**
      ii. **Uterine tubes**
      iii. **Ovarian ligament of the uterus** (which is a remnant of the gubernaculum in the embryo)
      iv. **Round ligament of the uterus** (which is a remnant of the gubernaculum in the embryo)
      v. **Epoophoron** (which is a remnant of the mesonephric tubules in the embryo)
      vi. **Paroöphoron** (which is a remnant of the mesonephric tubules in the embryo)
      vii. **Gartner duct** (which is a remnant of the mesonephric duct in the embryo)
      viii. **Ureter** (which lies at the base of the broad ligament posterior and inferior to the uterine artery). During a hysterectomy, the ureters may be inadvertently ligated along with the uterine artery due to their close anatomic relationship.
      ix. **Uterine artery, vein, and nerves** (which lie at the base of the broad ligament within the transverse cervical ligament)

E. POSITION OF THE UTERUS

1. The uterus is normally is an anteflexed and anteverted position, which places the uterus in a nearly horizontal position lying on the superior wall of the urinary bladder.
2. **Anteflexed** refers to the anterior bend of the uterus at the angle between the cervix and the body of the uterus.
3. **Anteverted** refers to the anterior bend of the uterus at the angle between the cervix and the vagina.

F. CLINICAL CONSIDERATIONS

1. **Endometrial adenocarcinoma (Figure 16-4)** is the most common gynecologic cancer in women and is linked to prolonged estrogen stimulation of the endometrium. Risk factors include exogenous estrogen treatment for menopause, obesity, diabetes, nulliparity, early menarche, and late menopause. This cancer grows in a diffuse or polypoid pattern and often involves multiple sites. The most common histologic variant is composed entirely of glandular cells (called pure endometrial adenocarcinoma). Clinical features include perimenopausal or postmenopausal women who complain of abnormal uterine bleeding. The photograph shows an opened uterine cavity to reveal a partially necrotic, polypoid endometrial cancer.

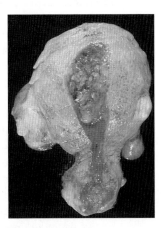

● **Figure 16-4 Endometrial adenocarcinoma.**

2. **Endometriosis (Figure 16-5)** is the presence of endometrial glandular tissue in abnormal locations outside of the uterus. The ectopic sites most frequently involved include the ovary (80% of the cases), uterine ligaments, rectovaginal septum, pouch of Douglas, pelvic peritoneum covering the uterus, uterine tubes, rectosigmoid colon, and bladder. Early foci of endometriosis on the ovary or peritoneal surface appear as red or bluish nodules ("**mulberry nodules**") about 1 to 5 mm in size. Since this ectopic endometrial tissue shows cyclic changes synchronous with the endometrium of the uterus (i.e., participates in the menstrual cycle), repeated bleedings lead to a deposition of hemosiderin forming "**gunpowder mark**" lesions. In the ovary, repeated bleedings may lead to the formation of large (15 cm) cysts containing inspissated chocolate-colored material ("**chocolate cysts**"). Endometriosis results in infertility, dysmenorrhea, and pelvic pain (most pronounced at the time of menstruation). The photograph shows an ovary with red and/or bluish nodules ("mulberry nodules").

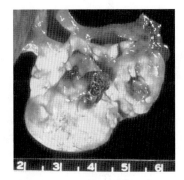

● **Figure 16-5 Endometriosis.**

3. **Uterine fibroids (leiomyoma) (Figure 16-6)** are a common benign neoplasm resulting from a proliferation of smooth muscle cells of the uterus, which may become calcified. The fibroids may be located within the myometrium of the uterus (intramural); beneath the endometrium (submucosa), where they may grow into the uterine cavity; or beneath the serosa (subserosal), where they may grow into the peritoneal cavity. This may result in infertility if the fibroids block the uterine tube or prevent implantation of the conceptus. Fibroids may be palpated as irregular, nodular masses protruding against the anterior abdominal wall. The radiograph shows a calcified mass just to the left of the midline. Calcifications in fibroids are often popcornlike in appearance. A very large fibrinoid may occupy the entire pelvic cavity or may even extend into the abdomen.

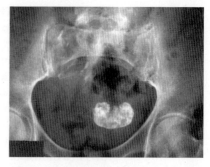

● **Figure 16-6 Uterine fibroids (leiomyoma).**

IV Cervix

A. The cervix is the lower part of the uterus that measures about 2.5 to 3.0 cm in length.

B. The cervix is divided into a **supravaginal portion** (lying above the vaginal vault) and a **vaginal portion (portio vaginalis)**, which protrudes into the vagina.

C. The junction between the cervix and uterus is at the **internal os.**

D. The cervical mucus produced during the proliferative phase of the menstrual cycle is **watery**, whereas the cervical mucus produced during the secretory phase of the menstrual cycle is **viscous.**

E. During childbirth, the cervix undergoes "cervical softening," where the connective tissue becomes pliable due to the action of **relaxin.**

V Ectocervix

A. The outer epithelial surface of the vaginal portion of the cervix (portio vaginalis) is called the **ectocervix.**

B. The epithelial surface lining the lumen of the **endocervical canal** is called the **endocervix.**

C. The **endocervical canal** connects the uterine cavity with the vaginal cavity and extends from the internal os to the **external os.**

D. At puberty, the simple columnar epithelium of the endocervical canal extends onto the ectocervix. However, exposure of the simple columnar epithelium to the acidic (pH = 3) environment of the vagina induces a transformation from columnar to squamous epithelium (i.e., **squamous metaplasia**) and the formation of a **transformation zone.**

E. The transformation zone is the site of **Nabothian cysts**, which develop as stratified squamous epithelium grows over the mucus-secreting simple columnar epithelium and entraps large amounts of mucous.

F. SQUAMOUS CELL CARCINOMA OF THE CERVIX (FIGURE 16-7). The transformation zone is the most common site of **squamous cell carcinoma of the cervix**, which is usually preceded by epithelial changes called **cervical intraepithelial neoplasias (CINs)** diagnosed by a Papanicolaou smear. **Human papillomavirus (HPV)** has also been linked as an important factor in cervical oncogenesis and is often tested for. Cervical carcinoma may spread to the side wall of the pelvis, where the ureters may become obstructed leading to hydronephrosis. The most common site of lymph node spread (i.e., sentinel nodes) is to the **obturator lymph nodes.** The computed tomography (CT) scan shows a mass (*large arrow*) immediately posterior to the urinary bladder. A small amount of gas is present within the mass (*small arrow*) secondary to necrosis. Note the indentation of the posterior margin of the urinary bladder.

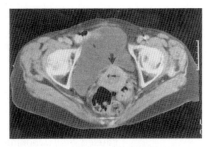

● **Figure 16-7 Squamous cell carcinoma of the cervix.**

VI Vagina (Figure 16-1)

A. GENERAL FEATURES

1. The vagina extends from the cervix to the vestibule of the vagina.
2. The vagina is the longest part of the birth canal, and its distention during childbirth is limited by the ischial spine and sacrospinous ligaments.
3. The vagina forms a recess around the cervix called the **fornix**. The fornix is divided into three regions:
    a. **Anterior fornix** is located anterior to the cervix and is related to the **vesicouterine pouch.** The urinary bladder is palpable through the anterior fornix during a digital examination.
    b. **Lateral fornices** are located lateral to the cervix.
    c. **Posterior fornix** is located posterior to the cervix and is related to the **rectouterine pouch (of Douglas).** The rectum, sacral promontory (S1 vertebral body), and coccyx are palpable through the posterior fornix during digital examination. The posterior fornix is a site for culdocentesis.

B. ARTERIAL SUPPLY

1. The arterial supply of the superior portion of the vagina is from the **vaginal branches of uterine artery,** which arises from the internal iliac artery.
2. The arterial supply of the middle and lower portions of the vagina is from the **internal pudendal artery,** which arises from the internal iliac artery.

C. VENOUS DRAINAGE. The venous drainage of the vagina is to the **vaginal venous plexus** (which is continuous with the uterine venous plexus), which empties into the internal iliac veins → inferior vena cava.

D. CLINICAL CONSIDERATIONS

1. **Culdocentesis** is a procedure where a needle is passed through the posterior fornix into the rectouterine pouch of the peritoneal cavity to obtain a fluid sample for analysis or to collect oocytes for in vitro fertilization. It provides diagnostic information for many gynecologic conditions (e.g., pelvic inflammatory disease, ectopic tubal pregnancy).
2. **Cystocele** is the herniation of the urinary bladder into the anterior wall of the vagina.

3. **Rectocele** is the herniation of the rectum into the posterior wall of the vagina.

4. **Bartholin cyst (Figure 16-8)** is caused by an obstruction of the duct from the greater vestibular glands of Bartholin. The diagram shows a Bartholin cyst (BC) on the left side of the vestibule of the vagina.

5. **Vaginitis** is a chronic infection most often caused by *Trichomonas vaginalis* (15% of cases), *Candida albicans* (25%), or *Gardnerella vaginalis* (30%). The vaginal epithelium is resistant to bacteria, fungal, and protozoan invasion so that the pathogens remain within the lumen of the vagina.

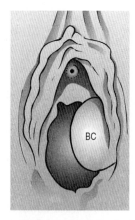

● **Figure 16-8 Bartholin cyst (BC).**

  a. *T. vaginalis* is a **flagellated protozoan**, which is sexually transmitted. It produces a vaginitis characterized by an inflammatory vaginal smear with numerous neutrophils, fiery-red appearance of the vaginal and cervical mucosa ("strawberry mucosa"), and a **thin, gray-white, frothy, purulent, malodorous discharge (pH >4.5)**. Postcoital bleeding is a common complaint. The organism is best seen in fresh preparations diluted with warm saline where the tumbling motility of the organism can be observed.

  b. *C. albicans* is a **yeast** that produces pseudohyphae and true hyphae in tissues. It produces superficial white patches or large fluffy membranes that easily detach leaving a red, irritated underlying surface and a **thick, white, "cottage cheese" discharge (pH <4.5)**. The organism can be observed on KOH preparations of the discharge.

  c. *G. vaginalis* is a **Gram-negative bacillus** bacterial infection generally called **bacterial vaginosis**, where higher levels than normal of the bacteria are present. It is not sexually transmitted. It produces a vaginitis characterized by no inflammatory vaginal smear, no changes in the mucosa, and a **thin, homogenous, somewhat adherent, fishy-odor discharge (pH >4.5)**. The discharge gives a positive amine test ("whiff test"; fishy amine smell) when mixed with KOH. A vaginal smear will show an increased number of bacteria and "clue cells," which are squamous cells with a clumped nucleus and a folded cytoplasm covered with bacteria.

Ⅶ External Genitalia (Figure 16-9)

A. LABIA MAJORA. The labia majora are two folds of hairy skin with underlying fat pads.

B. LABIA MINORA

1. The labia minora are two folds of hairless skin located medial to the labia majora that enclose the vestibule of the vagina.

2. Each labium minus is continuous anteriorly with the **prepuce of the clitoris** and the **frenulum of the clitoris.**

3. Each labium minus is continuous posteriorly with the **fourchette**, which connects the labia minora with the **vaginal introitus (entry).**

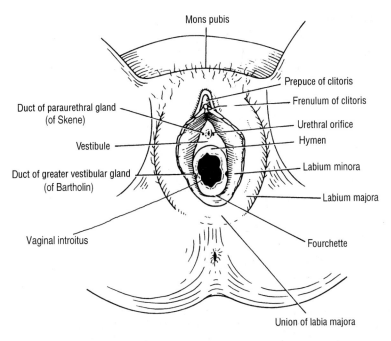

● **Figure 16-9 Diagram of the female genitalia.**

C. **VESTIBULE OF THE VAGINA**
 1. The vestibule of the vagina is the space between the labia minora.
 2. The vestibule contains the **urethral orifice; paraurethral glands (of Skene); vaginal introitus (entry)** incompletely covered by the **hymen; greater vestibular glands (of Bartholin);** and **lesser vestibular glands.**

D. **CLITORIS**
 1. Although the clitoris is homologous with the penis, the clitoris has *no* corpus spongiosum and does *not* transmit the urethra.
 2. The **body of the clitoris** is formed by two **corpora cavernosa,** which are continuous with the crura of the clitoris.
 3. The **glans of the clitoris** is formed by the fusion of the **vestibular bulbs.**

VIII Innervation of the Female Reproductive System
A. **PARASYMPATHETIC**
 1. Preganglionic neuronal cell bodies are located in the **gray matter of the S2–4 spinal cord.** Preganglionic axons form the **pelvic splanchnic nerves,** which interact with the **inferior hypogastric plexus.**
 2. Postganglionic neuronal cell bodies are located near or within the female viscera.
 3. Postganglionic axons terminate on smooth muscle and glands.

B. **SYMPATHETIC**
 1. Preganglionic neuronal cell bodies are located in the **intermediolateral cell column** of the spinal cord. Preganglionic axons form the **sacral splanchnic nerves.**
 2. Postganglionic neuronal cell bodies are located in the **inferior hypogastric plexus.**
 3. Postganglionic axons terminate on smooth muscle and glands.

Chapter 17

Male Reproductive System

① Testes (Figures 17-1 and 17-2)

A. GENERAL FEATURES

1. The testes are surrounded incompletely (medially, laterally, and anteriorly, but not posteriorly) by a sac of peritoneum called the **tunica vaginalis**.

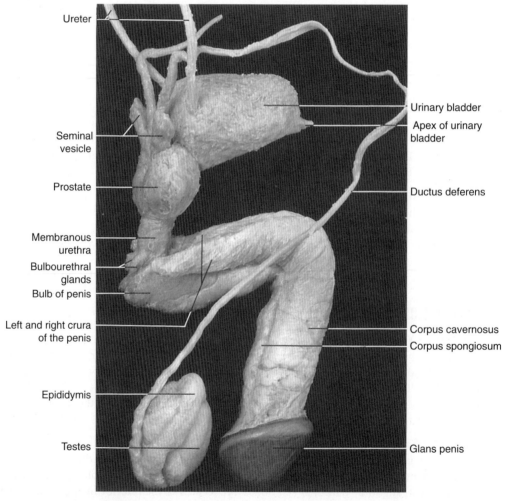

Ureter

Seminal vesicle

Prostate

Membranous urethra

Bulbourethral glands

Bulb of penis

Left and right crura of the penis

Epididymis

Testes

Urinary bladder

Apex of urinary bladder

Ductus deferens

Corpus cavernosus

Corpus spongiosum

Glans penis

● **Figure 17-1 Male reproductive system.** Note the pathway of sperm: seminiferous tubules → straight tubules → rete testes → efferent ductules → epididymis → ductus deferens → ejaculatory duct → prostatic urethra → membranous urethra → penile urethra.

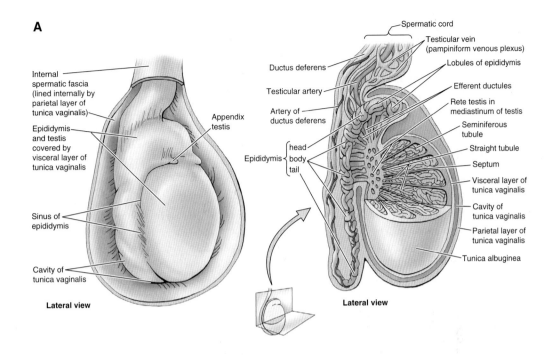

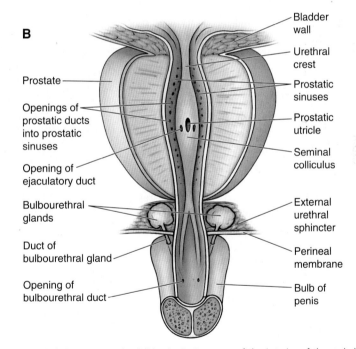

● **Figure 17-2 A:** Anatomy of the testes and epididymis. **B:** Anatomy of the interior of the male bladder and urethra.

2. Beneath the tunica vaginalis, the testes are surrounded by a thick connective tissue capsule called the **tunica albuginea** because of its whitish color.

3. Beneath the tunica albuginea, the testes are surrounded by a highly vascular layer of connective tissue called the **tunica vasculosa**.

4. The tunica albuginea projects connective tissue septae inward toward the mediastinum, which divides the testes into about 250 lobules, each of which contains one to four highly coiled **seminiferous tubules**. These septae converge toward the midline on the posterior surface, where they meet to form a ridgelike thickening called the **mediastinum**.

5. The testes contain the **seminiferous tubules, straight tubules, rete testes, efferent ductules,** and the **Leydig (interstitial) cells.**

B. ARTERIAL SUPPLY
1. The arterial blood supply of the testes is from the **testicular arteries,** which arise from the abdominal aorta just inferior to the renal arteries.
2. There is a rich collateral arterial blood supply from the internal iliac artery via the **artery of the ductus deferens,** inferior epigastric artery via the **cremasteric artery,** and femoral artery via the **external pudendal artery.** The collateral circulation is sufficient to allow ligation of the testicular artery during surgery.

C. VENOUS DRAINAGE
1. The venous drainage of the testes is to the **right testicular vein** (which empties into the inferior vena cava) and the **left testicular vein** (which empties into the left renal vein).
2. This is important in males, where the appearance of a **left-side testicular varicocele** may indicate occlusion of the left testicular vein and/or left renal vein due to a malignant tumor of the kidney.
3. The testicular veins are formed by the union of the veins of the **pampiniform plexus.**

D. CLINICAL CONSIDERATIONS
1. **Cryptorchidism (Figure 17-3)** occurs when the testes begin to descend along the normal pathway but fail to reach the scrotum (versus an **ectopic testes,** which descends along an abnormal pathway). The undescended testis is generally found within the **inguinal canal** or **abdominal cavity near the deep inguinal ring.** Bilateral cryptorchidism results in **sterility** since the cooler temperature of the scrotal sac is necessary for spermatogenesis. Cryptorchidism is associated with an increased incidence of cancer and torsion. The photograph shows that both testes have not descended into the scrotal sac. The undescended right testis is apparent (*arrow*).

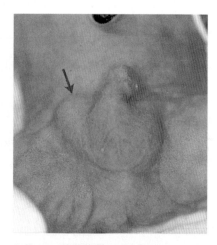

● Figure 17-3 Cryptorchidism.

2. **Hydrocele of testes (Figure 17-4)** occurs when a small patency of the processus vaginalis remains so that peritoneal fluid can flow into the tunica vaginalis surrounding the testes. The photograph shows a bilateral hydrocele.

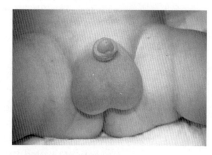

● Figure 17-4 Hydrocele.

3. **Varicocele (Figure 17-5)** is an abnormal dilatation of the pampiniform plexus and testicular vein and usually presents as a palpable "bag of worms" scrotal swelling. It most often occurs on the left side (90%) due to compression of the left testicular vein by the sigmoid colon, which contains stored feces and is often associated with infertility. The diagram shows the abnormal dilatation of the pampiniform plexus of veins (*arrow*).

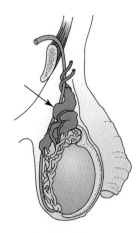

● **Figure 17-5 Varicocele.**

4. **Torsion (Figure 17-6)** is the rotation of the testes about the spermatic cord, usually toward the penis (i.e., medial rotation). An increased incidence occurs in men with testes in a horizontal position and a high attachment of the tunica vaginalis to the spermatic cord ("bell clapper deformity"). Torsion is a medical emergency since compression of the testicular vessels results in ischemic necrosis within 6 hours. The photograph shows the cut section of the testis from a man who experienced sudden excruciating scrotal pain. Note the diffuse hemorrhage and necrosis of the testis and adnexal structures.

● **Figure 17-6 Torsion.**

Ⅱ Epididymis (Figures 17-1 and 17-2)

A. The epididymis is a very long (6 meters) and highly coiled duct that is described as having a **head region**, **body region**, and **tail region** (which is continuous with the ductus deferens).

B. **SPERM MATURATION (i.e., MOTILITY) AND STORAGE** occur in the head and body of the epididymis.

C. The principal cells lining the epididymis have the following functions: continued resorption of testicular fluid that began in the efferent ductules; phagocytosis of degenerating sperm or spermatid residual bodies not phagocytosed by the Sertoli cells; and secretion of glycoproteins, which bind to the surface of the cell membrane of the sperm, sialic acid, and glycerophosphocholine (which inhibits capacitation, thus preventing sperm from fertilizing a secondary oocyte until the sperm enter the female reproductive tract).

D. In the tail region of the epididymis, the muscular coat consists of an **inner longitudinal layer, middle circular layer**, and **outer longitudinal layer of smooth muscle**. These three layers contract due to neural stimulation during sexual excitation and force sperm from the tail of the epididymis to the ductus deferens. This is the initial muscular component that contributes to the force of emission.

III Ductus Deferens (Figures 17-1 and 17-2)

A. GENERAL FEATURES

1. The ductus deferens begins at the inferior pole of the testes, ascends to enter the spermatic cord, transits the inguinal canal, enters the abdominal cavity by passing through the deep inguinal ring, crosses the external iliac artery and vein, and enters the pelvis.

2. The distal end of the ductus deferens enlarges to form the **ampulla**, where it is joined by a short duct from the seminal vesicle to form the **ejaculatory duct**.

3. The smooth muscular coat of the ductus deferens is similar to the tail region of the epididymis (i.e., **inner longitudinal layer, middle circular layer,** and **outer longitudinal layer of smooth muscle**) and contributes to the force of emission.

B. ARTERIAL SUPPLY.
The arterial supply of the ductus deferens is from the **artery of the ductus deferens**, which arises from the internal iliac artery and anastomoses with the testicular artery.

C. VENOUS DRAINAGE.
The venous drainage of the ductus deferens is to the **testicular vein** and the **distal pampiniform plexus**.

D. CLINICAL CONSIDERATION: VASECTOMY.
The scalpel will cut through the following layers in succession to gain access to the ductus deferens: skin → Colles fascia and dartos muscle → external spermatic fascia → cremasteric fascia and muscle → internal spermatic fascia → extraperitoneal fat. The tunica vaginalis is not cut.

IV Contents of the Spermatic Cord.
The contents of the spermatic cord include the following: ductus deferens; testicular artery; artery of the ductus deferens; cremasteric artery; pampiniform venous plexus; sympathetic and parasympathetic nerves, which form the testicular plexus of nerves; genitofemoral nerve; and lymphatics.

V Ejaculatory Duct (Figures 17-1 and 17-2)

A. The distal end of the ductus deferens enlarges to form the **ampulla**, where it is joined by a short duct from the seminal vesicle to form the **ejaculatory duct**.

B. The ejaculatory duct passes through the prostate gland and opens into the prostatic urethra at the **seminal colliculus** of the urethral crest.

C. The ejaculatory duct has no smooth muscular coat so it does not contribute to the force for emission.

VI Seminal Vesicles (Figures 17-1 and 17-2)

A. The seminal vesicles are highly coiled tubular diverticula that originate as evaginations of the ductus deferens distal to the ampulla.

B. Contraction of the smooth muscle of the seminal vesicle during emission will discharge seminal fluid into the ejaculatory duct.

C. The seminal fluid is a whitish-yellow viscous material that contains **fructose** (the principal metabolic substrate for sperm) and **other sugars, choline, proteins, amino acids, ascorbic acid, citric acid,** and **prostaglandins.**

D. Seminal fluid accounts for 70% of the volume of the ejaculated semen.

E. In **forensic medicine**, the presence of fructose (which is not produced elsewhere in the body) and choline crystals are used to determine the presence of semen.

Ⅶ Bulbourethral (BU) Glands of Cowper (Figures 17-1 and 17-2)

A. The BU glands are located in the deep perineal space embedded in the skeletal muscles of the urogenital diaphragm (i.e., deep transverse perineal muscle and sphincter urethrae muscle) and adjacent to the membranous urethrae.

B. The ducts of the BU glands open into the penile urethra.

C. The BU fluid is a clear, mucuslike, slippery fluid that contains **galactose, galactosamine, galacturonic acid, sialic acid,** and **methylpentose.**

D. This fluid makes up a major portion of the preseminal fluid (or pre-ejaculate fluid) and probably serves to lubricate the penile urethra.

Ⅷ Prostate Gland (Figures 17-1, 17-2, and 17-7)

A. GENERAL FEATURES

1. The prostate gland is located between the base of the urinary bladder and the urogenital diaphragm.

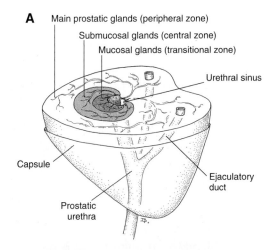

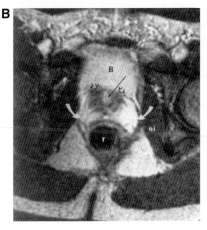

● **Figure 17-7 Prostate gland. A:** Diagram of the prostate gland indicating the relationship of the peripheral zone, central zone, and transitional (periurethral) zone to the prostatic urethra. **B:** Magnetic resonance image shows a high-intensity peripheral zone (*curved arrows*), the urethra (*long arrow*), and the low-intensity transitional zone (*open arrows*). B = bladder, oi = obturator internus muscle, r = rectum.

2. The anterior surface of the prostate is related to the retropubic space. The posterior surface of the prostate is related to the seminal vesicles and rectum. The prostate gland can be easily palpated by a digital examination via the rectum.

3. The prostate gland consists of five lobes: **right and left lateral lobes, right and left posterior lobes,** and a **middle lobe.**

4. The prostate gland is a collection of 30 to 50 compound tubuloalveolar glands that are arranged in three zones: the **peripheral zone** (contains the largest glands and highest number of glands), **central zone,** and **transitional (periurethral) zone.**

5. Prostatic fluid contains **citric acid, prostatic acid phosphatase (PAP), prostaglandins, fibrinogen,** and **prostatic specific antigen (PSA,** a serine protease that liquefies semen after ejaculation).

B. ARTERIAL SUPPLY. The arterial supply is from the **inferior vesical artery,** which arises from the internal iliac artery.

C. VENOUS DRAINAGE. The venous drainage follows two pathways:
1. The first pathway is to the **prostatic venous plexus → internal iliac veins → inferior vena cava (IVC).** This may explain the metastasis of prostatic cancer to the heart and lungs.
2. The second pathway is to the **prostatic venous plexus → vertebral venous plexus → cranial dural sinuses.** This may explain the metastasis of prostatic cancer to the vertebral column and brain.

D. CLINICAL CONSIDERATIONS
1. **Benign prostatic hyperplasia (BPH) (Figure 17-8)** is characterized by hyperplasia of the **transitional (periurethral) zone,** which generally involves the lateral and middle lobes and develops in all men. Hyperplasia of epithelial and fibromuscular stromal cells leads to the formation of soft, yellow-pink nodules. BPH compresses the prostatic urethra and obstructs urine flow. The hyperplasia may be due to increased sensitivity of prostate to **dihydrotestosterone (DHT).** BPH is *not* premalignant. Clinical signs include increased frequency of urination, nocturia, dysuria, difficulty starting and stopping urination, dribbling, and sense of incomplete emptying of bladder. Treatment may include 5α-reductase inhibitors (e.g., finasteride [**Proscar**]) to block conversion of testosterone to DHT and/or α-adrenergic antagonists (e.g., **terazosin, prazosin, doxazosin**) to inhibit prostate gland secretion. The intravenous pyelogram (IVP) radiograph shows an elevation of the base of the bladder by a smooth half-moon filling defect (*arrows*). This causes a deformity in the pathway of the ureter such that the ureters end in a hook ("fish-hooking" phenomena; *open arrows*).

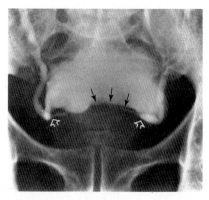

● **Figure 17-8 Benign prostatic hyperplasia.**

2. **Prostatic carcinoma (PC) (Figure 17-9)** is most commonly found in the **peripheral zone**, which generally involves the posterior lobes (which can be palpated upon a digital rectal exam). Neoplastic epithelial cells lead to the formation of yellow, firm, gritty tumors that invade nearby structures. Since PC begins in the peripheral zone, by the time urethral blockage occurs (i.e., patient complains of difficulty in urination), the carcinoma is in an advanced stage. The patient is usually asymptomatic until the advanced stages; clinical signs include indurated mass on digital rectal examination, obstructive uropathy, and low back or pelvic pain. **Prostatic intraepithelial neoplasia (PIN)** is frequently associated with PC. Serum **PSA levels** are diagnostic. Metastasis to bone (e.g., lumbar vertebrae, pelvis) is frequent. Treatment may include **leuprolide (Lupron)**, which is a gonadotropin-releasing hormone (GNRH) agonist that inhibits the release of follicle-stimulating hormone (FSH) and leuteinizing hormone (LH) when administered in a continuous fashion, thereby inhibiting secretion of testosterone; **cyproterone (Androcur)** or **flutamide (Eulexin)**, which are androgen receptor antagonists; radiation; and/or prostatectomy. The magnetic resonance image (MRI) shows a low-intensity prostate carcinoma (*large open arrow*) in the peripheral zone (*curved arrows*). The urethra (*long arrow*) and the low-intensity transitional zone (*small open arrow*) are apparent.

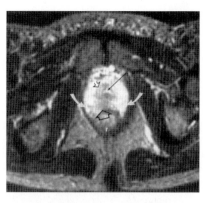

● **Figure 17-9 Prostatic carcinoma.** r, rectum.

IX Penis (Figure 17-10)

A. GENERAL FEATURES

1. The penis consists of three columns of erectile tissue bounded together by the **tunica albuginea**: one **corpus spongiosum** and two **corpora cavernosa**.
2. The **corpus spongiosum** begins as the **bulb of the penis** and ends as the **glans penis**. It is ventrally situated in the penis and transmits the urethra. During erection, the corpus spongiosum does not get as turgid as the corpora cavernosa.
3. The **corpora cavernosa** begin as the **crura of the penis** and end proximal to the **glans penis**. They are dorsally situated in the penis.
4. The **erectile tissue of the penis** found within the corpus spongiosum and corpora cavernosa consists of vascular channels that are lined by endothelium.
5. The penis is supported by the **suspensory ligament**, which arises from the linea alba and inserts into the deep fascia (of Buck).

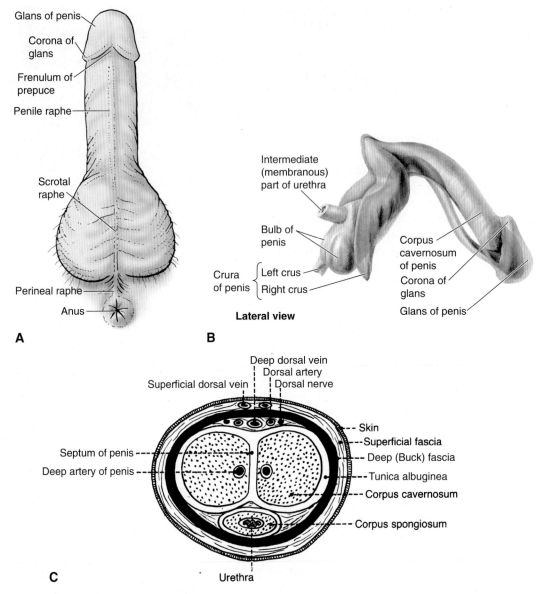

● **Figure 17-10 A:** Urethral surface of a circumcised penis (inferior view). **B:** Internal structure of the penis. **C:** Cross section of the penis.

B. ARTERIAL SUPPLY. The arterial supply is from the **deep artery of the penis** (involved in the erection of the penis) and **dorsal artery of the penis**, which arise from the internal pudendal artery.

C. VENOUS DRAINAGE. The venous drainage follows two pathways:
1. The first is to the **deep dorsal vein of the penis** → prostatic venous plexus → internal iliac vein → IVC.
2. The second is to the **superficial dorsal vein of the penis** → external pudendal vein → great saphenous vein → femoral vein → external iliac vein → IVC.

D. INNERVATION. The penis is innervated by the pudendal nerve via the **dorsal nerve of the penis.**

X Innervation of the Male Reproductive System

A. PARASYMPATHETIC

1. Preganglionic neuronal cell bodies are located in the **gray matter of the S2–4 spinal cord**. Preganglionic axons form the **pelvic splanchnic nerves**, which interact with the **inferior hypogastric plexus**.
2. Postganglionic neuronal cell bodies are located near or within the male viscera.
3. Postganglionic axons terminate on smooth muscle and glands.

B. SYMPATHETIC

1. Preganglionic neuronal cell bodies are located in the **intermediolateral cell column** of the spinal cord. Preganglionic axons form the **sacral splanchnic nerves**.
2. Postganglionic neuronal cell bodies are located in the **inferior hypogastric plexus**.
3. Postganglionic axons terminate on smooth muscle and glands.

XI Erection, Secretion, Emission, and Ejaculation

A. ERECTION. Erection of the penis is controlled by the parasympathetic nervous system via the **pelvic splanchnic nerves (S2–4)**, which dilate blood vessels supplying the erectile tissue. This engorges the corpora cavernosa and corpus spongiosum with blood; compresses the veins, which impedes venous return; and causes a full erection. The erection of the penis is also maintained by the somatic nervous system via the **perineal branch of the pudendal nerve**, which contracts the **bulbospongiosus muscles** and **ischiocavernosus muscles**. This compresses the erectile tissue of the bulb of the penis and the crura of the penis and helps to maintain the erection.

B. SECRETION. Secretion from the seminal vesicles, bulbourethral glands of Cowper, and prostate gland is controlled by the parasympathetic nervous system via the **pelvic splanchnic nerves (S2–4)**, which stimulate the secretory activity of these glands.

C. EMISSION. Emission from the penis is controlled by the sympathetic nervous system via the **L3 and L4 lumbar splanchnic nerves** and the **sacral splanchnic nerves**, which contracts the smooth muscle of the tail region of the epididymis, ductus deferens, seminal vesicle, and prostate gland, thus promoting movement of sperm and fluid; and contracts the internal urethral sphincter (i.e., smooth muscle), thus preventing reflux of sperm and fluid into the urinary bladder.

D. EJACULATION. Ejaculation from the penis is controlled by the somatic nervous system via the **pudendal nerve**, which contracts the **bulbospongiosus muscle** (i.e., skeletal muscle) to propel sperm and fluid and relaxes the **sphincter urethrae muscle** located within the deep perineal space (i.e., skeletal muscle; also called the external urethral sphincter).

Chapter 18

Pelvis

❶ **Bones of the Pelvis (Figure 18-1).** The bony pelvis is a basin-shaped ring of bone that consists of the following:

A. COXAL (HIP BONE). There are two coxal bones, and each coxal bone is formed by the fusion of the **ischium, ilium,** and **pubis,** which join at the acetabulum (an incomplete cup-shaped cavity) of the hip joint.

    1. Ilium
        a. The ilium forms the lateral part of the hip bone and joins the ischium and pubis to form the acetabulum and ala.
        b. The ilium is composed of the anterior-superior iliac spine, anterior-inferior iliac spine, posterior iliac spine, greater sciatic notch, iliac fossa, and gluteal lines.

    2. Ischium
        a. The ischium joins the ilium and superior ramus of the pubis to form the acetabulum.
        b. The ramus of the ischium joins the inferior pubic ramus to form the ischiopubic ramus.
        c. The ischium is composed of the ischial spine, ischial tuberosity, and lesser sciatic notch.

    3. Pubis
        a. The pubis forms the anterior part of the acetabulum and the anteromedial part of the hip bone.
        b. The pubis is composed of the body, superior ramus, and inferior ramus.

B. SACRUM
    1. The sacrum is formed by the fusion of the **S1–5 vertebrae** and is the posterior portion of the bony pelvis.
    2. The sacrum contains the **dorsal sacral foramina,** which transmit dorsal primary rami of sacral spinal nerves; **ventral sacral foramina,** which transmit ventral primary rami of sacral spinal nerves; and **sacral hiatus,** which is formed due to the failure of the laminae of the S5 vertebrae to fuse.
    3. The pedicles form the **sacral cornua,** which are important landmarks in locating the sacral hiatus for administration of caudal anesthesia.

C. COCCYX (TAIL BONE). The coccyx is formed by the fusion of the Co1–4 vertebrae.

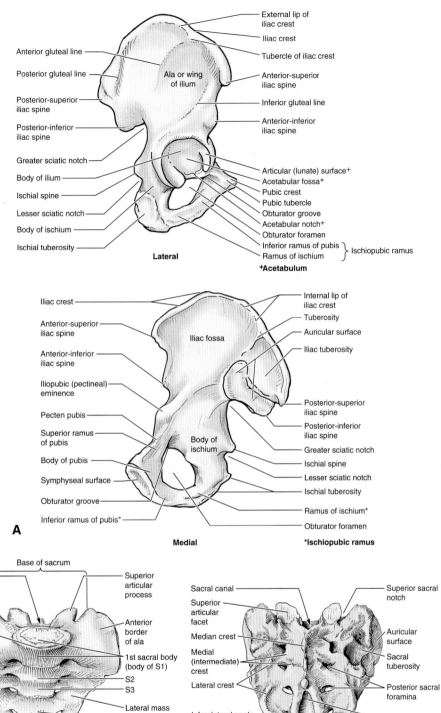

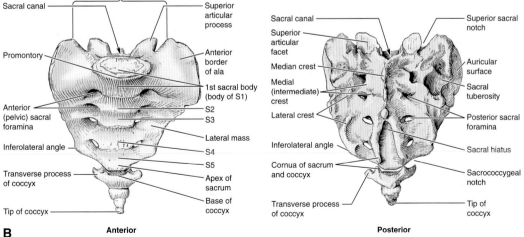

● **Figure 18-1 Bones of the pelvis. A:** The right coxal bone (lateral and medial views). **B:** The sacrum and coccyx (posterior and anterior views).

Greater and Lesser Sciatic Foramina. The **sacrotuberous ligament** (runs from the sacrum to the ischial tuberosity) and **sacrospinous ligament** (runs from the sacrum to the ischial spine) help define the borders of the foramina.

A. GREATER SCIATIC FORAMEN

1. The greater sciatic foramen is divided into the **suprapiriformis recess** and **infrapiriformis recess** by the piriformis muscle.

2. The greater sciatic foramen transmits the following important structures as they exit the pelvic cavity to enter the gluteal and thigh regions: **superior gluteal vein, artery, and nerve; piriformis muscle; inferior gluteal vein, artery, and nerve; sciatic nerve; internal pudendal vein and artery;** and **pudendal nerve.**

B. LESSER SCIATIC FORAMEN

1. The lesser sciatic foramen transmits the following important structures as they re-enter the pelvic cavity and proceed to the perineum: **internal pudendal vein and artery** and **pudendal nerve.**

2. Note that the internal pudendal vein, internal pudendal artery, and pudendal nerve exit the pelvic cavity via the greater sciatic foramen and then re-enter the pelvic cavity through the lesser sciatic foramen and proceed to the perineum.

Pelvic Inlet (Pelvic Brim) (Figure 18-2)

A. The pelvic inlet is defined by the **sacral promontory (S1 vertebra)** and the **linea terminalis.** The linea terminalis includes **the pubic crest, iliopectineal line,** and **arcuate line.**

B. The pelvic inlet divides the pelvic cavity into two parts: the **major (false) pelvic cavity,** which lies above the pelvic inlet between the iliac crests and is actually part of the abdominal cavity, and the **minor (true) pelvic cavity,** which lies below the pelvic inlet and extends to the pelvic outlet.

C. The pelvic inlet is **oval shaped in females** and **heart shaped in males.**

D. The measurements of the pelvic inlet include the following:

1. **True conjugate diameter** is the distance from the sacral promontory to the superior margin of the pubic symphysis. This diameter is measured radiographically on a lateral projection.

2. **Diagonal conjugate diameter** is the distance from the sacral promontory to the inferior margin of the pubic symphysis. This diameter is measured during an obstetric examination.

Pelvic Outlet

A. The pelvic outlet is defined by the **coccyx, ischial tuberosities, inferior pubic ramus,** and **pubic symphysis.**

B. The pelvic outlet is closed by the **pelvic diaphragm** and **urogenital diaphragm.**

C. The pelvic outlet is **diamond shaped in both females and males.**

D. The pelvic outlet is divided into the **anal triangle** and **urogenital triangle** by a line passing through the ischial tuberosities.

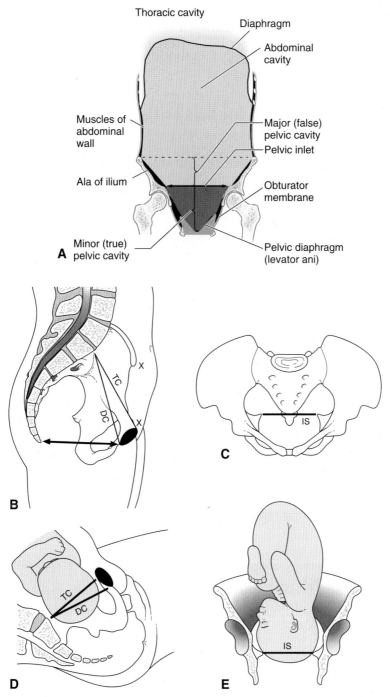

● **Figure 18-2 A:** Diagram shows the relationships of the thoracic, abdominal, and pelvic cavities. **B:** A lateral view of the pelvis. The diameter of the pelvic inlet is measured by the true conjugate (TC) diameter and the diagonal conjugate (DC) diameter. The opening of the pelvic outlet is shown (*line with arrows*) extending from the pubic symphysis to the coccyx. Note also that in the natural position of the bony pelvis, the anterior-superior iliac spine and the pubic tubercle lie in the same vertical plane (see X's). **C:** A superior view of the pelvis. The diameter of the pelvic outlet is measured by the transverse diameter (not shown) and the interspinous (IS) diameter. **D:** A lateral view of the pelvis. Note that during childbirth the fetal head must pass through the pelvic inlet. The TC and DC diameters measure the diameter of the pelvic inlet. **E:** A frontal view of the pelvis. Note that during childbirth the fetal head must pass through the pelvic outlet. The IS diameter measures the diameter of the pelvic outlet. The TC, DC, and IS diameters are important during childbirth, where the fetus must travel through the birth canal, which consists of the pelvic inlet → minor pelvis → cervix → vagina → pelvic outlet.

E. The measurements of the pelvic outlet include the following:
   1. **Transverse diameter** is the distance between the ischial tuberosities.
   2. **Interspinous diameter** is the distance between the ischial spines. The ischial spines may present a barrier to the fetus during childbirth if the interspinous diameter is less than 9.5 cm.

V Comparison of the Female and Male Pelvis (Table 18-1)

TABLE 18-1	COMPARISON OF FEMALE AND MALE PELVIS
Female Pelvis	**Male Pelvis**
Thin and light	Thick and heavy
Pelvic inlet is oval-shaped	Pelvic inlet is heart-shaped
Pelvic outlet is diamond-shaped	Pelvic outlet is diamond-shaped
Pelvic outlet is comparatively larger due to everted ischial tuberosities	Pelvic outlet is comparatively small
Major (false) pelvic cavity is shallow	Major (false) pelvic cavity is deep
Minor (true) pelvic cavity is wide and shallow; cylindrical	Minor (true) pelvic cavity is narrow and deep; tapering
Subpubic angle (pubic arch) is wide (>80 degrees)	Subpubic angle (pubic arch) is narrow (<70 degrees)
Greater sciatic notch is wide (~90 degrees)	Greater sciatic notch is narrow (~70 degrees); inverted V
Sacrum is short and wide	Sacrum is long and narrow
Obturator foramen is triangular shaped	Obturator foramen is round shaped

VI Muscles of the Pelvis (Figure 18-3).
The muscles of the pelvis include the **obturator internus muscle, piriformis muscle, coccygeus muscle,** and **levator ani muscles (iliococcygeus, pubococcygeus,** and **puborectalis muscles).**

VII Arterial Supply

A. **INTERNAL ILIAC ARTERY.** The internal iliac artery arises from the bifurcation of the common iliac artery. The internal iliac artery is commonly divided into an **anterior division** and a **posterior division.**
   1. **Anterior division** gives off the following branches:
      a. **Inferior gluteal artery** exits the pelvis via the infrapiriformis recess of the greater sciatic foramen (i.e., inferior to the piriformis muscle). This artery supplies the pelvic diaphragm, piriformis, quadratus femoris, uppermost hamstrings, gluteus maximus, and sciatic nerve.

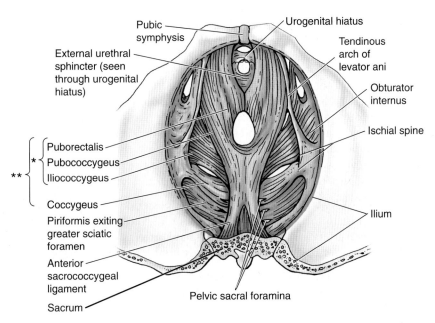

Pubic symphysis

Urogenital hiatus

Tendinous arch of levator ani

External urethral sphincter (seen through urogenital hiatus)

Obturator internus

Ischial spine

Puborectalis
Pubococcygeus
Iliococcygeus

Coccygeus
Piriformis exiting greater sciatic foramen

Ilium

Anterior sacrococcygeal ligament

Pelvic sacral foramina

Sacrum

● **Figure 18-3 Muscles of the pelvis.** Superior view of the muscles of the pelvis. * = levator ani, ** = pelvic diaphragm (floor).

b. **Internal pudendal artery** exits the pelvis via the infrapiriformis recess of the greater sciatic foramen (i.e., inferior to the piriformis muscle), enters the perineum via the lesser sciatic foramen, and courses to the urogenital triangle via the pudendal canal. This artery supplies the perineum (main artery of the perineum), including the skin and muscles of the anal triangle and the urogenital triangle and the erectile bodies.

c. **Umbilical artery** runs along the lateral pelvic wall and alongside the bladder for a short distance, then obliterates to form the **medial umbilical ligament**. The umbilical artery gives rise to the **superior vesical artery**, which supplies the superior part of the urinary bladder.

d. **Obturator artery** runs along the lateral pelvic wall and exits the pelvis via the obturator canal. This artery supplies the pelvic muscles, muscles of the medial compartment of the thigh, head of the femur, and ilium.

e. **Vaginal artery (female) or inferior vesical artery (male).** The **vaginal artery** in the female supplies the anterior and posterior walls of the vagina, vestibular bulb, and adjacent rectum. The **inferior vesical artery** in the male runs in the lateral ligament of the bladder and supplies the fundus of the bladder, prostate gland, seminal vesicle, ductus deferens, and lower part of the ureter.

f. **Uterine artery (female) or artery of the ductus deferens (male).** The **uterine artery** in the female runs medially in the base of the broad ligament to reach the junction of the cervix and body of the uterus and runs in front of and above the ureter near the lateral fornix of the vagina. This uterine artery supplies the uterus, ligaments of the uterus, uterine tube, ovary, cervix, and vagina. The artery of the ductus deferens in the male supplies the ductus deferens.

g. **Middle rectal artery** runs medially and descends in the pelvis. The middle rectal artery supplies the lower part of the rectum, upper part of the anal canal, prostate gland, and seminal vesicles.

2. **Posterior division** gives off the following branches:
   a. **Iliolumbar artery** ascends anterior to the sacroiliac joint and posterior to the psoas major muscle. This artery supplies the psoas major, iliacus, quadratus lumborum, and cauda equina in the vertebral canal.
   b. **Lateral sacral artery** runs medially in front of the sacral plexus and gives rise to branches that enter the anterior sacral foramina and then emerge from the posterior sacral foramina. This artery supplies the meninges, roots of the sacral nerves, and muscles and skin overlying the sacrum.
   c. **Superior gluteal artery** exits the pelvis via the suprapiriformis recess of the greater sciatic foramen (i.e., superior to the piriformis muscle). This artery supplies the piriformis, gluteal muscles, and tensor fascia lata.

B. MEDIAN SACRAL ARTERY arises from the posterior aspect of the abdominal aorta and runs close to the midline over the L4 and L5 vertebrae, sacrum, and coccyx. The median sacral artery gives rise to **medial sacral arteries**. This median sacral artery supplies the posterior part of the rectum, lower lumbar vertebrae, sacrum, and coccyx.

C. SUPERIOR RECTAL ARTERY is a continuation of the inferior mesenteric artery and descends into the pelvis between the layers of the sigmoid mesocolon. This artery supplies the superior part of the rectum.

D. OVARIAN ARTERY (FEMALE) OR TESTICULAR ARTERY (MALE). The ovarian artery in the female arises from the abdominal aorta and reaches the ovary through the suspensory ligament of the ovary. This artery supplies the ureter, ovary, and ampulla of the uterine tube. The testicular artery in the male arises from the abdominal aorta and then runs in the inguinal canal to enter the scrotum. This artery supplies the ureter, testis, and epididymis.

Venous Drainage

A. PELVIC VENOUS PLEXUSES. The pelvic venous plexuses within the minor (true) pelvic cavity are formed by intercommunicating veins surrounding the pelvic viscera and include the **rectal venous plexus, vesical venous plexus, prostatic venous plexus, uterine venous plexus,** and **vaginal venous plexus.** These pelvic venous plexuses drain venous blood via a number of different pathways, as follows:
   1. Pelvic venous plexuses → internal iliac veins, which join the external iliac veins to form the common iliac veins → common iliac veins, which join to form the inferior vena cava (IVC). This is the main venous drainage pathway.
   2. Pelvic venous plexuses → median sacral vein → common iliac vein → inferior vena cava
   3. Pelvic venous plexuses → ovarian veins → inferior vena cava
   4. Pelvic venous plexuses → superior rectal vein → inferior mesenteric vein → portal vein
   5. Pelvic venous plexuses → lateral sacral veins → internal vertebral venous plexus → cranial dural sinuses

Nerves (Figure 18-4)

A. SACRAL PLEXUS. The components of the sacral plexus include:
   1. **Rami** are the L4–5 (lumbosacral trunk) and **S1–4 ventral primary rami** of spinal nerves.
   2. **Divisions (anterior and posterior)** are formed by rami dividing into anterior and posterior divisions.

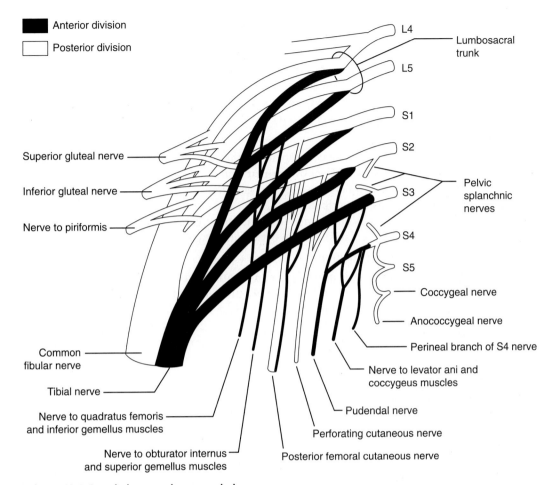

■ Anterior division
□ Posterior division

L4

Lumbosacral trunk

L5

S1

Superior gluteal nerve

S2

Inferior gluteal nerve

S3

Pelvic splanchnic nerves

Nerve to piriformis

S4

S5

Coccygeal nerve

Anococcygeal nerve

Perineal branch of S4 nerve

Common fibular nerve

Nerve to levator ani and coccygeus muscles

Tibial nerve

Pudendal nerve

Nerve to quadratus femoris and inferior gemellus muscles

Perforating cutaneous nerve

Nerve to obturator internus and superior gemellus muscles

Posterior femoral cutaneous nerve

● **Figure 18-4 Sacral plexus and coccygeal plexus.**

3. **Branches.** The major terminal branches are:
   a. **Superior gluteal nerve (L4–S1)** innervates the gluteus medius, gluteus minimus, and tensor fascia lata muscles.
   b. **Inferior gluteal nerve (L5–S2)** innervates the gluteus maximus muscle.
   c. **Nerve to piriformis (S1, S2)** innervates the piriformis muscle.
   d. **Common fibular nerve (L4, L5, S1, S2)**
   e. **Tibial nerve (L4, L5, S1–3).** The tibial nerve and common fibular nerve comprise the **sciatic nerve** (see Chapter 21).
   f. **Nerve to the quadratus femoris and inferior gemellus (L5–S1)** innervates the quadratus femoris and inferior gemellus muscles.
   g. **Nerve to the obturator internus and superior gemellus (L5–S2)** innervates the obturator internus and superior gemellus muscles.
   h. **Posterior femoral cutaneous nerve (S1–3)** innervates the skin of the buttock, thigh, and calf (sensory). This nerve gives rise to the **inferior cluneal nerves** and **perineal branches.**
   i. **Perforating cutaneous nerve (S2, S3)** innervates the skin in the perineal area.

j.   **Pudendal nerve (S2–4)** passes through the greater sciatic foramen, crosses the ischial spine, and enters the perineum with the internal pudendal artery through the pudendal canal. This nerve gives rise to the **inferior rectal nerve, perineal nerve,** and **dorsal nerve of the penis (or clitoris).**

k.   **Nerve to the levator ani and coccygeus (S3, S4)** innervates the levator ani muscles and the coccygeus muscle.

l.   **Perineal branch of spinal nerve (S4)** innervates the skin of the perineum (sensory).

B.   COCCYGEAL PLEXUS. The components of the coccygeal plexus include:

1.   **Rami** are the **S4 and S5 ventral primary rami** of spinal nerves.

2.   **Coccygeal nerve** innervates the coccygeus muscle, part of the levator ani muscles, and the sacrococcygeal joint.

3.   **Branches.** There is one branch from the coccygeal plexus called the **anococcygeal nerve,** which innervates the skin between the tip of the coccyx and the anus.

C.   AUTONOMIC COMPONENTS

1.   **Superior hypogastric plexus** is a continuation of the intermesenteric plexus from the inferior mesenteric ganglion below the aortic bifurcation and receives the L3 and L4 lumbar splanchnic nerves. This plexus contains ganglionic neuronal cell bodies upon which preganglionic sympathetic axons of the L3 and L4 lumbar splanchnic nerves synapse on. The superior hypogastric plexus descends anterior to the L5 vertebra and ends by dividing into the **right hypogastric nerve** and **left hypogastric nerve.**

2.   **Right and left hypogastric nerves** descend on either side lateral to the rectum and join the right or left inferior hypogastric plexus, respectively.

3.   **Right and left inferior hypogastric plexuses** are located against the posterolateral pelvic wall lateral to the rectum, vagina, and base of the bladder. The right and left inferior hypogastric plexuses are formed by the union of the **right or left hypogastric nerves, sacral splanchnic nerves (L5 and S1–3),** and **pelvic splanchnic nerves (S2–4).** This plexus contains ganglionic neuronal cell bodies upon which preganglionic sympathetic axons of the sacral splanchnic nerves (L5 and S1–3) synapse on.

4.   **Sacral sympathetic trunk** is a continuation of the paravertebral sympathetic chain ganglia in the pelvis. The sacral trunks descend on the inner surface of the sacrum medial to the sacral foramina and converge to form the small median **ganglion impar** anterior to the coccyx.

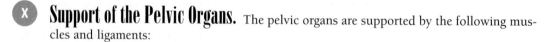 **Support of the Pelvic Organs.** The pelvic organs are supported by the following muscles and ligaments:

A.   PELVIC DIAPHRAGM (FLOOR). The pelvic diaphragm is composed of the following muscles:

1.   **Coccygeus muscle**

2.   **Levator ani muscles,** which consist of:

a.   Iliococcygeus

b.   Pubococcygeus

c.   Puborectalis. This muscle forms a U-shaped sling around the anorectal junction causing a 90-degree perineal flexure. This muscle is important in maintaining fecal continence.

B. **UROGENITAL DIAPHRAGM.** The urogenital diaphragm is composed of the following muscles:
   1. **Deep transverse perineal muscle**
   2. **Sphincter urethra muscle**

C. **TRANSVERSE CERVICAL LIGAMENT (CARDINAL LIGAMENT OF MACKENRODT)** is a condensation of endopelvic fascia that extends laterally from the cervix to the side wall of the pelvis.

D. **UTEROSACRAL LIGAMENT** is a condensation of endopelvic fascia that extends posteriorly from the cervix to the sacrum.

E. **PUBOCERVICAL LIGAMENT** is a condensation of endopelvic fascia that extends anteriorly from the cervix to the pubic symphysis.

 XI **Clinical Considerations**

A. **PELVIC RELAXATION** is the weakening or loss of support of pelvic organs due to damage of the pelvic diaphragm, urogenital diaphragm, transverse cervical ligament (cardinal ligament of Mackenrodt), uterosacral ligament, and/or pubocervical ligament. This may result in **cystocele** (prolapse of urinary bladder into the anterior vaginal wall), **rectocele** (prolapse of rectum into posterior wall of vagina), or **uterine prolapse** (prolapse of uterus into vaginal vault). It is caused by multiple childbirths; birth trauma; increased intra-abdominal pressure due to obesity, heavy lifting, or chronic cough; or menopausal loss of muscle tone. Clinical signs include a heavy sensation in the lower abdomen that exacerbates upon heavy lifting or prolonged standing, increased frequency of urination with burning sensation due to urine stagnation and bacterial proliferation, and urine leakage with coughing or sneezing (i.e., stress incontinence).

B. **THE PELVIC RING.** The pelvic ring consists of the sacrum and the two coxal bones that have resilient articulations where small degrees of movement are possible between the sacroiliac (SI) joint and the pubic symphysis. The **sacrum is the keystone of the femoral-sacral arch** that supports the vertebral column over the legs. The **anterior and posterior SI ligaments** attach the upper sacrum to the ilium. The **sacrotuberous ligament** and the **sacrospinous ligament** attach the lower sacrum to the ischium. The functional stability of the pelvic ring depends on these ligaments.

C. **PUDENDAL NERVE BLOCK (FIGURE 18-5)** provides perineal anesthesia during forceps childbirth delivery by anesthetizing the pudendal nerve. A 1% lidocaine solution is injected transvaginally or just lateral to the labia majora **around the tip of the ischial spine** and **through the sacrospinous ligament**. The pain of childbirth is transmitted by the pudendal nerve through sensory fibers of S2–5 spinal nerves. The pudendal nerve passes out of the pelvic cavity through the greater sciatic foramen, travels around the posterior surface of the ischial spine, and re-enters the pelvic cavity through the lesser sciatic foramen. The pudendal nerve travels within the fascia of the obturator internus muscle (called the **pudendal canal of Alcock**) and divides into the **inferior rectal nerve**, **perineal nerve**, and **dorsal nerve of the penis (or clitoris)**. To obtain a complete anesthesia of the perineal region, the **ilioinguinal nerve** (which branches into the **anterior labial nerves**), **genitofemoral nerve**, and **perineal branch of the posterior femoral cutaneous nerve** are anesthetized.

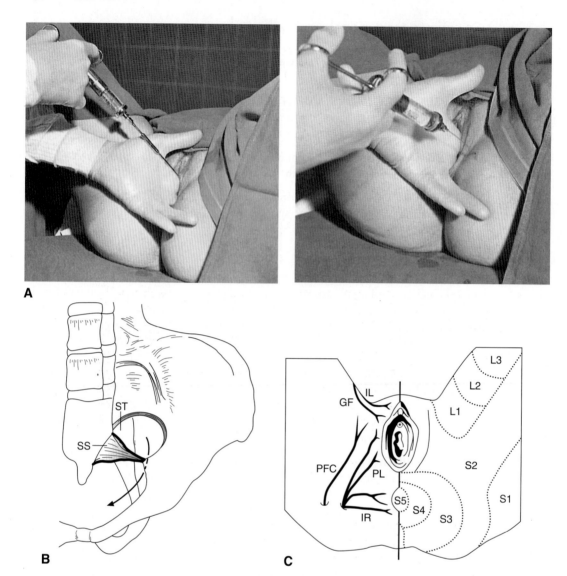

A

B

C

● **Figure 18-5 A:** Photographs of the clinical administration of a pudendal nerve block both transvaginally and lateral to the labia majora. The ischial spine is a good anatomic landmark. **B:** Diagram indicating the path of the pudendal nerve (*curved arrow*) as it passes out of the pelvic cavity through the greater sciatic foramen (posterior to the ischial spine) and returns to the pelvic cavity through the lesser sciatic foramen as it proceeds to the perineum. SS = sacrospinous ligament, ST = sacrotuberous ligament. **C:** Diagram of the perineum in the lithotomy position. The posterior labial nerves (PL) and inferior rectal nerves (IR), which are terminal branches of the pudendal nerve, are shown. In addition, the ilioinguinal nerve (IL), genitofemoral nerve (GF), and perineal branch of the posterior femoral cutaneous nerve (PFC) are indicated.

XII Radiology

A. ANTEROPOSTERIOR RADIOGRAPH OF FEMALE PELVIS (FIGURE 18-6)

A

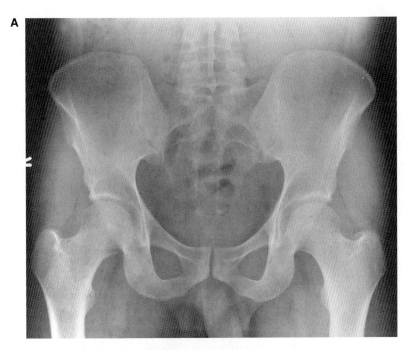

B

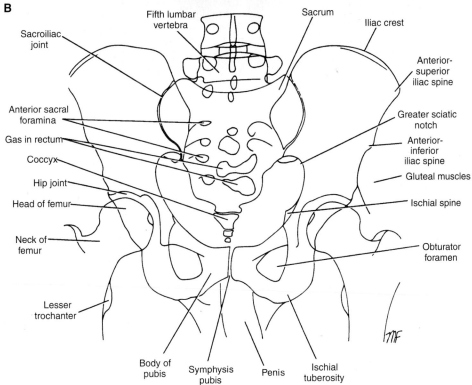

● **Figure 18-6 A:** Anteroposterior radiograph of the male pelvis. **B:** Diagrammatic representation of the radiograph in A.

Chapter 19

Perineum

① Perineum (Figure 19-1)

A. The perineum is a part of the pelvic outlet located inferior to the pelvic diaphragm.

B. The perineum is diamond shaped and can be divided by a line passing through the ischial tuberosities into two triangles: the **urogenital (UG) triangle** and the **anal triangle**.

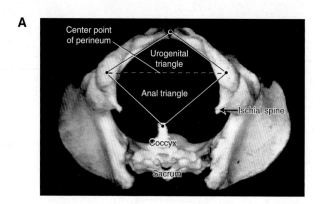

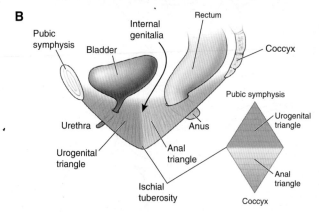

● **Figure 19-1 Perineum. A:** Osseous boundaries of the perineum. The diamond-shaped perineum extends from the pubic symphysis to the coccyx. Note that a transverse line joining the anterior ends of the ischial tuberosities divides the perineum into two unequal triangular areas, the urogenital triangle anteriorly and the anal triangle posteriorly. The midpoint of the transverse line indicates the site of the perineal body (central perineal tendon). **B:** Lateral diagram of the perineum. The lateral diagram of the perineum shows that the urogenital triangle and the anal triangle do not occupy the same plane.

 **Urogenital (UG) Triangle.** The UG triangle is composed of the following:

A. DEEP PERINEAL SPACE

1. The deep perineal space is a space that lies between the **superior fascia of the UG diaphragm** and the **inferior fascia of the UG diaphragm (perineal membrane)**.

2. This space contains a number of structures that completely occupy it. The anatomic structures found within the deep perineal space of the male and female are indicated in **Table 19-1**.

3. One of those structures is the **UG diaphragm**, which consists of the **deep transverse perineal muscle** and the **sphincter urethrae muscle**.

B. SUPERFICIAL PERINEAL SPACE

1. The superficial perineal space is a space that lies between the **inferior fascia of the UG diaphragm (perineal membrane)** and **superficial perineal fascia (Colles fascia)**.

2. The anatomic structures found within the superficial perineal space of the male and female are indicated in **Table 19-1**.

TABLE 19-1	**STRUCTURE WITHIN THE DEEP AND SUPERFICIAL PERINEAL SPACES**
Male	**Female**
Structures within the Deep Perineal Space	
Membranous urethra	Urethra
	Vagina
Urogenital (UG) diaphragm	UG diaphragm
Deep transverse perineal muscle	Deep transverse perineal muscle
Sphincter urethrae muscle	Sphincter urethrae muscle
Branches of internal pudendal artery	Branches of internal pudendal artery
Artery of the penis	Artery of the clitoris
Branches of pudendal nerve	Branches of pudendal nerve
Dorsal nerve of the penis	Dorsal nerve of the clitoris
Bulbourethral glands (of Cowper)	No glands
Structures within the Superficial Perineal Space	
Penile (spongy) urethra	Urethra
	Vestibule of the vagina
Bulbospongiosus muscle	Bulbospongiosus muscle
Ischiocavernosus muscle	Ischiocavernosus muscle
Superficial transverse perineal muscle	Superficial transverse perineal muscle
Branches of internal pudendal artery	Branches of the internal pudendal artery
Perineal artery → posterior scrotal arteries	Perineal artery → posterior labial arteries
Dorsal artery of the penis	Dorsal artery of the clitoris
Deep artery of the penis	Deep artery of the clitoris
Branches of pudendal nerve	Branches of pudendal nerve
Perineal nerve → posterior scrotal nerves	Perineal nerve → posterior labial nerves
Dorsal nerve of the penis	Dorsal nerve of the clitoris
Bulb of the penis	Vestibular bulb
Crura of the penis	Crura of the clitoris
Perineal body	Perineal body
	Round ligament of the uterus
Duct of the bulbourethral gland	Greater vestibular glands (of Bartholin)

C.   **CLINICAL CONSIDERATION.** Episiotomy is an incision of the perineum made in order to enlarge the vaginal opening during childbirth. There are two types of episiotomies.

   1.   **Median episiotomy** starts at the **frenulum of the labia minora** and proceeds directly downward cutting through the **skin** → **vaginal wall** → **perineal body** → **superficial transverse perineal muscle.** The external anal sphincter muscle may be inadvertently cut.

   2.   **Mediolateral episiotomy** starts at the frenulum of the labia minora and proceeds at a 45-degree angle cutting through the **skin** → **vaginal wall** → **bulbospongiosus muscle.** This procedure has a higher risk of bleeding in comparison to a median episiotomy but creates more room than a median episiotomy.

Ⅲ **Anal Triangle.** The anal triangle is composed of the following:

A.   **ISCHIORECTAL FOSSA**

   1.   The ischiorectal fossa is located on either side of the anorectum and is separated from the pelvic cavity by the levator ani muscle.

   2.   This fossa contains ischiorectal fat, inferior rectal nerves, inferior rectal artery and vein, perineal branches of the posterior femoral cutaneous nerve, and the pudendal (Alcock) canal, which transmits the pudendal nerve and the internal pudendal artery and vein.

B.   **MUSCLES OF THE ANAL TRIANGLE.** The muscles of the anal triangle include the obturator internus, external anal sphincter, levator ani, and coccygeus muscles.

IV Muscles of the Male and Female Perineum (Figure 19-2)

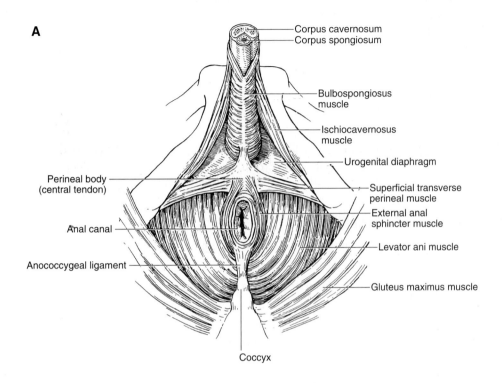

A

- Corpus cavernosum
- Corpus spongiosum
- Bulbospongiosus muscle
- Ischiocavernosus muscle
- Urogenital diaphragm
- Superficial transverse perineal muscle
- External anal sphincter muscle
- Levator ani muscle
- Gluteus maximus muscle
- Perineal body (central tendon)
- Anal canal
- Anococcygeal ligament
- Coccyx

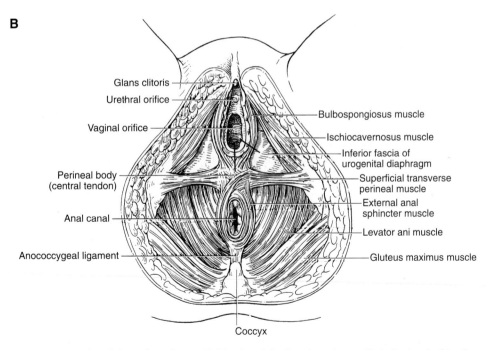

B

- Glans clitoris
- Urethral orifice
- Vaginal orifice
- Bulbospongiosus muscle
- Ischiocavernosus muscle
- Inferior fascia of urogenital diaphragm
- Superficial transverse perineal muscle
- External anal sphincter muscle
- Levator ani muscle
- Gluteus maximus muscle
- Perineal body (central tendon)
- Anal canal
- Anococcygeal ligament
- Coccyx

● **Figure 19-2 A:** Muscles of the male perineum. **B:** Muscles of the female perineum. Note the two incision lines for a median and mediolateral episiotomy (*thick black lines*).

Chapter 20

Upper Limb

I **Bones.** The bones of the upper limb include the clavicle, scapula, humerus, radius, ulna, carpal bones (scaphoid, lunate, triquetrum, pisiform, trapezium, trapezoid, capitate, and hamate), metacarpals, and phalanges (proximal, middle, and distal).

II **Muscles (See Appendix 1)**

A. **ANTERIOR AXIOAPPENDICULAR MUSCLES** include the pectoralis major, pectoralis minor, subclavius, and serratus anterior.

B. **POSTERIOR AXIOAPPENDICULAR AND SCAPULOHUMERAL MUSCLES** include the trapezius, latissimus dorsi, levator scapulae, rhomboid major and minor, deltoid, supraspinatus, infraspinatus, teres minor, teres major, and subscapularis.

C. **MUSCLES OF THE ANTERIOR (FLEXOR) COMPARTMENT OF THE ARM** include the biceps brachii, brachialis, and coracobrachialis.

D. **MUSCLES OF THE POSTERIOR (EXTENSOR) COMPARTMENT OF THE ARM** include the triceps and anconeus.

E. **MUSCLES OF THE ANTERIOR (FLEXOR) COMPARTMENT OF THE FOREARM** include the pronator teres, flexor carpi radialis, palmaris longus, flexor carpi ulnaris, flexor digitorum superficialis, flexor digitorum profundus, flexor pollicis longus, and pronator quadratus.

F. **MUSCLES OF THE POSTERIOR (EXTENSOR) COMPARTMENT OF THE FOREARM** include the brachioradialis, extensor carpi radialis longus, extensor carpi radialis brevis, extensor digitorum, extensor digiti minimi, extensor carpi ulnaris, supinator, extensor indicis, abductor pollicis longus, extensor pollicis longus, and extensor pollicis brevis.

G. **INTRINSIC MUSCLES OF THE HAND** include the opponens pollicis, abductor pollicis brevis, flexor pollicis brevis, adductor pollicis, abductor digiti minimi, flexor digiti minimi brevis, opponens digiti minimi, lumbricals (first through fourth), dorsal interossei (first through fourth), and palmar interossei (first through third).

III **Arterial Supply (Figure 20-1)**

A. **SUBCLAVIAN ARTERY** extends from the **arch of the aorta** to the **lateral border of the first rib.** The subclavian artery gives off the following branches:

1. **Internal thoracic artery** is continuous with the **superior epigastric artery**, which anastomoses with the **inferior epigastric artery** (a branch of the external iliac artery). This may provide a route of collateral circulation if the abdominal aorta is blocked (e.g., postductal coarctation of the aorta).

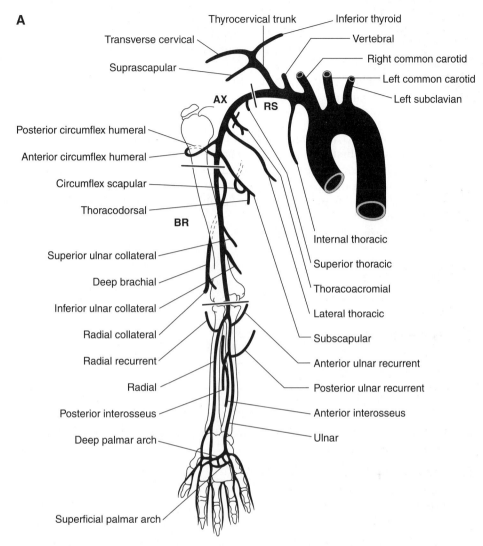

A

Thyrocervical trunk
Inferior thyroid
Transverse cervical
Vertebral
Suprascapular
Right common carotid
Left common carotid
Left subclavian
AX
RS
Posterior circumflex humeral
Anterior circumflex humeral
Circumflex scapular
Thoracodorsal
BR
Internal thoracic
Superior ulnar collateral
Superior thoracic
Deep brachial
Thoracoacromial
Inferior ulnar collateral
Lateral thoracic
Radial collateral
Subscapular
Radial recurrent
Anterior ulnar recurrent
Radial
Posterior ulnar recurrent
Posterior interosseus
Anterior interosseus
Deep palmar arch
Ulnar
Superficial palmar arch

B

Transverse cervical
Suprascapular
Clavicle
Thyrocervical trunk
First rib
Circumflex scapular
Subscapular

C

Anterior interosseous
Radial artery
Ulnar artery
Palmar carpal branch
Palmar carpal branch
Superficial palmar branch
Deep carpal branch
Deep palmar branch
Superficial palmar arch
Deep palmar arch
Princeps pollicis
Palmar metacarpal
Common palmar digital
Proper palmar digital
Radialis indicis

● **Figure 20-1 A:** Arterial supply of the upper limb. The lines from proximal to distal indicate the lateral border of the first rib, inferior border of the teres major muscle, and the cubital fossa, respectively. **B:** Diagram of the collateral circulation around the shoulder. **C:** Arterial supply of the hand.

2. **Vertebral artery**
3. **Thyrocervical trunk** has three branches:
   a. **Suprascapular artery**, which participates in collateral circulation around the shoulder
   b. **Transverse cervical artery**, which participates in collateral circulation around the shoulder
   c. **Inferior thyroid artery**

B. **AXILLARY ARTERY** is a continuation of the subclavian artery and extends from the lateral border of the first rib to the **inferior border of the teres major muscle.** The tendon of the pectoralis minor muscle crosses the axillary artery anteriorly and divides the axillary artery into three distinct parts (i.e., the first part is medial, the second part is posterior, and the third part lateral to the muscle). The axillary artery gives off the following branches:
   1. **First Part**
      a. **Superior thoracic artery**
   2. **Second Part**
      a. **Thoracoacromial artery** is a short, wide trunk that divides into four branches: acromial, deltoid, pectoral, and clavicular.
      b. **Lateral thoracic artery**
   3. **Third Part**
      a. **Anterior humeral circumflex artery**
      b. **Posterior humeral circumflex artery**
      c. **Subscapular artery**, which gives off the **circumflex scapular artery** and the **thoracodorsal artery**

C. **BRACHIAL ARTERY** is a continuation of the axillary artery and extends from the **inferior border of the teres major muscle** to the **cubital fossa,** where it ends in the cubital fossa opposite the neck of the radius. The brachial artery gives off the following branches:
   1. **Deep brachial artery**
      a. A fracture of the humerus at midshaft may damage the **deep brachial artery and radial nerve** as they travel together on the posterior aspect of the humerus in the radial groove.
      b. The deep brachial artery ends by dividing into the **middle collateral artery** and **radial collateral artery.**
   2. **Superior ulnar collateral artery** runs with the ulnar nerve posterior to the medial epicondyle and anastomoses with the posterior ulnar recurrent artery to participate in collateral circulation around the elbow.
   3. **Inferior ulnar collateral artery** anastomoses with the anterior ulnar recurrent artery to participate in collateral circulation around the elbow.
   4. **Radial artery** gives off the following branches:
      a. **Recurrent radial artery** anastomoses with the radial collateral artery.
      b. **Palmar carpal branch**
      c. **Dorsal carpal branch**
      d. **Superficial palmar branch** completes the superficial palmar arch.
      e. **Princeps pollicis artery** divides into two **proper digital arteries** for each side of the thumb.
      f. **Radialis indicis artery**
      g. **Deep palmar arch** is the main termination of the radial artery and anastomoses with the deep palmar branch of the ulnar artery. It gives rise to three **palmar metacarpal arteries,** which join the common palmar digital arteries from the superficial arch.
   5. **Ulnar artery** gives off the following branches:
      a. **Anterior ulnar recurrent artery**
      b. **Posterior ulnar recurrent artery**

    c. **Common interosseous artery,** which divides into the **anterior interosseous artery** and **posterior interosseous artery.** The posterior interosseous artery gives rise to the **recurrent interosseous artery.**

    d. **Palmar carpal branch**

    e. **Dorsal carpal branch**

    f. **Deep palmar branch** completes the deep palmar arch.

    g. **Superficial palmar arch** is the main termination of the ulnar artery and anastomoses with the superficial palmar branch of the radial artery. It gives rise to three **common palmar digital arteries,** each of which divides into **proper palmar digital arteries,** which run distally to supply the adjacent sides of the fingers.

D. **COLLATERAL CIRCULATION** exists in the upper limb in the following regions:

  1. **Collateral circulation around the shoulder**

    a. Thyrocervical trunk → transverse cervical artery → circumflex scapular artery → subscapular artery → axillary artery

    b. Thyrocervical trunk → suprascapular artery → circumflex scapular artery → subscapular artery → axillary artery

  2. **Collateral circulation around the elbow** involves the following pathways:

    a. Superior ulnar collateral artery → posterior ulnar recurrent artery

    b. Inferior ulnar collateral artery → anterior ulnar recurrent artery

    c. Middle collateral artery → recurrent interosseus artery

    d. Radial collateral artery → recurrent radial artery

  3. **Collateral circulation in the hand** involves the following pathway:

    a. Superficial palmar arch → deep palmar arch

E. **CLINICAL CONSIDERATIONS**

  1. **Subclavian steal syndrome** refers to retrograde flow in the vertebral artery due to an ipsilateral subclavian artery stenosis. The subclavian artery stenosis results in lower pressure in the distal subclavian artery. As a result, blood flows from the contralateral vertebral artery to the basilar artery and then in a retrograde direction down the ipsilateral vertebral artery away from the brainstem. Although this may have deleterious neurologic effects, the reversed vertebral artery blood flow serves as an important collateral circulation for the arm in the setting of a significant stenosis or occlusion of the subclavian artery. The most common cause for a subclavian steal syndrome is atherosclerosis. Subclavian steal is more common on the left side probably due to a more acute origin of the subclavian artery, which results in increased turbulence and accelerated atherosclerosis.

  2. **Placement of Ligatures.** A surgical ligature may be placed on the subclavian artery or axillary artery **between the thyrocervical trunk** and **subscapular artery.** A surgical ligature may also be placed on the brachial artery **distal to the inferior ulnar collateral artery.** A surgical ligature may *not* be placed on the axillary artery between the **subscapular artery** and the **deep brachial artery.**

  3. In order to control profuse bleeding due to trauma of the axilla (e.g., a stab or bullet wound), the third part of the axillary artery may be compressed against the humerus in the inferior part of the lateral wall of the axilla. If compression is required more proximally, the first part of the axillary artery may be compressed at its origin by downward pressure in the angle between the clavicle and the inferior attachment of the sternocleidomastoid muscle.

  4. **Percutaneous arterial catheterization** employs the brachial artery (if the femoral artery approach is unavailable). The **left brachial artery** is preferred because approaching from the left side allows access to the descending aorta without crossing the right brachiocephalic trunk and left common carotid arteries, thereby reducing the risk of stroke.

5. **Blood Pressure.** The brachial artery is used to measure blood pressure by inflating a cuff around the arm, which compresses and occludes the brachial artery against the humerus. A stethoscope is placed over the cubital fossa and the air in the cuff is gradually released. The first audible sound indicates systolic pressure. The point at which the pulse can no longer be heard indicates the diastolic pressure. In order to control profuse bleeding due to trauma, the brachial artery may be compressed near the middle of the arm medial to the humerus.

6. **Access for chronic hemodialysis** most commonly uses the **radial artery and the cephalic vein**, which establishes an arteriovenous fistula between the two vessels.

7. **The Allen test** is a test for occlusion of either the ulnar or radial artery. For example, blood is forced out of the hand by making a tight fist and then the physician compresses the ulnar artery. If blood fails to return to the palm and fingers after the fist is opened, then the uncompressed radial artery is occluded.

8. **Deep Laceration.** The deep palmar arch lies posterior to the tendons of the flexor digitorum superficialis and flexor digitorum profundus muscles. Therefore, a deep laceration at the metacarpal-carpal (MC) joint that cuts the deep palmar arch will also compromise flexion of the fingers.

9. **Laceration of the Palmar Arches.** This results in profuse bleeding due to the collateral circulation between the superficial and palmar arches. It is usually not sufficient to ligate either the ulnar or radial artery. It may be necessary to compress the brachial artery proximal to the elbow to prevent blood from reaching both the ulnar and radial arteries.

10. **Raynaud Syndrome.** This is an idiopathic condition characterized by intermittent bilateral attacks of ischemia of the fingers with cyanosis, paresthesia, and pain. This may also be brought about by cold temperature or emotional stimuli. Since arteries are innervated by postganglionic sympathetic neurons, a cervicodorsal **presynaptic sympathectomy** may be performed to dilate the digital arteries to the fingers.

IV Venous Drainage

A. **SUPERFICIAL VEINS OF THE UPPER LIMB.** The **dorsal venous network** located on the dorsum of the hand gives rise to the cephalic vein and basilic vein. The **palmar venous network** located on the palm of the hand gives rise to the median antebrachial vein.

1. **Cephalic Vein**
   a. The cephalic vein courses along the anterolateral surface of the forearm and arm and then between the deltoid and pectoralis major muscles along the deltopectoral groove and enters the clavipectoral triangle.
   b. The cephalic vein pierces the costocoracoid membrane and empties into the axillary vein.

2. **Basilic Vein**
   a. **The basilic vein** courses along the medial side of the forearm and arm.
   b. The basilic vein pierces the brachial fascia and merges with the venae comitantes of the axillary artery to form the axillary vein.

3. **Median Cubital Vein**
   a. The median cubital vein connects the cephalic vein to the basilic vein over the cubital fossa.
   b. The median cubital vein lies superficial to the bicipital aponeurosis and is used for intravenous injections, blood transfusions, and withdrawal.

4. **Median Antebrachial Vein**
   a. The median antebrachial vein courses on the anterior aspect of the forearm and empties into the basilic vein of the median cubital vein.

B. **DEEP VEINS OF THE UPPER LIMB.** The deep veins follow the arterial pattern of the arm leading finally to the **axillary vein.**

C. **COMMUNICATING VENOUS SYSTEM.** The communicating venous system is a network of **perforating veins** that connect the superficial veins with the deep veins.

Cutaneous Nerves of the Upper Limb (Figure 20-2). The cutaneous nerves of

the upper limb include the supraclavicular nerve, medial brachial cutaneous nerve, medial antebrachial cutaneous nerve, lateral brachial cutaneous nerve, lateral antebrachial cutaneous nerve, posterior brachial and antebrachial cutaneous nerves, intercostobrachial nerve, median nerve, ulnar nerve, and superficial branch of the radial nerve.

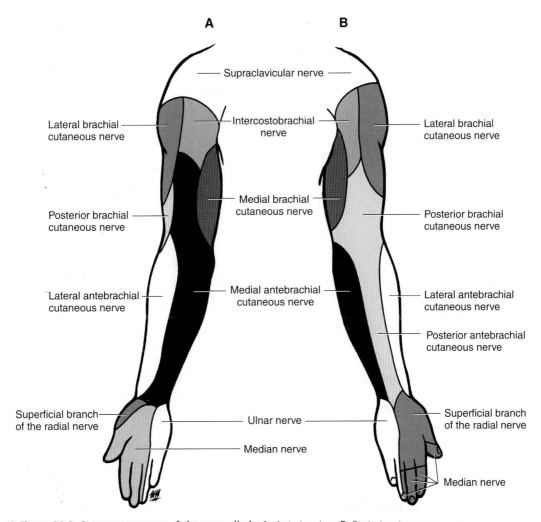

● **Figure 20-2 Cutaneous nerves of the upper limb. A:** Anterior view. **B:** Posterior view.

VI Brachial Plexus (Figures 20-3 and 20-4).

The components of the brachial plexus include:

A. RAMI are the **C5–T1 ventral primary rami** of spinal nerves and are located between the **anterior scalene** and **middle scalene muscles.**

A

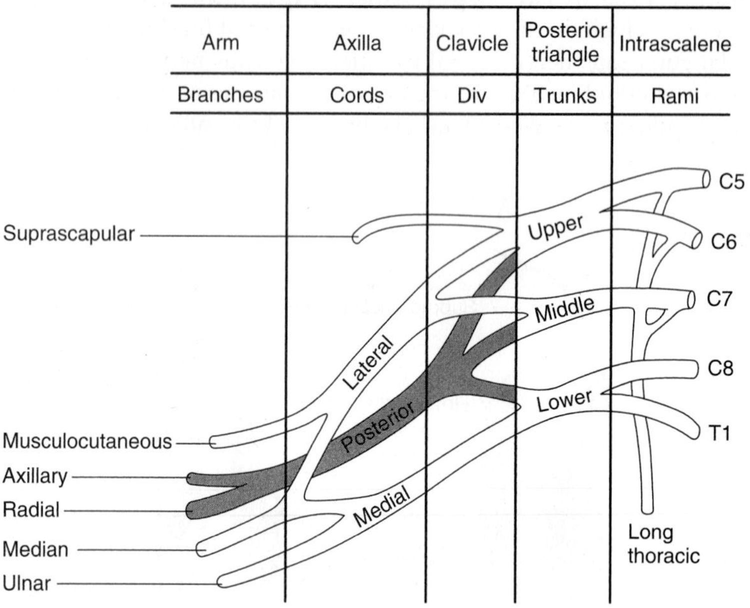

Arm
Axilla
Clavicle
Posterior triangle
Intrascalene
Branches
Cords
Div
Trunks
Rami
C5
C6
C7
C8
T1
Suprascapular
Upper
Middle
Lower
Lateral
Posterior
Medial
Musculocutaneous
Axillary
Radial
Median
Ulnar
Long thoracic

Injury	Description of injury	Nerves damaged	Clinical sign
Erb-Duchenne (C5 and C6, upper trunk)	Violent stretch between the head and shoulder (i.e., adduction traction of the arm and hyper-extension of the neck)	Musculocutaneous Suprascapular Axillary Phrenic	Pronated and medially rotated arm ("waiter's tip hand") Ipsilateral paralysis of diaphragm
Klumpke (C8 and T1, lower trunk)	Sudden upward pull of the arm (i.e., abduction injury)	Median Ulnar Sympathetics of T1 spinal nerve	Loss of function of the wrist and hand Horner syndrome

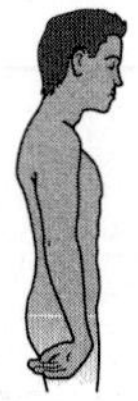

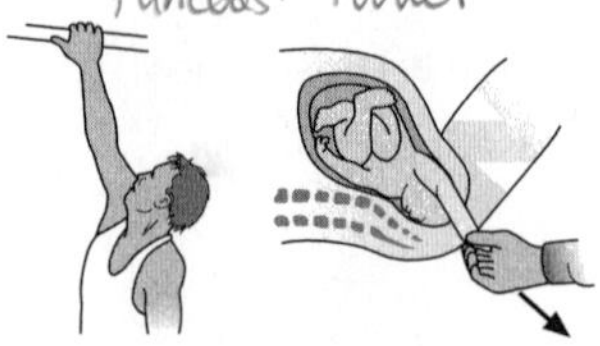

B

● **Figure 20-3 A:** Diagram of the brachial plexus shows the rami, trunks, divisions, cords, and five major terminal branches along with their respective anatomic position. For example, during surgery in the posterior triangle of the neck, the trunks of the brachial plexus may be damaged. The posterior divisions, cords, and branches are shaded. The suprascapular nerve and long thoracic nerve are also shown. **B:** Diagram indicating the Erb-Duchenne injury and Klumpke injury to the brachial plexus.

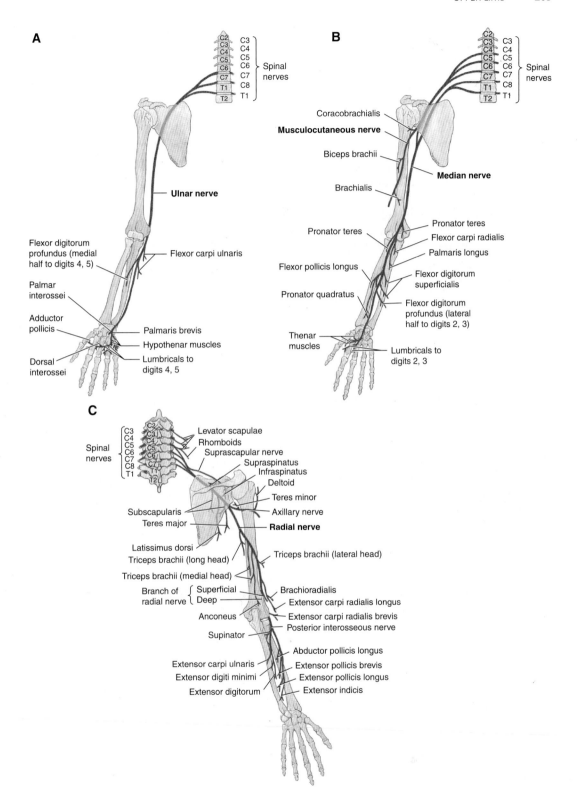

A: Anterior view: ulnar nerve.

C3
C4
C5
C6
C7
C8
T1 } Spinal nerves

Ulnar nerve

Flexor digitorum profundus (medial half to digits 4, 5)

Flexor carpi ulnaris

Palmar interossei

Adductor pollicis

Palmaris brevis

Hypothenar muscles

Dorsal interossei

Lumbricals to digits 4, 5

B: Anterior view: median and musculocutaneous nerves.

C3
C4
C5
C6
C7
C8
T1 } Spinal nerves

Coracobrachialis

Musculocutaneous nerve

Biceps brachii

Median nerve

Brachialis

Pronator teres

Pronator teres
Flexor carpi radialis
Palmaris longus

Flexor pollicis longus

Flexor digitorum superficialis

Pronator quadratus

Flexor digitorum profundus (lateral half to digits 2, 3)

Thenar muscles

Lumbricals to digits 2, 3

C: Posterior view: radial nerve.

C3
C4
C5
C6
C7
C8
T1 } Spinal nerves

Levator scapulae
Rhomboids
Suprascapular nerve
Supraspinatus
Infraspinatus
Deltoid
Teres minor
Axillary nerve

Subscapularis
Teres major

Radial nerve

Latissimus dorsi
Triceps brachii (long head)

Triceps brachii (lateral head)

Triceps brachii (medial head)

Branch of radial nerve { Superficial
Deep

Brachioradialis
Extensor carpi radialis longus

Anconeus

Extensor carpi radialis brevis
Posterior interosseous nerve

Supinator

Abductor pollicis longus

Extensor carpi ulnaris
Extensor digiti minimi

Extensor pollicis brevis
Extensor pollicis longus
Extensor indicis

Extensor digitorum

● **Figure 20-4 Innervation of the upper limb muscles. A:** Anterior view: ulnar nerve. **B:** Anterior view: median and musculocutaneous nerves. **C:** Posterior view: radial nerve.

B. **TRUNKS (UPPER, MIDDLE, LOWER)** are formed by the joining of rami and are located in the **posterior triangle of the neck.**

C. **DIVISIONS (THREE ANTERIOR AND THREE POSTERIOR)** are formed by trunks dividing into anterior and posterior divisions, are located **deep to the clavicle,** and are named according to their relationship to the **axillary artery.**

D. **CORDS (LATERAL, MEDIAL, POSTERIOR)** are formed by joining of the anterior and posterior divisions and are located in the **axilla deep to the pectoralis minor muscle.**

E. **BRANCHES.** The five major terminal branches are:
1. **Musculocutaneous Nerve (C5–7)**
2. **Axillary Nerve (C5, C6)**
3. **Radial Nerve (C5–8, T1)**
4. **Median Nerve (C5–8, T1)**
5. **Ulnar Nerve (C8, T1)**

F. **CLINICAL CONSIDERATION: INJURIES TO THE BRACHIAL PLEXUS**
1. **Erb-Duchenne or Upper Trunk Injury**
   a. This injury involves the **C5 and C6** ventral primary rami and is caused by a violent stretch between the head and shoulder (i.e., adduction traction of the arm with hyperextension of the neck).
   b. This damages the **musculocutaneous nerve** (innervates the biceps brachii and brachialis muscles), **suprascapular nerve** (innervates the infraspinatus muscle), **axillary nerve** (innervates the teres minor muscle), and **phrenic nerve** (innervates the diaphragm).
   c. Clinical signs include:
      i. The arm is pronated and medially rotated ("**waiter's tip hand**"). This occurs because the biceps brachii muscle (which is a supinator of the forearm) is weakened so that the pronator muscles dominate and the infraspinatus muscle (which is a lateral rotator of the arm) is weakened so that the medial rotator muscles dominate.
      ii. **Ipsilateral paralysis of the diaphragm** due to involvement of the C5 component of the phrenic nerve
2. **Klumpke or Lower Trunk Injury**
   a. This injury involves the **C8 and T1** ventral primary rami and is caused by a sudden pull upward of the arm (i.e., abduction injury).
   b. This damages the **median nerve, ulnar nerve** (both of which innervate muscles of the forearm and hand), and **sympathetics of the T1 spinal nerve.**
   c. Clinical signs include:
      i. **Loss of function of the wrist and hand**
      ii. **Horner syndrome,** in which **miosis** (constriction of the pupil due to paralysis of the dilator pupillae muscle), **ptosis** (drooping of the eyelid due to paralysis of the superior tarsal muscle), and **hemianhydrosis** (loss of sweating on one side) occur if the cervical sympathetic ganglia are also injured.

VII **Nerve Lesions.** The nerve lesions are summarized in Table 20-1.

TABLE 20-1		NERVE LESIONS	
Nerve Injury	**Injury Description**	**Impairments**	**Clinical Aspects**
Long thoracic nerve	Stab wound Mastectomy	Abduction of arm past horizontal is compromised	Test: Push against a wall causes winging of scapula
Axillary nerve	Surgical neck fracture of humerus Anterior dislocation of shoulder joint	Abduction of arm to horizontal is compromised Sensory loss on lateral side of upper arm	Test: Abduct arm to horizontal and ask patient to hold position against a downward pull
Radial nerve	Midshaft fracture of humerus Badly fitted crutch Arm draped over a chair	Extension of wrist and digits is lost Supination is compromised Sensory loss on posterior arm, posterior forearm, and lateral aspect of dorsum of hand	Wrist drop
Median nerve at elbow	Supracondylar fracture of humerus	Flexion of wrist is weakened Hand will deviate to ulnar side on flexion Flexion of index and middle fingers at DIP, PIP, and MP joints is lost Abduction, opposition, and flexion of thumb are lost Sensory loss on palmar and dorsal aspects of the index, middle, and half of the ring fingers and palmar aspect of thumb	Ape hand Benediction hand
Median nerve at wrist	Slashing of wrist Carpal tunnel syndrome	Flexion of index and middle fingers at MP joint is weakened Abduction and opposition of thumb are lost Sensory loss same as at elbow	Test: Make an O with thumb and index finger
Ulnar nerve at elbow	Fracture of medial epicondyle of humerus	Hand will deviate to radial side upon flexion Flexion of ring and little finger at DIP is lost Flexion at MP joint and extension at DIP and PIP joints of ring and little finger are lost Adduction and abduction of fingers are lost Adduction of thumb is lost Little finger movements are lost Sensory loss on palmar and dorsal aspects of half of ring finger and little finger	Claw hand
Ulnar nerve at wrist	Slashing of wrist	Flexion at MP joint and extension at DIP and PIP joints of ring and little finger are lost Adduction and abduction of fingers are lost Adduction of thumb is lost Little finger movements are lost Sensory loss same as at elbow	Test: Hold paper between middle and ring fingers

DIP = distal interphalangeal, MIP = middle interphalangeal, MP = metacarpophalangeal, PIP = proximal interphalangeal.

Shoulder Region (Figure 20-5)

A. THE AXILLA

1. The axilla is a pyramid-shaped region located between the upper thoracic wall and the arm.
2. The medial wall of the axilla is the upper ribs and the serratus anterior muscle.
3. The lateral wall of the axilla is the humerus.
4. The posterior wall of the axilla is the subscapularis, teres major, and latissimus dorsi muscles.
5. The anterior wall of the axilla is the pectoralis major and pectoralis minor muscles.
6. The base of the axilla is the axillary fascia.
7. The apex of the axilla is the space between the clavicle, scapula, and rib 1.

B. SPACES

1. **Quadrangular Space**
   a. The quadrangular space transmits the **axillary nerve** and **posterior humeral circumflex artery**.
   b. The quadrangular space is bounded superiorly by the teres major and subscapularis muscles, inferiorly by the teres major muscle, medially by the long head of the triceps, and laterally by the surgical neck of the humerus.
2. **Upper Triangular Space**
   a. The upper triangular space transmits the **circumflex scapular artery**.
   b. The upper triangular space is bounded superiorly by the teres minor muscle, inferiorly by the teres major muscle, and laterally by the long head of the triceps.
3. **Lower Triangular Space**
   a. The lower triangular space transmits the **radial nerve** and **deep brachial artery**.
   b. The lower triangular space is bounded superiorly by the teres major muscle, medially by the long head of the triceps, and laterally by the medial head of the triceps.

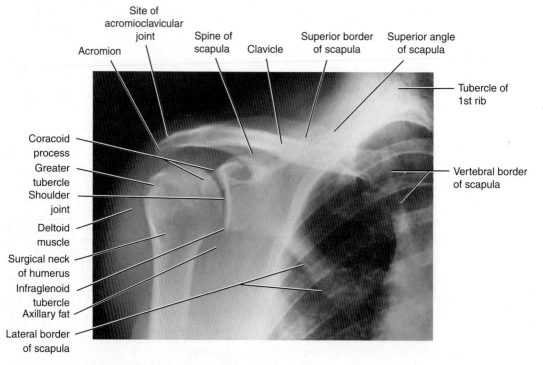

● **Figure 20-5** Anteroposterior radiograph of the shoulder region.

C. GLENOHUMERAL JOINT

1. General Features

a. The glenohumeral joint is the articulation of the head of the humerus with the glenoid fossa of the scapula.

b. This joint has two prominent bursae: the **subacromial bursa** (which separates the tendon of the supraspinatus muscle from the deltoid muscle) and the **subscapular bursa** (which separates the scapular fossa and the tendon of the subscapularis muscle).

c. The "**rotator cuff**" (along with the **tendon of the long head of the biceps brachii muscle**) contributes to the stability of the glenohumeral joint by holding the head of the humerus against the glenoid surface of the scapula.

d. The rotator cuff is formed by the tendons of the following muscles (SITS acronym):

    i. **S**ubscapularis muscle, innervated by the subscapular nerve

    ii. **I**nfraspinatus muscle, innervated by the suprascapular nerve

    iii. **T**eres minor muscle, innervated by the axillary nerve

    iv. **S**upraspinatus muscle, innervated by the suprascapular nerve

2. Clinical Considerations

a. **Anterior-inferior dislocation of the humerus** ("**shoulder dislocation**"; Figure 20-6) is the most common direction of a shoulder dislocation. The head of the humerus lies anterior and inferior to the **coracoid process** of the scapula and may damage the **axillary nerve** or **axillary artery**. The dislocation occurs due to the shallowness of the glenoid fossa. Impaction of the anterior-inferior surface of the glenoid labrum on the posterolateral aspect of the humeral head after it dislocates may cause a depressed humeral head fracture called the **Hill-Sachs lesion**. Clinical signs include loss of normal round contour of the shoulder, a palpable depression under the acromion, and the head of humerus being palpable in the axilla. The anteroposterior (AP) radiograph shows an anterior dislocation of the shoulder. The humeral head is displaced out of the glenoid fossa (GF) inferior to the coracoid process (*) of the scapula.

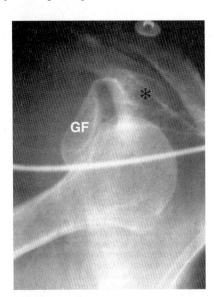

● **Figure 20-6 Anterior dislocation of the shoulder.**

b. **Rotator cuff injury (also called subacromial bursitis or painful arc syndrome; Figure 20-7). Rotator cuff tendinitis** most commonly involves the **tendon of the supraspinatus muscle** and the **subacromial bursa**. It presents in middle-aged men with pain upon lifting the arm above the head. **Acute rotator cuff tear** presents as acute onset of pain with an inability to lift the arm above the head after a traumatic event. The most common rotator cuff

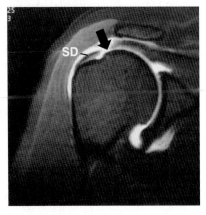

● **Figure 20-7 Complete tear of the rotator cuff.** SD = subdeltoid bursa.

tear is an isolated tear of the **supraspinatus tendon** at its insertion at the greater tuberosity. Partial thickness tears are more common than full thickness tears. The magnetic resonance (MR) arthrogram shows a complete tear of the rotator cuff (*arrow*). The bright contrast medium flows from the glenohumeral joint space into the subdeltoid bursa (SD), which does not occur in a normal, intact rotator cuff.

D. ACROMIOCLAVICULAR JOINT (FIGURE 20-8)
1. **General Features**
   a. The acromioclavicular joint is the articulation of the lateral end of the clavicle with the acromion of the scapula.
   b. This joint is stabilized by **the coracoacromial ligament, coracoclavicular ligament** (subdivided into the **conoid** and **trapezoid ligaments**), and the **acromioclavicular ligament.**

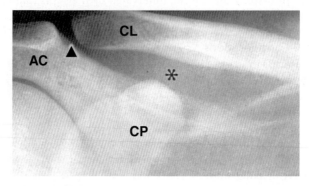

● **Figure 20-8 Acromioclavicular joint.** The anteroposterior radiograph shows a normal adult shoulder. The acromioclavicular ligament (*arrowhead*) and the coracoclavicular ligament (*) are shown. AC = acromion, CL = clavicle, CP = coracoid process.

2. **Clinical Considerations**
   a. **Acromioclavicular subluxation** ("shoulder separation"; **Figure 20-9**) is a common injury due to a downward blow at the tip of the shoulder. There are three grades of shoulder separation: **grade I**, where there is no ligament tearing and no abnormal joint spaces (i.e., minor sprain); **grade II**, where the acromioclavicular ligament is torn and the acromioclavicular space is 50% wider than the normal, contralateral shoulder; and **grade III**, where the coracoclavicular ligament and acromioclavicular ligament are torn, and the coracoclavicular space and acromioclavicular space are 50% wider than the normal contralateral shoulder. Clinical signs include the following: the injured arm hangs noticeably lower than the normal arm; there is a noticeable bulge at the tip of the shoulder due to upward

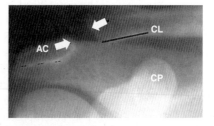

● **Figure 20-9 Acromioclavicular subluxation ("shoulder separation").** AC = acromion, CL = clavicle, CP = coracoid process.

displacement of the clavicle; pushing down on the lateral end of the clavicle and releasing causes a rebound ("piano key sign"); and radiography with a 10-pound weight shows a marked separation of the acromion from the clavicle in a grade II and III. The AP radiograph (weight bearing) shows a grade II acromioclavicular separation. The acromion (AC; *dashed line*) is more inferior to the clavicle (CL; *solid line*). The acromioclavicular space (*arrows*) is abnormally widened but the coracoclavicular space in normal.

b. **Fracture of the clavicle (Figure 20-10)** most commonly occurs at the middle one third of the clavicle. This fracture results in the upward displacement of the proximal fragment due to the pull of the sternocleidomastoid muscle and downward displacement of the distal fragment due to the pull of the deltoid muscle and gravity. The subclavian artery, the subclavian vein, and divisions of the brachial plexus, which are located deep to the clavicle, may be put in jeopardy. The AP radiograph shows a fracture of the middle one third of the clavicle (*thick arrow*). Note the upward displacement of the proximal fragment (*small arrow*) and the downward displacement of the distal fragment (*small arrow*).

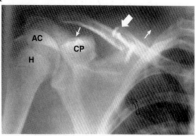

● **Figure 20-10 Fracture of the clavicle.** AC = acromion, CP = coracoid process, H = head of the humerus.

IX Elbow Region (Figure 20-11)

A. ELBOW JOINT consists of three articulations among the humerus, ulnar, and radial bones.

1. **Humeroulnar Joint**
   a. The humeroulnar joint is reinforced by the **ulnar collateral ligament.**
   b. The actions of flexion and extension of the forearm occur at this joint.
   c. A tear of the ulnar collateral ligament will permit abnormal **abduction** of the forearm.

2. **Humeroradial Joint**
   a. The humeroradial joint is reinforced by the **radial collateral ligament.**
   b. The actions of flexion and extension of the forearm occur at this joint.
   c. A tear of the radial collateral ligament will permit abnormal **adduction** of the forearm.

3. **Radioulnar Joint**
   a. The radioulnar joint is reinforced by the **annular ligament.**
   b. The actions of pronation and supination of the forearm occur at this joint.

B. CLINICAL CONSIDERATIONS

1. **Nursemaid's Elbow.** A severe distal traction of the radius (e.g., a parent yanking the arm of a child) can cause **subluxation of the head of the radius** from its encirclement by the **annular ligament.** The reduction of nursemaid's elbow involves applying direct pressure posteriorly on the head of the radius while simultaneously supinating and extending the forearm. This manipulation effectively "screws" the head of the radius into the annular ligament. Clinical signs include a child presenting with a flexed and pronated forearm held close to the body.

A

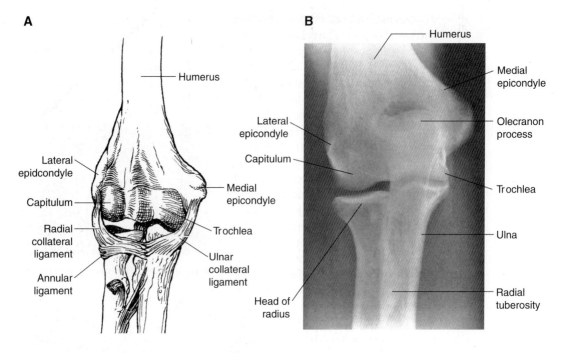

B

C

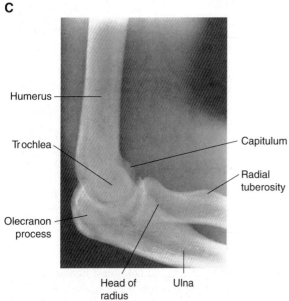

● **Figure 20-11 Normal elbow joint. A:** Diagram of the elbow joint. Note the location of the ligaments that support the elbow joint. **B:** Anteroposterior radiograph of the right elbow joint. **C:** Lateral radiograph of the right elbow joint.

2. **Lateral epicondylitis (tennis elbow)** is inflammation of the **common extensor tendon** of the wrist where it originates on the lateral epicondyle of the humerus.

3. **Medial epicondylitis (golfer's elbow)** is inflammation of the **common flexor tendon** of the wrist where it originates on the medial epicondyle of the humerus.

4. **Tommy John surgery** (named after a famous Chicago White Sox baseball pitcher) replaces or augments a torn ulnar collateral ligament. When this ligament is torn, it is impossible to throw a ball with force and speed. A replacement tendon is taken from the hamstring muscle and wrapped in a figure-eight pattern through holes drilled in the humerus and ulnar bones.

5. **Supracondylar fracture of the humerus (Figure 20-12)** places the contents of the cubital fossa in jeopardy, specifically the median nerve (see Table 20-1) and brachial artery. The contents of the cubital fossa include the **median nerve, brachial artery, biceps brachii tendon, median cubital vein** (superficial to the bicipital aponeurosis), and **radial nerve** (lying deep to the brachioradialis muscle). The lateral radiograph of a supracondylar fracture of the humerus shows a fracture site (*arrow*) with posterior displacement of the distal fragment (*) as well as the radius and ulna. The displacement of the humerus places the contents of the cubital fossa in jeopardy, specifically the median nerve and brachial artery.

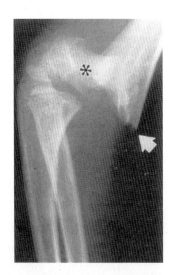

● Figure 20-12 Supracondylar fracture.

6. **Little leaguer's elbow (Figure 20-13)** is the avulsion of the medial epicondyle by violent or multiple contractions of the flexor forearm muscles (e.g., strenuous or repeated throwing of a ball). The AP radiograph shows avulsion of the medial epicondyle (*arrowhead*) and soft tissue swelling on the medial side of the elbow (*).

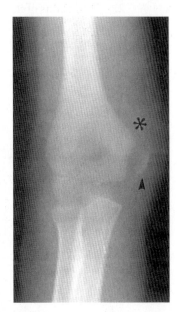

● Figure 20-13 Little leaguer's elbow.

7. **Dislocation of the elbow (Figure 20-14)** is most commonly a posterior dislocation of the radius and ulna with respect to the distal end of the humerus. Depending on the magnitude and direction of the dislocating force, fractures of the distal humerus, coronoid process of the ulna, or radial head may occur. The lateral radiograph shows a posterior dislocation of the elbow with a small bony fragment (*arrow*) arising from the tip of the coronoid process interposed between the trochlea and the base of the coronoid process.

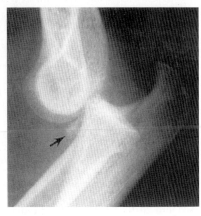

● Figure 20-14 Posterior dislocation of the elbow.

8. **Fracture of the olecranon (Figure 20-15)** may result from a fall on the forearm with the elbow flexed, in which case the fracture is transverse, or may result from a fall directly on the olecranon process itself, in which case the fracture is comminuted. The lateral radiograph shows a comminuted fracture of the olecranon process with proximal retraction of the proximal fragments (*) due to the unopposed action of the triceps tendon.

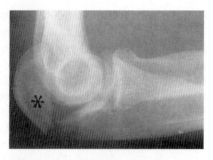

● Figure 20-15 Fracture of the olecranon.

ⓧ Wrist and Hand Region (Figure 20-16)

A. WRIST JOINT (RADIOCARPAL JOINT)

1. The wrist joint is the articulation of the concave distal end of the radius with the scaphoid and lunate carpal bones.
2. The actions of flexion/extension and abduction/adduction of the hand occur at this joint.
3. The ulnar bone plays a minor role at the wrist joint.

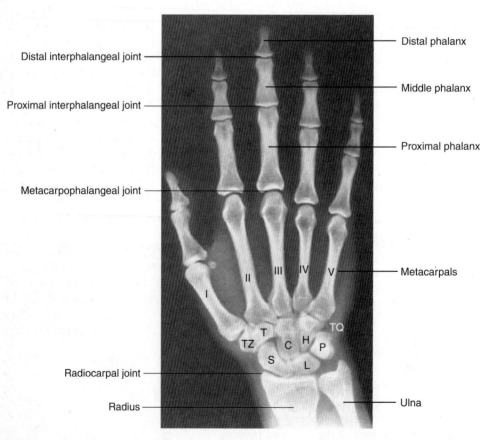

● **Figure 20-16 Anteroposterior radiograph of the hand and wrist region.** I = first metacarpal, II = second metacarpal, III = third metacarpal, IV = fourth metacarpal, V = fifth metacarpal, C = capitate, H = hamate, L = lunate, P = pisiform, S = scaphoid, T = trapezoid, TQ = triquetrum, TZ = trapezium.

B. METACARPOPHALANGEAL (MP) JOINT
1. The MP joint is the joint between the metacarpals and the proximal phalanx.
2. The action of flexion at the MP joint is accomplished by the flexor digitorum superficialis, flexor digitorum profundus, and lumbrical muscles.
3. The action of **ad**duction at the MP joint is accomplished by the **p**almar interosseus muscles (PAD acronym).
4. The action of **ab**duction at the MP joint is accomplished by the **d**orsal interosseus muscles (DAB acronym).

C. PROXIMAL INTERPHALANGEAL (PIP) JOINT
1. The PIP joint is the joint between the proximal phalanx and middle phalanx.
2. The action of flexion at the PIP joint is accomplished primarily by the flexor digitorum superficialis muscle.

D. DISTAL INTERPHALANGEAL (DIP) JOINT
1. The DIP joint is the joint between the middle phalanx and distal phalanx.
2. The action of flexion at the DIP joint is accomplished primarily by the flexor digitorum profundus muscle.

E. CLINICAL CONSIDERATIONS
1. **Carpal tunnel syndrome** is a tendosynovitis due to repetitive hand movements (e.g., data entry) that compresses the **median nerve** within the carpal tunnel. The flexor retinaculum (composed of the **volar carpal ligament** and **transverse carpal ligament**) is attached to the palmar surface of the carpal bones and forms the **carpal tunnel**. The structures that pass through the carpal tunnel include the **flexor digitorum superficialis tendons, flexor digitorum profundus tendons, flexor pollicis longus tendon**, and **median nerve**. No arteries pass through the carpal tunnel. Clinical signs include sensory loss on the palmar and dorsal aspects of the index, middle, and half of the ring fingers and palmar aspect of the thumb, and flattening of the thenar eminence ("ape hand"); tapping of the palmaris longus tendon produces a tingling sensation (Tinel test), and forced flexion of the wrist reproduces symptoms, while extension of the wrist alleviates symptoms (Phalen test).
2. **Slashing of the Wrist ("Suicide Cuts").** A deep laceration on the radial side of the wrist may cut the following structures: **radial artery, median nerve, flexor carpi radialis tendon**, and **palmaris longus tendon**. A deep laceration on the ulnar side of the wrist may cut the following structures: **ulnar artery, ulnar nerve**, and **flexor carpi ulnaris tendon**.
3. **Dupuytren contracture** is a thickening and contracture of the palmar aponeurosis that results in the progressive flexion of the fingers (usually more pronounced in the ring finger and little finger). This is highly correlated with coronary artery disease, possibly due to vasospasm caused by sympathetic innervation of the vasculature within the T1 component of the ulnar nerve.
4. **Volkmann ischemic contracture** is a contracture of the forearm muscles commonly due to a supracondylar fracture of the humerus where the brachial artery goes into spasm, thereby reducing the blood flow. This may also occur due to an overly tight cast or compartment syndrome, where muscles are subjected to increased pressure due to edema or hemorrhage.

5. **Fracture of the Scaphoid (Figure 20-17).** The scaphoid bone is the most commonly fractured carpal bone. The scaphoid bone articulates with the distal end of the radius at the radiocarpal joint. A fracture of the scaphoid is associated with **osteonecrosis** of the scaphoid bone (proximal fragment) because the blood supply to the scaphoid bone flows from distal to proximal. Clinical signs include tenderness in the "**anatomic snuff box**" (formed by the tendons of the extensor pollicis longus, extensor pollicis brevis, and abductor pollicis longus) because the scaphoid lies in the floor of the snuff box; the radiograph may be negative for several weeks until bone resorption occurs. The radiograph shows a scaphoid fracture (*arrow*). Note that the proximal part of the scaphoid is prone to osteonecrosis.

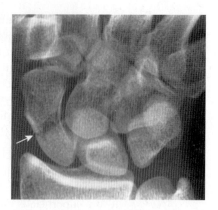

● Figure 20-17 Scaphoid fracture.

6. **Colles fracture (Figure 20-18)** is a fracture of the distal portion of the radius where the distal fragment of the radius is displaced posteriorly ("dinner fork deformity"). This occurs when a person falls on an outstretched hand with the wrist extended. A Colles fracture is commonly accompanied by a fracture of the ulnar styloid process. The lateral radiograph shows a Colles fracture (*large arrow*). Note that the distal fragment of the radius together with the bones of the wrist and hand are displaced posteriorly, rotated, and impacted in the typical "dinner fork deformity."

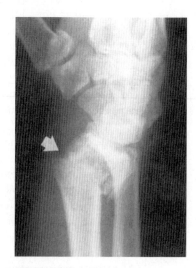

● Figure 20-18 Colles fracture.

7. **Gamekeeper's thumb (Figure 20-19)** is a disruption of the ulnar collateral ligament of the MP joint of the thumb often associated with an avulsion fracture at the base of the proximal phalanx of the thumb. This occurs in skiing falls where the thumb gets entangled with the ski pole. The radiograph shows a gamekeeper's thumb with an avulsion fracture (*arrow*) at the base of the proximal phalanx of the thumb associated with the ulnar collateral ligament.

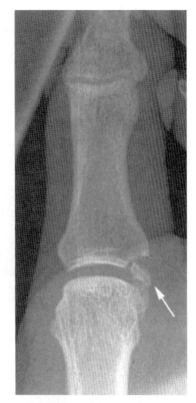

● Figure 20-19 Gamekeeper's thumb.

8. **Boxer fracture (Figure 20-20)** is a fracture at the head of the fifth metacarpal (i.e., little finger). This occurs when a closed fist is used to hit something hard. Clinical signs include pain on the ulnar side of the hand and depression of the head of the fifth metacarpal; attempts to flex the little finger elicit pain. The radiograph shows a fracture at the head of the fifth metacarpal (i.e., little finger; *arrow*).

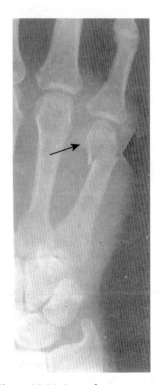

● Figure 20-20 Boxer fracture.

XI Cross-sectional Anatomy of Right Arm and Right Forearm (Figure 20-21)

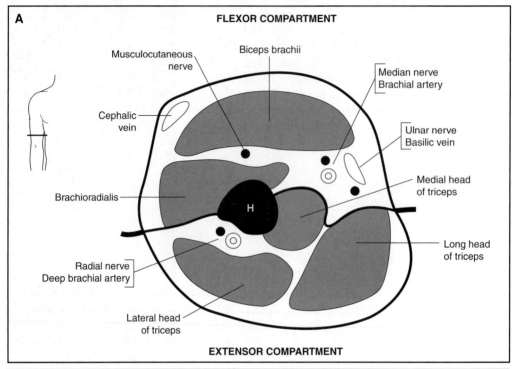

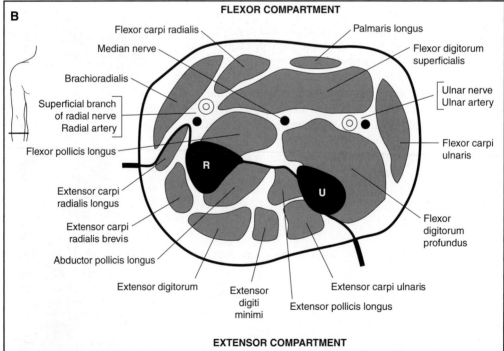

● **Figure 20-21 A:** Cross section through the right brachium (arm). *Black line* divides the flexor (anterior) compartment from the extensor (posterior) compartment. Note the radial nerve within the extensor compartment traveling with the deep brachial artery, the median nerve traveling with the brachial artery, and the ulnar nerve near the basilic vein. H, humerus. **B:** Cross section through the right antebrachium (forearm). *Black line* divides the flexor (anterior) compartment from the extensor (posterior) compartment. Note the location of the ulnar artery, ulnar nerve, median nerve, radial artery, and the superficial branch of the radial nerve within the flexor compartment. R, radius; U, ulna.

Case Study 20-1

A 40-year-old male seaman comes to your office complaining that "I've had some pain in my left arm that comes and goes for about three years. But, now the pain has gotten a lot worse." He tells you that "the pain is like a dull ache and spreads out into the shoulder, neck, and into my left arm and hand. My pinky finger and ring finger get numb and tingle." After some discussion, you learn that he has recently started a new job on a large cargo container ship that requires quite a bit of upper arm movement. What is the most likely diagnosis?

Relevant Physical Findings

- Tenderness in the left supraclavicular space
- Wasting of the left thenar eminence
- Pulling the arm down increases the pain.

Relevant Lab Findings

- Anteroposterior radiograph shows a cervical rib.

Diagnosis

Thoracic Outlet Syndrome

- Thoracic outlet syndrome (TOS) refers to compression of the neurovascular structures at the superior aperture of the thorax. The brachial plexus, subclavian vein, and subclavian artery are affected. Neurologic symptoms occur in 95% of cases and include pain, especially in the medial aspect of the arm, forearm, and ulnar 1.5 digits; paresthesias, often nocturnal, awakening patient with pain or numbness; loss of dexterity; cold intolerance; supraclavicular tenderness; diminished sensation to light touch; and weakness (usually subtle) in the affected limb.
- There are three major causes of TOS: anatomic (e.g., cervical ribs), trauma/repetitive activities, and neurovascular entrapment at the costoclavicular space.
- In this case, the medial cord of the brachial plexus (C8 and T1) is involved.

Chapter 21

Lower Limb

I   **Bones.** The bones of the lower limb include the hip (coxal) bone formed by the fusion of the ilium, ischium, and pubis; femur; patella; tibia; fibula; tarsal bones (talus, calcaneus, navicular, cuboid, and three cuneiform bones); metatarsals; and phalanges (proximal, middle, and distal).

II   **Muscles (See Appendix 2)**

    A. MUSCLES OF THE GLUTEAL REGION (ABDUCTORS AND ROTATORS OF THE THIGH) include the gluteus maximus, gluteus medius, gluteus minimus, tensor of fascia lata, piriformis, obturator internus, superior and inferior gemelli, and quadratus femoris.

    B. MUSCLES OF THE ANTERIOR COMPARTMENT OF THE THIGH (FLEXORS OF THE HIP JOINT AND EXTENSORS OF THE KNEE JOINT) include the pectineus, psoas major, psoas minor, iliacus, sartorius, rectus femoris, vastus lateralis, vastus medialis, and vastus intermedius.

    C. MUSCLES OF THE MEDIAL COMPARTMENT OF THE THIGH (ADDUCTORS OF THE THIGH) include the adductor longus, adductor brevis, adductor magnus, gracilis, and obturator externus.

    D. MUSCLES OF THE POSTERIOR COMPARTMENT OF THE THIGH (EXTENSORS OF THE HIP JOINT AND FLEXORS OF THE KNEE JOINT) include the semitendinosus, semimembranosus, and biceps femoris.

    E. MUSCLES OF THE ANTERIOR AND LATERAL COMPARTMENTS OF THE LEG include the tibialis anterior, extensor digitorum longus, extensor hallucis longus, fibularis tertius, fibularis longus, and fibularis brevis.

    F. MUSCLES OF THE POSTERIOR COMPARTMENT OF THE LEG include the gastrocnemius, soleus, plantaris, popliteus, flexor hallucis longus, flexor digitorum longus, and tibialis posterior.

    G. MUSCLES OF THE FOOT include the first layer: abductor hallucis, flexor digitorum brevis, and abductor digiti minimi; second layer: quadratus plantae and lumbricales; third layer: flexor hallucis brevis, adductor hallucis, and flexor digiti minimi brevis; fourth layer: plantar interossei and dorsal interossei; and dorsum of the foot: extensor digitorum brevis and extensor hallucis brevis.

III   **Arterial Supply (Figure 21-1)**

    A. SUPERIOR GLUTEAL ARTERY is a branch of the internal iliac artery and enters the buttock through the greater sciatic foramen above the piriformis muscle. This artery anastomoses with the lateral circumflex, medial circumflex, and inferior gluteal artery.

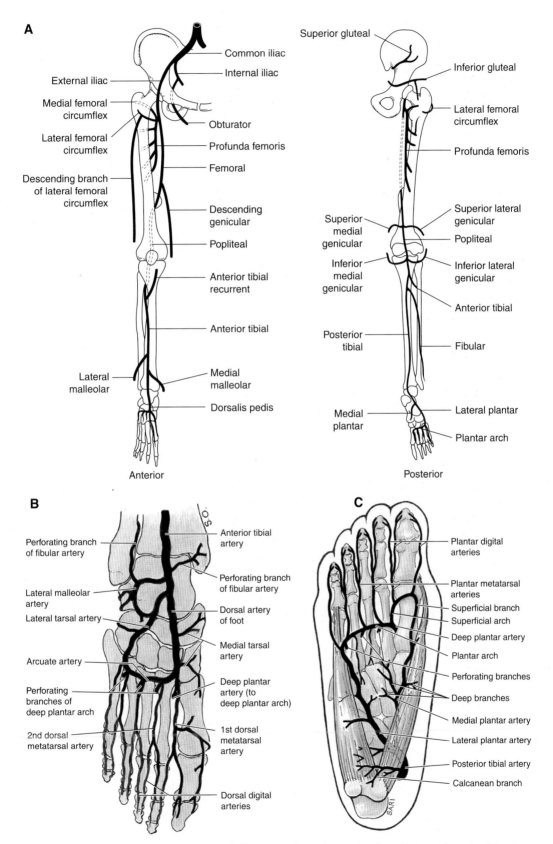

A:

Common iliac
Internal iliac
External iliac
Medial femoral circumflex
Lateral femoral circumflex
Obturator
Profunda femoris
Descending branch of lateral femoral circumflex
Femoral
Descending genicular
Popliteal
Anterior tibial recurrent
Anterior tibial
Lateral malleolar
Medial malleolar
Dorsalis pedis

Superior gluteal
Inferior gluteal
Lateral femoral circumflex
Profunda femoris
Superior medial genicular
Superior lateral genicular
Popliteal
Inferior medial genicular
Inferior lateral genicular
Anterior tibial
Posterior tibial
Fibular
Medial plantar
Lateral plantar
Plantar arch

Anterior

Posterior

B:

Perforating branch of fibular artery
Anterior tibial artery
Perforating branch of fibular artery
Lateral malleolar artery
Lateral tarsal artery
Dorsal artery of foot
Medial tarsal artery
Arcuate artery
Perforating branches of deep plantar arch
Deep plantar artery (to deep plantar arch)
2nd dorsal metatarsal artery
1st dorsal metatarsal artery
Dorsal digital arteries

C:

Plantar digital arteries
Plantar metatarsal arteries
Superficial branch
Superficial arch
Deep plantar artery
Plantar arch
Perforating branches
Deep branches
Medial plantar artery
Lateral plantar artery
Posterior tibial artery
Calcanean branch

● **Figure 21-1 Arterial supply of the lower limb. A:** Anterior and posterior views. **B:** Arterial supply of the dorsum of the foot. **C:** Arterial supply of the sole of the foot.

B. **INFERIOR GLUTEAL ARTERY** is a branch of the internal iliac artery and enters the buttock through the greater sciatic foramen below the piriformis muscle. This artery participates in the cruciate anastomosis and also anastomoses with the superior gluteal artery, internal pudendal artery, and obturator artery.

C. **OBTURATOR ARTERY** is a continuation of the internal iliac artery and passes through the obturator foramen close to the femoral ring, where it may complicate surgical repair of a femoral hernia. The obturator artery gives off the following branches:
   1. **Muscular branches to the adductor muscles**
   2. **Artery of the ligamentum teres (artery to the head of the femur).** This artery is of considerable importance in *children* because it supplies the head of the femur **proximal** to the epiphyseal growth plate. After the epiphyseal growth plate closes in the adult, this artery plays an insignificant role in supplying blood to the head of the femur.

D. **FEMORAL ARTERY** is a continuation of the external iliac artery distal to the inguinal ligament and enters the **femoral triangle** posterior to the inguinal ligament and midway between the anterior-superior iliac spine and the symphysis pubis. At this location the **femoral pulse** can be palpated, arterial blood can be obtained for **blood gas measurements**, or **percutaneous arterial catheterization** can be performed. The femoral artery is commonly used for percutaneous arterial catheterization because it is superficial and easily palpated, and hemostasis can be achieved by applying pressure over the head of the femur. The preferred entry site is **below the inguinal ligament** at the level of the **midfemoral head** (a site that is confirmed by fluoroscopy). If the femoral artery is punctured above the inguinal ligament or below the femoral head, control of hemostasis is difficult or impossible. The femoral artery gives off the following branches:
   1. **Superficial Epigastric Artery**
   2. **Superficial Circumflex Iliac Artery**
   3. **Superficial External Pudendal Artery**
   4. **Deep External Pudendal Artery**
   5. **Descending Genicular Artery**
   6. **Profunda femoris (deep femoral) artery** branches into the following:
      a. **Four perforating arteries** supply the adductor magnus and the hamstring muscles. The first perforating artery participates in the cruciate anastomosis with the inferior gluteal artery and the medial and lateral circumflex arteries.
      b. **Medial circumflex artery** participates in the cruciate anastomosis and provides the main blood supply to the head and neck of the femur in the adult.
      c. **Lateral circumflex artery** participates in the cruciate anastomosis and also sends a **descending branch of the lateral circumflex artery** to participate in the genicular anastomosis around the knee joint.

E. **POPLITEAL ARTERY** is a continuation of the femoral artery at the adductor hiatus in the adductor magnus muscle and extends through the popliteal fossa, where the **popliteal pulse** can be palpated against the popliteus muscle with the leg flexed. The popliteal artery gives off the following branches:
   1. **Genicular arteries** participate in the genicular anastomosis around the knee joint and supply the capsule and ligaments of the knee joint. There are four genicular arteries: **superior lateral, inferior lateral, superior medial,** and **inferior medial.**
   2. **Anterior tibial artery** descends on the anterior surface of the interosseus membrane with the **deep fibular nerve** and terminates as the dorsalis pedis artery. The anterior tibial artery gives off the following branches:
      a. **Anterior tibial recurrent artery**
      b. **Medial malleolar artery**
      c. **Lateral malleolar artery**

    d. **Dorsalis pedis artery.** The dorsalis pedis artery lies between the extensor hallucis longus and extensor digitorum longus tendons midway between the medial and lateral malleolus, where the **dorsal pedal pulse** can be palpated. The dorsalis pedis artery gives off the following branches:

        i. **Lateral tarsal artery** anastomoses with the arcuate artery.

        ii. **Arcuate artery** runs laterally across the bases of the lateral four metatarsals and gives rise to the **second, third, and fourth dorsal metatarsal arteries.** The dorsal metatarsal arteries branch into two **dorsal digital arteries.**

        iii. **First dorsal metatarsal artery**

        iv. **Deep plantar artery** enters the sole of the foot and joins the lateral plantar artery to form the **plantar arch.**

  3. **Posterior tibial artery** passes behind the medial malleolus with the **tibial nerve,** where it can be palpated. The posterior tibial artery gives off the following branches:

    a. **Fibular artery** passes behind the lateral malleolus, gives rise to the **posterior lateral malleolar artery,** and ends in branches around the ankle and heel.

    b. **Medial plantar artery** gives rise to a **superficial branch,** which forms three superficial digital branches, and a **deep branch,** which supplies the big toe.

    c. **Lateral plantar artery** arches medially across the foot to form the **plantar arch** in conjunction with the deep plantar artery (from the dorsalis pedis artery). The plantar arch gives rise to four **plantar metatarsal arteries** and three **perforating branches,** which anastomose with the arcuate artery. The plantar metatarsal arteries branch into two **plantar digital arteries.**

F. COLLATERAL CIRCULATION

  1. **Around the hip joint (cruciate anastomosis)** involves the following arteries:

    a. Inferior gluteal artery (a branch of the internal iliac artery)

    b. Medial femoral circumflex artery

    c. Lateral femoral circumflex artery

    d. First perforating branch of profundus femoris artery

  2. **Around the head of the femur (trochanteric anastomosis)** involves the following arteries:

    a. Superior gluteal artery

    b. Inferior gluteal artery

    c. Medial femoral circumflex artery

    d. Lateral femoral circumflex artery

  3. **Around the knee joint (genicular anastomosis)** maintains blood supply to the leg during full flexion and involves the following arteries:

    a. Superior lateral genicular artery

    b. Inferior lateral genicular artery

    c. Superior medial genicular artery

    d. Inferior medial genicular artery

    e. Descending genicular artery (from the femoral artery)

    f. Descending branch of the lateral femoral circumflex artery

    g. Anterior tibial recurrent artery

G. CLINICAL CONSIDERATIONS

  1. **Placement of Ligatures.** In emergency situations, the femoral artery can be ligated anywhere along its course in the anterior compartment of the thigh without risking total loss of blood supply to the lower limb distal to the ligature site. However, sudden occlusion of the femoral artery by ligature or embolism is usually followed by gangrene. In general, collateral circulation in the lower limb is not as robust as in the upper limb.

2. **Acute arterial occlusion** is most commonly caused by an **embolism** or **thrombosis**. This occlusion most frequently occurs where the femoral artery gives off the profunda femoris artery. Clinical signs include pain, paralysis, paresthesia, pallor, poikiloderma, and pulselessness (i.e., the 6 P's). This may lead to loss of lower limb due to muscle and nerve damage (both are very sensitive to anoxia) within 4 to 8 hours if prompt treatment does not occur.

3. **Chronic arterial occlusive disease** is most commonly caused by **atherosclerosis**. This disease most frequently involves the femoral artery near the adductor hiatus and popliteal artery (i.e., femoropopliteal in 50% of the cases), anterior tibial artery, posterior tibial artery, and fibular artery (i.e., tibiofibular in diabetic patients). Clinical signs include **intermittent claudication**, whose key feature is profound fatigue or aching upon exertion (never by sitting or standing for prolonged periods), which is relieved by short periods of rest (5 to 10 minutes); and **ischemic rest pain**, which features pain across the distal foot and toes that usually occurs at night (patient awakens from sleep); pain is exacerbated by elevation and relieved by a dependent position (patient sleeps with leg over the side of the bed).

4. **Compartment syndrome** is an increase in the interstitial fluid pressure within an osseofascial compartment of sufficient magnitude (30 mm Hg or greater) to compromise microcirculation (ischemia), leading to muscle and nerve damage. This syndrome most frequently occurs in the anterior compartment of the thigh due to crush injuries (e.g., car accidents) involving the **femoral artery** and **femoral nerve**, and the anterior compartment of the leg due to tibial fractures involving the **anterior tibial artery** and **deep fibular nerve**. Clinical signs include swollen, tense compartment; pain upon passive stretching of the tendons within the compartment; and pink color, warmth, and presence of a pulse over the involved compartment.

IV Venous Drainage

A. SUPERFICIAL VEINS OF THE LOWER LIMB

1. **Great saphenous vein** (has 10 to 12 valves) is formed by the union of the dorsal vein of the big toe and the dorsal venous arch of the foot. The great saphenous vein passes anterior to the medial malleolus (travels with the **saphenous nerve**), where it is accessible for venous puncture or catheter insertion and passes posterior to the medial condyle of the femur. The great saphenous vein anastomoses with the lesser saphenous vein. The great saphenous vein courses along the medial aspect of the leg and thigh and finally empties into the femoral vein within the femoral triangle.

2. **Small saphenous vein** is formed by the union of the dorsal vein of the little toe and the dorsal venous arch of the foot. The small saphenous vein passes posterior to the lateral malleolus (travels with the **sural nerve**). The small saphenous vein courses along the lateral border of the calcaneal tendon, ascends between the heads of the gastrocnemius muscle, and finally empties into the popliteal vein within the popliteal fossa.

B. DEEP VEINS OF THE LOWER LIMB follow the arterial pattern of the leg leading finally to the **femoral vein**.

C. COMMUNICATING VENOUS SYSTEM is a network of **perforating veins** that penetrate the deep fascia and connect the superficial veins (which contain valves) with the deep veins. This allows flow of blood only from the **superficial veins → deep veins** and enables muscular contractions to propel blood toward the heart against gravity. Incompetent valves allow backflow of blood into the superficial veins (superficial veins ← deep veins), causing dilation of the superficial veins and leading to **varicose veins**.

D. CLINICAL CONSIDERATION. Deep venous thrombosis (DVT) is a blood clot (thrombus) within the deep veins of the lower limb (most commonly), which may lead to a pulmonary embolus. DVT is usually caused by venous stasis (e.g., prolonged immobilization, congestive heart failure, obesity), hypercoagulation (e.g., oral contraceptive use, pregnancy), or endothelial damage. The nidus of DVT is stagnant blood behind the cusp of a venous valve (i.e., the venous sinus). Treatment includes intravenous heparin for 5 to 7 days followed by Coumadin for 3 months (Coumadin is contraindicated in pregnant women because it is teratogenic).

Ⓥ Cutaneous Nerves of the Lower Limb (Figure 21-2). The cutaneous nerves of the lower limb include the superior, middle, and inferior clunial nerves; genitofemoral nerve; iliohypogastric nerve; lateral femoral cutaneous nerve; posterior femoral cutaneous nerve; cutaneous branch of the obturator nerve; anterior cutaneous branches of the femoral nerve; saphenous nerve (travels with the great saphenous vein); lateral sural cutaneous nerve; medial sural cutaneous nerve; sural nerve (formed by the union of the lateral and medial sural cutaneous nerves); superficial fibular nerve; deep fibular nerve; calcaneal nerves; medial plantar nerve; and lateral plantar nerve.

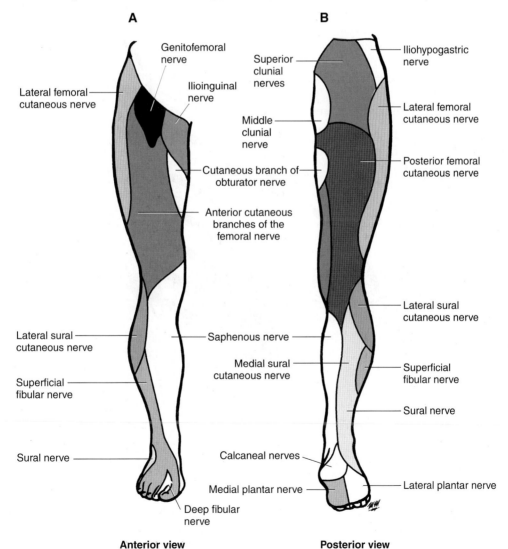

Anterior view   **Posterior view**

● **Figure 21-2 Cutaneous nerves of the lower limb. A:** Anterior view. **B:** Posterior view.

VI Lumbosacral Plexus (Figures 21-3 and 21-4). The components of the lumbosacral plexus include:

A. RAMI are the L1–5 and S1–4 ventral primary rami of spinal nerves.

B. DIVISIONS (ANTERIOR AND POSTERIOR) are formed by rami dividing into anterior and posterior divisions.

C. BRANCHES. The six major terminal branches are:
1. **Femoral Nerve (L2–4)**
2. **Obturator Nerve (L2–4)**
3. **Superior Gluteal Nerve (L4–5, S1)**
4. **Inferior Gluteal Nerve (L5, S1, S2)**

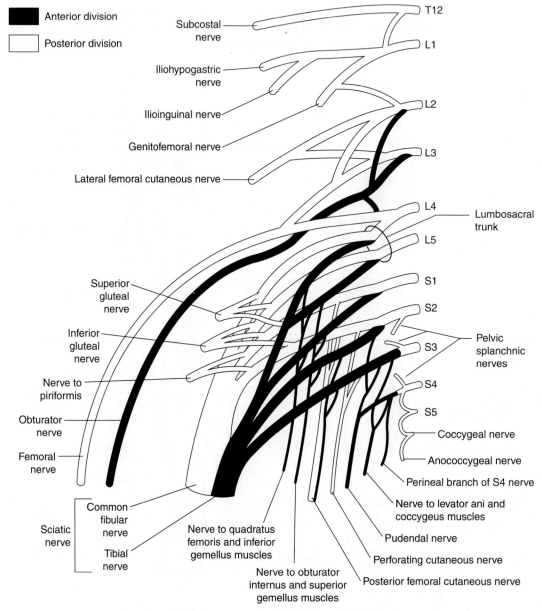

● **Figure 21-3 Diagram of the lumbosacral plexus showing the rami, divisions, and six major terminal branches.** The anterior divisions and branches are shown in black.

A

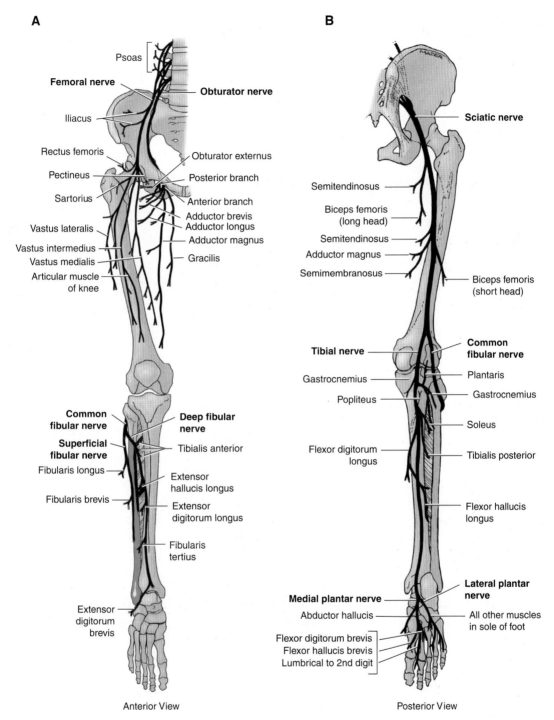

Psoas

Femoral nerve

Iliacus

Rectus femoris

Pectineus

Sartorius

Vastus lateralis

Vastus intermedius

Vastus medialis

Articular muscle
of knee

Obturator nerve

Obturator externus

Posterior branch

Anterior branch

Adductor brevis
Adductor longus

Adductor magnus

Gracilis

**Common
fibular nerve**

**Superficial
fibular nerve**

Fibularis longus

Fibularis brevis

**Deep fibular
nerve**

Tibialis anterior

Extensor
hallucis longus

Extensor
digitorum longus

Fibularis
tertius

Extensor
digitorum
brevis

Anterior View

B

Sciatic nerve

Semitendinosus

Biceps femoris
(long head)

Semitendinosus

Adductor magnus

Semimembranosus

Biceps femoris
(short head)

Tibial nerve

Gastrocnemius

Popliteus

Flexor digitorum
longus

**Common
fibular nerve**

Plantaris

Gastrocnemius

Soleus

Tibialis posterior

Flexor hallucis
longus

**Lateral plantar
nerve**

Medial plantar nerve

Abductor hallucis

Flexor digitorum brevis
Flexor hallucis brevis
Lumbrical to 2nd digit

All other muscles
in sole of foot

Posterior View

● **Figure 21-4 Innervation of the lower limb muscles. A:** Anterior view. Femoral nerve, obturator nerve, common
fibular nerve, and deep fibular nerve and their branches are shown. **B:** Posterior view. Sciatic nerve, tibial nerve, common fibular nerve, medial plantar nerve, and lateral plantar nerve and their branches are shown.

    5.   **Common fibular nerve (L4, L5, S1, S2)** divides into the:
       a.   Superficial fibular nerve
       b.   Deep fibular nerve
    6.   **Tibial nerve (L4, L5, S–3).** The tibial nerve and common fibular nerve comprise
       the **sciatic nerve.**

D. CLINICAL CONSIDERATIONS. The most common condition affecting the lumbosacral plexus is the herniation of intervertebral discs.

VII Nerve Lesions (Table 21-1)

TABLE 21-1		NERVE LESIONS	
Nerve Injury	**Injury Description**	**Impairments**	**Clinical Aspects**
Femoral nerve	Trauma at femoral triangle Pelvic fracture	Flexion of thigh is weakened Extension of leg is lost Sensory loss on anterior thigh and medial leg	Loss of knee jerk reflex Anesthesia on anterior thigh
Obturator nerve	Anterior hip dislocation Radical retropubic prostatectomy	Adduction of thigh is lost Sensory loss on medial thigh	
Superior gluteal nerve	Surgery Posterior hip dislocation Poliomyelitis	Gluteus medius and minimus function is lost Ability to pull pelvis down and abduction of thigh are lost	Gluteus medius limp or "waddling gait" Positive Trendelenburg sign Contralateral
Inferior gluteal nerve	Surgery Posterior hip dislocation	Gluteus maximus function is lost Ability to rise from a seated position, climb stairs, or jump is lost	Patient will lean the body trunk backward at heel strike
Common fibular nerve	Blow to lateral aspect of leg Fracture of neck of fibula	Eversion of foot is lost Dorsiflexion of foot is lost Extension of toes is lost Sensory loss on anterolateral leg and dorsum of foot	Patient will present with foot plantar flexed ("foot drop") and inverted Patient cannot stand on heels "Foot slap"
Tibial nerve at popliteal fossa	Trauma at popliteal fossa	Inversion of foot is weakened Plantar flexion of foot is lost Flexion of toes is lost Sensory loss on sole of foot	Patient will present with foot dorsiflexed and everted Patient cannot stand on toes

VIII **Hip and Gluteal Region (Figure 21-5).** The **piriformis muscle** is the landmark of the gluteal region. The superior gluteal vessels and nerve emerge superior to the piriformis muscle, whereas the inferior gluteal vessels and nerve emerge inferior to it. Gluteal intramuscular injections can be safely made in the **superolateral portion** of the buttock.

A. HIP JOINT is the articulation of the head of the femur with the lunate surface of the acetabulum and the acetabular labrum. The hip joint is supported by the following ligaments:

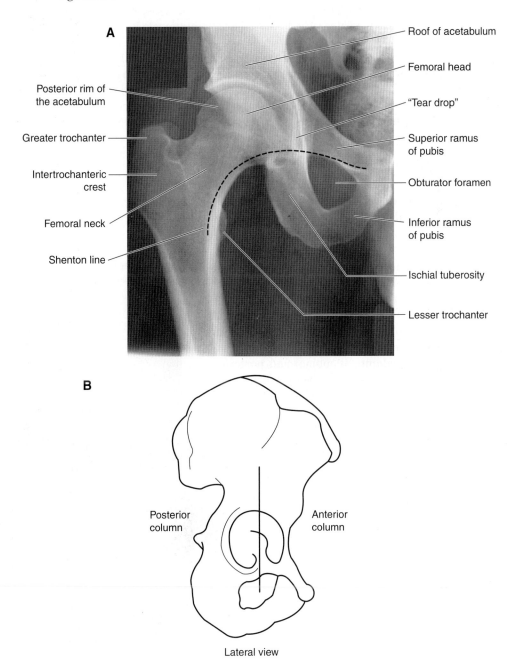

● Figure 21-5 Hip and gluteal region. **A:** Anteroposterior radiograph of the right hip region. Shenton's line is a radiology term describing a curved line drawn along the medial border of the femur and superior border of the obturator foramen. The "teardrop" appearance is caused by the superimposition of structures at the inferior margin of the acetabulum. **B:** Diagram of the hip (coxal) bone (lateral view) showing the anterior column and posterior columns of the acetabulum.

1. **Iliofemoral ligament (Y ligament of Bigelow)** is the largest ligament and reinforces the hip joint anteriorly.
2. **Pubofemoral ligament** reinforces the hip joint inferiorly.
3. **Ischiofemoral ligament** is the thinnest ligament and reinforces the hip joint posteriorly.
4. **Ligamentum teres** plays only a minor role in stability but carries the **artery to the head of the femur.**

B. **FEMORAL TRIANGLE.** The hip joint is related to the **femoral triangle**, whose boundaries are the inguinal ligament (superiorly), sartorius muscle (laterally), and adductor longus muscle (medially). The floor of the femoral triangle is the pectineus and the iliopsoas muscles. The roof of the femoral triangle is the fascia lata. The femoral triangle contains the following structures listed in a medial-to-lateral direction:
   1. **Femoral canal** (most medial structure) containing lymphatics and lymph nodes. The femoral canal is important clinically because this can be a path for herniation of abdominal contents. The femoral canal is within the femoral sheath.
   2. **Femoral Vein.** The **great saphenous vein** joins the femoral vein within the femoral triangle just below and lateral to the pubic tubercle. This is an important site where a great saphenous vein cutdown can be performed. The femoral vein is within the femoral sheath.
   3. **Femoral Artery.** The femoral artery is within the femoral sheath.
   4. **Femoral Nerve** (most lateral structure). The femoral nerve is *not* within the femoral sheath.

B. **CLINICAL CONSIDERATIONS**
   1. **Acetabular fractures (Figure 21-6)** most commonly occur in high-energy motor vehicle accidents or falls, where indirect forces are transmitted through the femoral head to the acetabulum. The acetabulum is nestled under an arch formed by the **anterior column** (iliopubic) and **posterior column** (ilioischial). The top anteroposterior (AP) radiograph shows an anterior column fracture of the acetabulum (*arrows*). The bottom AP radiograph shows a posterior rim fracture of the acetabulum (*arrow*).

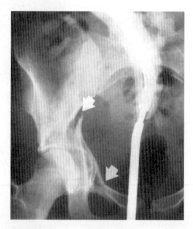

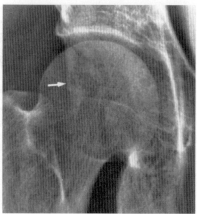

● Figure 21-6 Acetabular fractures.

2. **Anterior dislocations of the hip joint (Figure 21-7)** are not as common as posterior dislocations. The head of the femur comes to lie anterior to the iliofemoral ligament. The lower limb is **externally rotated** and **abducted**. The **femoral artery** may be damaged so that the lower limb may become cyanotic. The AP radiograph shows an anterior dislocation of the hip joint. Note that the femoral head lies anterior to the obturator foramen. The lower limb is externally rotated and abducted.

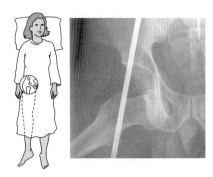

● **Figure 21-7 Anterior dislocation of hip joint.**

3. **Posterior Dislocation of the Hip Joint (Figure 21-8).** The hip joint is most commonly dislocated in a posterior direction due to a severe trauma (e.g., car accident where flexed knee hits the dashboard). The head of the femur comes to lie just posterior to the iliofemoral ligament and the posterior rim of the acetabulum may also be fractured. The lower limb is **internally rotated, adducted**, and **shorter** than the normal limb. Avascular necrosis of the femoral head may occur if the medial and lateral circumflex arteries are compromised. In addition, the **sciatic nerve** may be damaged. The AP radiograph shows a posterior dislocation of the hip joint. This type of dislocation is most common in car accidents, whereby the lower limb is internally rotated, adducted, and shorter than the normal limb.

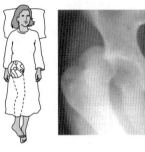

● **Figure 21-8 Posterior dislocation of hip joint.**

4. **Femoral neck fracture (Figure 21-9)** most commonly occurs in elderly women with osteoporosis just distal to the femoral head (i.e., subcapital location). The lower limb is **externally rotated** and **shorter** than the normal limb. Avascular necrosis of the femoral head may occur if the medial and lateral circumflex arteries are compromised. The AP radiograph shows a femoral neck fracture (subcapital). This type of fracture is most common in elderly women with osteoporosis, whereby the lower limb is externally rotated and shorter than the normal limb.

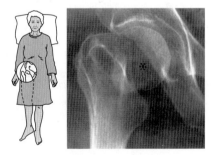

● **Figure 21-9 Femoral neck fracture.** Asterisk indicates the head of the femur.

5. **Legg-Perthes disease (Figure 21-10)** is an idiopathic avascular necrosis of the head of the femur that possibly occurs when the medial and lateral circumflex arteries gradually replace the artery to the head of the femur as the main blood supply to the head of the femur. It most

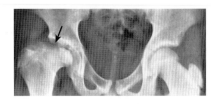

● **Figure 21-10 Legg-Perthes disease.**

commonly occurs unilaterally in Caucasian boys who present with hip pain, slight external rotation, and a limp. This disease has three major phases: **initial phase**, **degenerative phase**, and **regenerative phase**. The AP radiograph shows Legg-Perthes disease. This 8-year-old boy complained of pain in the right hip and demonstrated a limp. Note that the femoral head on the right side (*arrow*) is almost completely absent. Compare to the normal femoral head on the left side.

Ⅸ Knee Region (Figure 21-11)

A. KNEE (FEMOROTIBIAL) JOINT is the articulation of the medial and lateral condyles of the femur with the medial and lateral condyles of the tibia. The knee joint is supported by the following ligaments:

1. **Patellar ligament** is struck to elicit the knee-jerk reflex. The reflex is blocked by damage to the femoral nerve, which supplies the quadriceps muscle or damage to spinal cord segments L2–4.

2. **Medial (tibial) collateral ligament** extends from the medial epicondyle of the femur to the shaft of the tibia and prevents **abduction** at the knee joint. A torn medial collateral ligament can be recognized by abnormal passive abduction of the extended leg.

3. **Lateral (fibular) collateral ligament** extends from the lateral epicondyle of the femur to the head of the fibula and prevents **adduction** at the knee joint. A torn lateral collateral ligament can be recognized by the abnormal passive adduction of the extended leg.

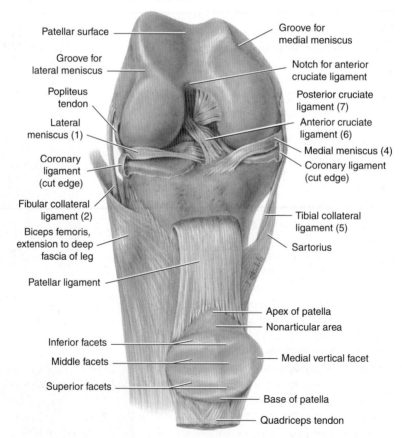

● **Figure 21-11** Diagram of the flexed knee joint with the patella reflected showing the articular surfaces and ligaments of the knee joint.

4. **Anterior cruciate ligament** extends from the *anterior* aspect of the tibia to the lateral condyle of the femur and prevents *anterior* movement of the tibia in reference to the femur. A torn anterior cruciate ligament can be recognized by abnormal passive *anterior* displacement of the tibia called an **anterior drawer sign.** A **hyperextension injury** at the knee joint will stretch the anterior cruciate ligament.

5. **Posterior cruciate ligament** extends from the *posterior* aspect of the tibia to the medial condyle of the femur and prevents *posterior* movement of the tibia in reference to the femur. A torn posterior cruciate ligament can be recognized by abnormal passive *posterior* displacement of the tibia called a **posterior drawer sign.** A **hyperflexion injury** at the knee joint will stretch the posterior cruciate ligament.

B. The knee joints contain menisci, which include:
1. **Medial meniscus** is a C-shaped fibrocartilage that is attached to the medial collateral ligament and is easily torn because it is not very mobile.
2. **Lateral meniscus** is an O-shaped fibrocartilage. Lateral meniscus tears are most commonly associated with anterior cruciate ligament tears.

C. The knee joint is related to the popliteal fossa, which contains the following structures: **tibial nerve, common fibular nerve, popliteal artery, popliteal vein,** and **small saphenous vein.**

D. **CLINICAL CONSIDERATION.** The "terrible triad of O'Donoghue" (Figure 21-12) is the result of fixation of a semiflexed leg receiving a violent blow on the lateral side (e.g., football "clipping") causing abduction and lateral rotation that damages the following structures: the **anterior cruciate ligament** is torn; the **medial meniscus** is torn as a result of its attachment to the medial collateral ligament; and the **medial collateral ligament** is torn due to excessive abduction of the knee joint. The diagram shows an injured left knee ("terrible triad") due to a violent blow to the lateral side of the knee. The *curved arrows* indicate the direction of movement at the knee joint (abduction and lateral rotation). Note the torn anterior cruciate ligament (6), torn medial meniscus (4), and torn medial collateral ligament (5).

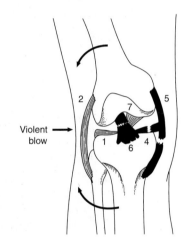

● **Figure 21-12 Terrible triad.** *1* = lateral meniscus, *2* = lateral (fibular) collateral ligament, *3* = head of fibula, *4* = medial meniscus, *5* = medial (tibial) collateral ligament, *6* = anterior cruciate ligament, *7* = posterior cruciate ligament.

X Ankle and Foot Region (Figure 21-13)

A. **ANKLE (TALOCRURAL) JOINT** is the articulation of the inferior surface of the tibia with the trochlea of the talus, where **dorsiflexion** and **plantar flexion** of the foot occur. The ankle joint is supported by the following ligaments:
1. **Medial (deltoid) ligament** extends from the medial malleolus of the tibia to the talus, navicular, and calcaneus bones. The medial ligament consists of the **anterior tibiotalar ligament, posterior tibiotalar ligament, tibionavicular ligament,** and **tibiocalcaneal ligament.**
2. **Lateral ligament** extends from the lateral malleolus of the fibula to the talus and calcaneus bones. The lateral ligament consists of the **anterior talofibular ligament, posterior talofibular ligament,** and **calcaneofibular ligament.**

A

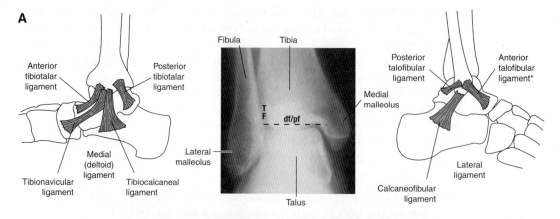

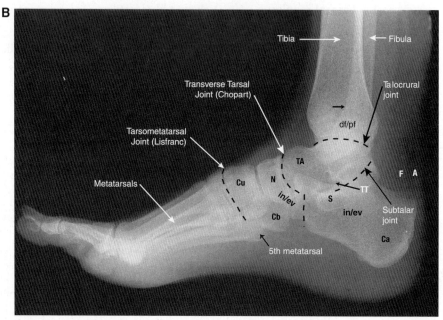

● **Figure 21-13 Ankle joint. A:** Anteroposterior radiograph of the right ankle. The *dotted line* indicates the talocrural joint where dorsiflexion and plantar flexion (df/pf) occur. The diagrams show the components of the medial (deltoid) ligament and lateral ligament that provide support for the talocrural joint. The anterior talofibular ligament (*) is most commonly injured in an ankle sprain. **B:** Lateral radiograph of the right ankle. The talocrural joint is shown where dorsiflexion and plantar flexion (df/pf) occur. The subtalar joint is shown where inversion and eversion (in/ev) occur. The transverse tarsal joint (Chopart joint) is shown where inversion and eversion (in/ev) also occur. The tarsometatarsal joint (Lisfranc joint) is also shown. → = superimposed tibia and fibula, A = Achilles tendon, Cb = cuboid, Cu = cuneiforms, F = fat, N = navicular, S = sustentaculum tali, TA = talus, TT = tarsal tunnel.

B. The ankle joint contains the medial malleolus, which is related to the following structures:

1. Anterior relationships include the **saphenous nerve** and **great saphenous vein** (an excellent location for a great saphenous vein cutdown).

2. Posterior relationships include the **flexor hallucis longus tendon, flexor digitorum longus tendon, tibial posterior tendon, posterior tibial artery,** and **tibial nerve.**

C. SUBTALAR JOINT is the articulation of the talus and the calcaneus where **inversion** and **eversion** of the foot occur.

D. TRANSVERSE TARSAL JOINT (CHOPART JOINT) is actually two joints: the **talonavicular** joint and the **calcaneocuboid** joint. It is the joint where **inversion** and **eversion** of the foot also occur.

E. TARSOMETATARSAL JOINT (LISFRANC JOINT) is the articulation of the tarsal bones with the metatarsals.

F. CLINICAL CONSIDERATIONS

1. **Inversion injury (most common ankle injury) (Figure 21-14)** occurs when the foot is forcibly **inverted** and results in the following: a stretch or tear of the lateral ligament (most commonly the **anterior talofibular ligament**); fracture of the fibula; and **avulsion of the tuberosity of the fifth metatarsal** (called a **Jones fracture**), where the **fibularis brevis muscle** attaches (depending on the severity of the injury). The diagram shows structures damaged due to an inversion injury to the right ankle. The AP radiograph shows a bimalleolar (both malleoli are fractured) fracture-dislocation caused by an inversion injury. The impaction force of the dislocated talus striking against the medial malleolus has resulted in an oblique fracture (*arrowheads*). The distance between the medial malleolus and the medial surface of the talus (*asterisk*) indicates that the medial (deltoid) ligament is disrupted. The force transmitted through the lateral ligament has caused a fracture of the lateral malleolus (*solid white arrow*). The normal relationship between the proximal fibular fracture and tibia (*open arrow*) indicates that the distal tibiofibular ligaments and the interosseous membrane are intact.

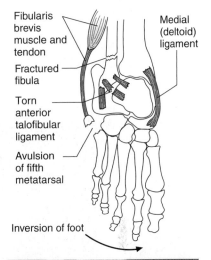

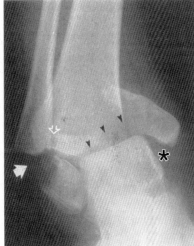

● **Figure 21-14 Inversion injury of the right ankle.**

2. **Eversion injury (Pott fracture; Figure 21-15)** occurs when the foot is forcibly **everted** and results in the following: **avulsion of the medial malleolus** (the medial ligament is so strong that instead of tearing, it avulses the medial malleolus) and **fracture of the fibula** due to the lateral movement of the talus. The diagram shows structures damaged due to an eversion injury to the right ankle. The AP radiograph shows an avulsion fracture of the medial malleolus (*white arrow*) and fracture of the fibula (*black arrow*). The presence of a fracture medial malleolus indicates that the medial (deltoid) ligament is intact. The widening of the tibiofibular joint (*open arrow*) indicates that the anterior and posterior distal tibiofibular ligaments are disrupted, as is the interosseous membrane.

3. **Ski boot injury** usually results in the fracture of the distal portions of the tibia and fibula.

4. **Calcaneal fracture (lover's fracture)** occurs when a person jumps from a great height (e.g., from a second story bedroom window, hence the name). A calcaneal fracture usually involves the **subtalar joint** and is usually associated with fractures of the **lumbar vertebrae** and **neck of the femur.**

5. **Lisfranc injury** occurs when bikers get their foot caught in the pedal clips or as a result of a high-energy car accident. A Lisfranc injury results in the fracture or dislocation at the tarsometatarsal joint (Lisfranc joint).

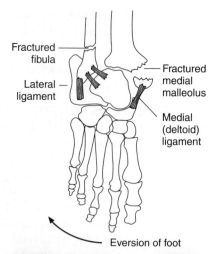

Fractured fibula

Lateral ligament

Fractured medial malleolus

Medial (deltoid) ligament

Eversion of foot

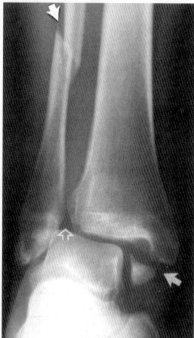

● **Figure 21-15 Eversion injury of the right ankle.**

XI Cross-sectional Anatomy of Right Thigh and Right Leg (Figure 21-16)

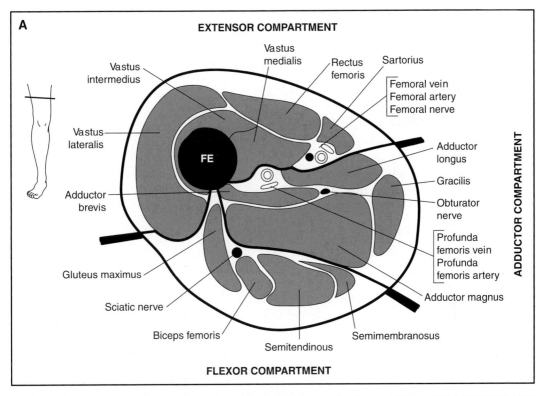

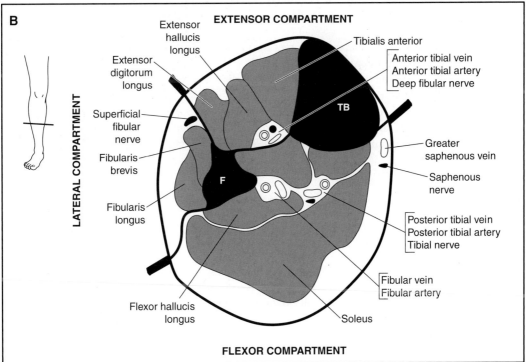

● **Figure 21-16 Cross-sectional anatomy of the thigh and leg. A:** A cross section through the right thigh. *Black lines* divide the extensor (anterior) compartment, flexor (posterior) compartment, and adductor (medial) compartments. **B:** A cross section through the right leg. *Black lines* divide the extensor, flexor, and lateral compartments.

Case Study 21-1

A 22-year-old man comes to the emergency room late at night complaining of severe pain in his left leg. After some discussion, you learn that earlier that day he was kicked very hard in the left leg while playing soccer. He tells you that "I tried to keep playing because this was an important game but I had to come out of the game because the pain was too much. When I was showering after the game, I noticed the color of my left foot was a little dark." What is the most likely diagnosis?

Relevant Physical Exam Findings

- Anterolateral surface of the left leg is swollen, very hard, and tender.
- Dorsalis pedis pulse is absent over the left foot.
- Temperature of his left foot is noticeably less than that of his right foot.
- Patient cannot extend the toes of the left foot.
- When attempting to walk patient drags his left foot.

Relevant Lab Findings

- Radiograph showed no fracture of any bone in the lower limb.

Diagnosis

Anterior Compartment Syndrome

- The direct trauma to the left leg resulted in hemorrhage and swelling inside the anterior compartment of the left leg. This swelling increased pressure on the nerves, veins, and arteries inside the compartment.
- Muscles and nerves can survive up to 4 hours of ischemia without irreversible damage. After 4 hours of ischemia, nerves will show irreversible damage.
- Prolonged compression of the nerves destroys their ability to function properly. The initial numbness, severe pain, and tenderness were due to increased pressure on the common fibular nerve.
- The eventual loss of extension of the toes and foot drop were also due to increased pressure on the common fibular nerve.
- The darkened foot color and loss of dorsalis pedis pulse were due to increased pressure on the anterior tibial artery.
- A fasciotomy is usually performed to relieve the symptoms. Two medial and lateral incisions are used to completely release all surrounding compartments. Skin incisions are left open to keep intracompartmental pressure low. Closure is performed 48 to 72 hours later.

Case Study 21-2

A mother brings her 7-year-old son into your office saying that "my son keeps complaining of pain in his left knee and sometimes the left hip. And it is like he won't put any weight on his left leg." She also tells you that "he fell down a couple of days ago while playing and came home crying. I also brought him to the emergency room about 8 months ago after a fall. They gave him some pain medication and everything was fine until now. It's like the same thing is happening all over again." After some discussion with the mother you learn that she observed symptoms over the past 2 months where her son showed signs of a worsening limp, decreased mobility, and increasing pain in the left knee and hip without any trauma. What is the most likely diagnosis?

Relevant Physical Exam Findings

- The boy is markedly obese.
- The left knee has a decreased range of motion and is tender to palpation.
- The left hip has a decreased range of motion and pain is elicited upon passive internal rotation when the boy is lying down.
- The left ankle and foot examinations are unremarkable.
- The boy refuses to bear weight on the left leg and demonstrates an antalgic gait.

Relevant Lab Findings

- Frontal radiograph shows loss of the normal spherical shape of the femoral head; the contour of the femoral head is flat. The flattening is most apparent along the lateral aspect of the femoral head.
- Frog-leg radiograph shows a prominent widening of the joint space and a peripheral fracture of the femoral epiphysis (crescent sign).

Diagnosis

Legg-Perthes Disease

- Legg-Perthes disease is a pediatric disorder that affects mainly Caucasian boys between 3 and 12 years of age.
- Osteonecrosis of the femoral head is the eventual outcome of this disease.
- The cause seems to be a vascular insult to the femoral epiphysis leading to infarction of the trabecular bone and structural collapse of the femoral head.
- The etiology of Legg-Perthes disease is not well understood and remains idiopathic in nature.

Case Study 21-3

A 14-year-old girl comes to your office complaining that "I've had a lot of pain in my right shin for about a month. It hurts just under the kneecap." She also tells you that she was recently on a wilderness vacation and did some pretty strenuous hiking up some steep rocky cliffs. She has also begun conditioning for the basketball season and has started playing a lot of practice scrimmages. She also says, "I take a lot of ibuprofen to stop the pain and it works pretty good. When I take a few days off from my conditioning program the pain goes away, but the pain always comes back when I start conditioning again. Doc, I have got to get in shape for the basketball season." What is the most likely diagnosis?

Relevant Physical Exam Findings

- Pain over the proximal tibial tuberosity at the insertion of the patellar ligament
- A visible soft tissue swelling at the insertion of the patellar ligament

Relevant Lab Findings

- Radiograph showed mild separation of small ossicle from the developing ossification center at the tibial tuberosity.

Diagnosis

Osgood-Schlatter Disease

- Osgood-Schlatter disease is one of the most common causes of knee pain in active adolescents.
- Osgood-Schlatter disease is thought to be caused by repeated traction on the developing ossification center at the tibial tuberosity, which leads to multiple subacute fractures or tendinous inflammation.
- Osgood-Schlatter disease is a benign, self-limiting disease that usually disappears within a year. Pain may persist for 2 to 3 years until the tibial growth plate closes in late adolescence.
- Persistent complaints of pain may be due to bony ossicle formation in the patellar ligament.

Chapter 22

Head

I **Skull.** The skull can be divided into two parts: the neurocranium and viscerocranium.

A. NEUROCRANIUM. The neurocranium consists of the flat bones of the skull (i.e., cranial vault) and the base of the skull, which include the following eight bones: **frontal bone, occipital bone, ethmoid bone, sphenoid bone, paired parietal bones,** and **paired temporal bones.**

B. VISCEROCRANIUM. The viscerocranium consists of the bones of the face that develop from the pharyngeal arches in embryologic development, which include the following 14 bones: **mandible, vomer, paired lacrimal bones, paired nasal bones, paired palatine bones, paired inferior turbinate bones, paired maxillary bones,** and **paired zygomatic bones.**

C. SUTURES
1. During fetal life and infancy, the flat bones of the skull are separated by dense connective tissue (fibrous joints) called **sutures.**
2. There are five sutures: **frontal suture, sagittal suture, lambdoid suture, coronal suture,** and **squamous suture.**
3. Sutures allow the flat bones of the skull to deform during childbirth (called **molding**) and to expand during childhood as the brain grows.
4. Molding may exert considerable tension at the "obstetric hinge" (junction of the squamous and lateral parts of the occipital bone) such that the **great cerebral vein (of Galen)** is ruptured during childbirth.

D. FONTANELLES
1. Fontanelles are large fibrous areas where several sutures meet.
2. There are six fontanelles: **anterior fontanelle, posterior fontanelle, two sphenoid fontanelles,** and two **mastoid fontanelles.**
3. The anterior fontanelle is the largest fontanelle and readily palpable in the infant. It pulsates because of the underlying cerebral arteries and can be used to obtain a blood sample from the underlying **superior sagittal sinus.**
4. The anterior fontanelle and the mastoid fontanelles close at about 2 years of age, when the main growth of the brain ceases.
5. The posterior fontanelle and the sphenoid fontanelles close at about 6 months of age.

E. FORAMINA OF THE SKULL (FIGURE 22-1)
1. The floor of the cranial cavity can be divided into the **anterior cranial fossa, middle cranial fossa,** and **posterior cranial fossa,** all of which contain foramina and fissures through which blood vessels and cranial nerves are transmitted.
2. In addition, the **falx cerebri** and **tentorium cerebelli** divide the interior of the skull into compartments, which becomes clinically important when increased intracranial pressure in one compartment causes the brain to "herniate" or shift to a lower-pressure compartment.

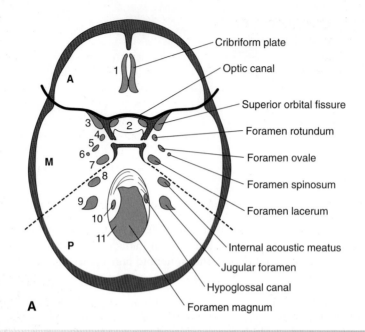

A

Foramen	Structures Transmitted
Anterior Cranial Fossa	
1. Cribriform plate	CN I; Discharge of CSF from the nose **(rhinorrhea)** will result from fracture of cribriform plate and dural tear
Foramen cecum	Emissary vein
Anterior and posterior ethmoidal foramina	Anterior and posterior ethmoidal nerves and arteries
Middle Cranial Fossa	
2. Optic canal	CN II, ophthalmic artery, central artery and vein of retina
3. Superior orbital fissure	CN III, CN IV, CN V$_1$, CN VI, ophthalmic veins
4. Foramen rotundum	CN V$_2$
5. Foramen ovale	CN V$_3$, lesser petrosal nerve, accessory meningeal artery
6. Foramen spinosum	Middle meningeal artery; **epidural hemorrhage** will result from a fracture in this area
7. Foramen lacerum	Empty
Carotid canal	Internal carotid artery and sympathetic carotid plexus
Hiatus of facial canal	Greater petrosal nerve
Posterior Cranial Fossa	
8. Internal acoustic meatus	CN VII, CN VIII, labyrinthine artery; discharge of CSF from *external* acoustic meatus **(otorrhea)** will result from fracture of mastoid process and dural tear
9. Jugular foramen	CN IX, CN X, CN XI, sigmoid sinus; **mass in jugular foramen** will result in difficulty swallowing (dysphagia) and speaking (dysarthria), uvula paralysis, and inability to shrug shoulders
10. Hypoglossal canal	CN XII
11. Foramen magnum	Medulla of the brainstem, CN XI, vertebral arteries
Condyloid foramen	Emissary vein
Mastoid foramen	Branch of occipital artery to the dura, emissary vein

CSF = cerebrospinal fluid.

B

● **Figure 22-1 Foramina of the skull. A:** Base of the skull (interior aspect) showing the various foramina within the anterior (A), middle (M), and posterior (P) cranial fossae are numbered. In a clinical vignette question, first identify the clinical features mentioned in the question; then, match the clinical features with the appropriate structures transmitted and the foramen (table); and finally, identify the foramen in the figure. Common clinical situations are indicated in the table (e.g., rhinorrhea, epidural hemorrhage, otorrhea, and mass in jugular foramen). **B:** Table of foramen and structures.

F. CLINICAL CONSIDERATIONS

1. Temporal Bone Formation

a. **Mastoid process.** This portion of the temporal bone is absent at birth, which leaves the **facial nerve (cranial nerve [CN] VII)** relatively unprotected as it emerges from the stylomastoid foramen. In a difficult delivery, forceps may damage CN VII. The mastoid process forms by 2 years of age.

b. **Petrosquamous fissure.** The petrous and squamous portions of the temporal bone are separated by the petrosquamous fissure, which opens directly into the mastoid antrum of the middle ear. This fissure, which may remain open until 20 years of age, provides a route for the spread of infection from the middle ear to the meninges.

2. Spheno-occipital joint is a site of growth up to about 20 years of age.

3. Fracture of the pterion may result in a rupture of the anterior branches of the middle meningeal artery and result in a life-threatening epidural hemorrhage.

4. Abnormalities in Skull Shape. These may result from failure of cranial sutures to form or from premature closure of sutures (**craniosynostoses**).

a. **Crouzon syndrome (CR) (Figure 22-2)** is one of eight *FGFR*-related craniosynostosis syndromes, which include Pfeiffer syndrome, Apert syndrome, Beare-Stevenson syndrome, *FGFR2*-related isolated coronal synostosis, Jackson-Weiss syndrome, Crouzon syndrome with acanthosis nigricans, and Muenke syndrome. CR is an autosomal dominant genetic disorder caused by a missense mutation in the *FGFR2* gene on **chromosome 10q25-q26** for **fibroblast growth factor receptor 2.** These missense mutations result in **constitutive activation of FGFR2** (i.e., a **gain-of-function mutation**), which indicates that *FGFR2* normally inhibits bone growth. Clinical findings include premature craniosynostosis, midface hypoplasia with shallow orbits, ocular proptosis, mandibular prognathism, normal extremities, progressive hydrocephalus, and no mental retardation. The photograph shows a young boy with Crouzon syndrome.

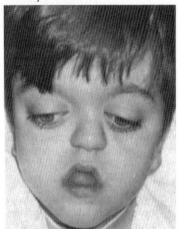

● Figure 22-2 Crouzon syndrome.

5. **Foramen magnum herniation (or Arnold-Chiari malformation; Figure 22-3)** is a congenital malformation and occurs when the cerebellar vermis, cerebellar tonsils, and medulla herniate through the foramen magnum along with cerebral aqueductal stenosis and breaking of the tectal plate. This results in stretching of CN IX, CN X, and CN XII and compression of medulla. Clinical findings include spastic dysphonia, difficulty in swallowing, laryngeal stridor (vibrating sound heard during respiration as a result of obstructed airway), diminished gag reflex, apnea, vocal cord paralysis, and hydrocephalus due to aqueductal stenosis. The magnetic resonance image (MRI) of the Arnold-Chiari malformation shows a herniation of the brainstem and cerebellum (*arrows*) through the foramen magnum. Note the presence of a syrinx (S; an abnormal cavity) in the cervical spinal cord.

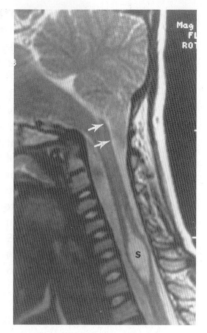

● **Figure 22-3** Foramen magnum herniation (Arnold-Chiari malformation).

II   Scalp.

The scalp is composed of five layers. The first three layers constitute the **scalp proper**, which moves as a unit. The five layers include the following:

A. **SKIN.** The thin skin has many sweat glands, sebaceous glands, and hair follicles.

B. **CONNECTIVE TISSUE.** The connective tissue forms a thick, vascularized subcutaneous layer.

C. **APONEUROSIS (GALEA APONEUROTICA).** The aponeurosis is a broad, strong tendinous sheet that covers the calvaria and serves as attachment for the occipitofrontalis muscle, temporoparietalis muscle, and superior auricular muscle.

D. **LOOSE CONNECTIVE TISSUE.** The loose connective tissue allows free movement of the scalp proper over the cranium. This layer contains the **emissary veins** through which infection can spread easily throughout the scalp or into the intracranial sinuses.

E. **PERICRANIUM.** The pericranium forms the **periosteum** of the neurocranium.

III   Meninges

A. **DURA MATER.** The cranial dura mater is a two-layered membrane consisting of the **external periosteal layer** (i.e., the **endosteum** of the neurocranium) and the **internal meningeal layer,** which is continuous with the dura of the vertebral canal and forms dural infoldings or reflections that divide the cranial cavity into compartments.

1. **Dural Infolding or Reflections**
   a. **Falx cerebri** extends between the cerebral hemispheres and contains the **inferior sagittal sinus** and **superior sagittal sinus.**
   b. **Falx cerebelli** extends between the cerebellar hemispheres.
   c. **Tentorium cerebelli** supports the occipital lobes of the cerebral hemispheres and covers the cerebellum. It encloses the **transverse sinus** and the **superior petrosal sinus.**
   d. **Diaphragma sellae** forms the roof of the sella turcica covering the hypophysis.

    2. Vasculature of the Dura
-   a.   The arterial supply of the dura mater is by the **middle meningeal artery** (a branch of the maxillary artery), which branches into an anterior branch and a posterior branch.
-   b.   The venous drainage of the dura mater is by **middle meningeal veins**, which drain into the **pterygoid plexus.**

    3. Innervation of the Dura
-   a.   Most of the meningeal dura is innervated by meningeal branches from CN V_1, CN V_2, and CN V_3.
-   b.   The meningeal dura of the posterior cranial fossa is innervated by the C1, C2, and C3 spinal nerves and by CN X.

B. ARACHNOID

1. The arachnoid is a filmy, transparent layer that is connected to the pia mater by **arachnoid trabeculae.**
2. The arachnoid is separated from the pia mater by the **subarachnoid space,** which contains cerebrospinal fluid (CSF) and enlarges at several locations to from **subarachnoid cisterns.**
3. The arachnoid projects **arachnoid villi** (collections of which are called **arachnoid granulations**) into the cranial venous sinuses, which serve as sites where CSF diffuses into the venous blood.

C. PIA MATER

1. The pia mater is a shiny, delicate layer that is closely applied to the brain and follows all the contours of the brain.
2. The cerebral arteries that run in the subarachnoid space penetrate the pia mater as they enter the brain, whereby the pia mater is reflected onto the surface of the cerebral artery continuous with the tunica adventitia.

D. CLINICAL CONSIDERATIONS

1. **Headaches.** The dura mater is sensitive to pain. If the dura is irritated or stretched (e.g., after a lumbar puncture to remove CSF), a headache results where pain is referred to regions supplied by CN V.
2. **Bacterial Meningitis.** Meningitis is inflammation of the pia arachnoid area of the brain, spinal cord, or both. Bacterial meningitis is caused by group B *streptococci* (e.g., *Streptococcus agalactiae*), *Escherichia coli,* and *Listeria monocytogenes* in newborns (<1 month old); *Streptococcus pneumoniae* in older infants and young children (1 to 23 months); *Neisseria meningitidis* in young adults (2 to 18 years of age); and *S. pneumoniae* in older adults (19 years of age and older). CSF findings include numerous neutrophils, decreased glucose level, and increased protein level. Clinical findings include fever, headache, nuchal rigidity, and Kernig sign (the patient can easily and completely extend the lower limb when in the dorsal decubitus position but not when in the sitting posture or when lying with thigh flexed upon the abdomen).
3. **Viral Meningitis (Aseptic Meningitis).** Viral meningitis is caused by mumps, echovirus, Coxsackie virus, Epstein-Barr virus, and herpes simplex type 2. CSF findings include numerous lymphocytes, normal glucose levels, and moderately increased protein levels. Clinical findings include fever, headache, nuchal rigidity, and Kernig sign.

IV   Muscles of the Head

A. MUSCLES OF THE FACE AND SCALP include the occipitofrontalis, orbicularis oculi, corrugator supercilii, procerus, levator labii superioris alaeque nasi, orbicularis oris levator labii superioris, zygomaticus minor, buccinator, zygomaticus major, levator anguli oris, risorius, depressor anguli oris, depressor labii inferioris, mentalis, and platysma. All these muscles are innervated by CN VII.

B. MUSCLES OF MASTICATION include the temporal, masseter, lateral pterygoid, and medial pterygoid.

C. MUSCLES OF THE SOFT PALATE include the tensor veli palatini, levator veli palatini, palatoglossus, palatopharyngeus, and musculus uvulae.

D. MUSCLES OF THE TONGUE include the genioglossus, hyoglossus, styloglossus, and palatoglossus.

V   Arterial Supply (Figures 22-4 and 22-5)

A. BRANCHES OF THE ARCH OF THE AORTA (see Chapter 5, Figure 5-17)
    1. Brachiocephalic artery
        a.   Right subclavian artery, which gives rise to the right vertebral artery
        b.   Right common carotid artery
    2. Left common carotid artery
    3. Left subclavian artery, which gives rise to the left vertebral artery

B. EXTERNAL CAROTID ARTERY (Figure 22-4) has eight branches in the neck, the more important of which include the **superior thyroid artery, lingual artery, facial artery, occipital artery, maxillary artery,** and **superficial temporal artery.** The maxillary artery enters the infratemporal fossa by passing posterior to the neck of the mandible and branches into the:
    1. Middle meningeal artery, which supplies the periosteal **dura mater** in the cranium. Skull fractures in the area of the **pterion** (junction of the parietal, frontal, temporal, and sphenoid bones) may sever the middle meningeal artery, resulting in an **epidural hemorrhage.**
    2. Inferior alveolar artery

C. INTERNAL CAROTID ARTERY (Figure 22-5) has no branches in the neck and forms the anterior circulation of the circle of Willis. The internal carotid artery has a number of important branches, which include:
    1. Ophthalmic artery enters the orbit with the optic nerve (CN II) and branches into the **central artery of the retina.** Occlusion results in **monocular blindness.**
    2. Anterior cerebral artery (ACA) supplies the motor cortex and sensory cortex for the leg. Occlusion results in **contralateral paralysis** and **contralateral anesthesia** of the leg.
    3. Middle Cerebral Artery (MCA). Occlusion of the main stem of the MCA results in **contralateral hemiplegia, contralateral hemianesthesia, homonymous hemianopia,** and **aphasia** if the dominant hemisphere is involved.

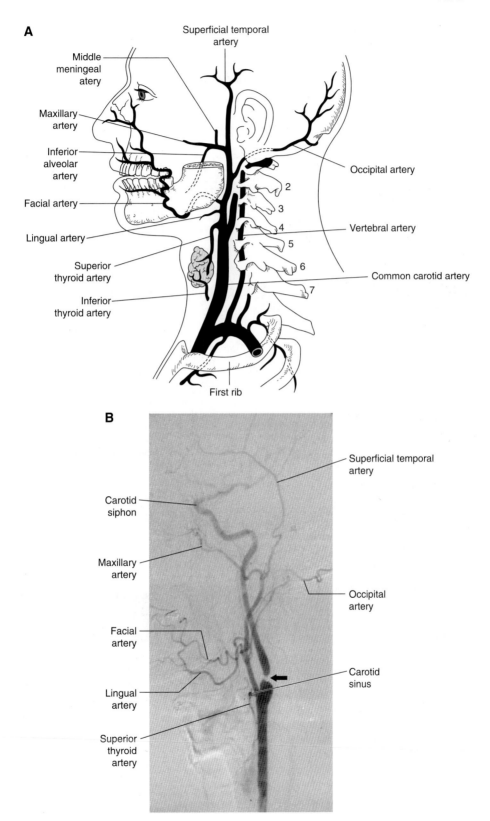

A

Middle meningeal atery

Superficial temporal artery

Maxillary artery

Inferior alveolar artery

Facial artery

Lingual artery

Superior thyroid artery

Inferior thyroid artery

Occipital artery

2
3
4
5
6
7

Vertebral artery

Common carotid artery

First rib

B

Carotid siphon

Maxillary artery

Facial artery

Lingual artery

Superior thyroid artery

Superficial temporal artery

Occipital artery

Carotid sinus

● **Figure 22-4 Blood supply of the head. A:** A diagram of the arterial supply of the head and neck region. **B:** Lateral arteriogram (digital subtraction) of the head and neck region with a blocked internal carotid artery (*arrow*). The most common location of atherosclerosis in the carotid artery system is at the **bifurcation of the common carotid artery.** Carotid artery plaques are usually ulcerated plaques.

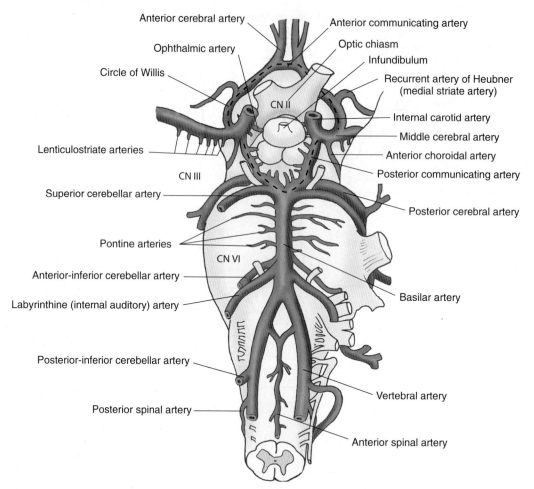

● **Figure 22-5 The internal carotid artery and vertebrobasilar system.** The basilar artery is formed by the confluence of the two vertebral arteries. Note the cerebral arterial circle (circle of Willis; marked by the *thick dashed black
line*). Note the close relationship of cranial nerve (CN) II, CN III, and CN VI to the vasculature.

a. **Lenticulostriate arteries (deep branches or lateral striate)** supply the
basal ganglia and the internal capsule.
Occlusion results in the **classic "para-
lytic stroke"** (Figure 22-6) with primarily a **contralateral hemiplegia** due to
destruction of descending motor fibers
in the posterior limb of the internal
capsule; **contralateral hemianesthesia**
may occur if ascending sensory thala-
mocortical fibers in the internal capsule are also destroyed. **Cerebrovascu-
lar disorders ("strokes")** are most
commonly cerebral infarcts due to occlusion of cerebral vessels by thrombosis or embolism, not by hemorrhage.
Strokes are characterized by a relatively
abrupt onset of a focal neurologic

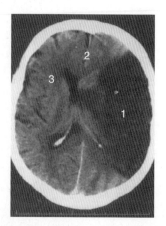

● **Figure 22-6 Middle cerebral artery territory stroke.** *1* = ischemic brain parenchyma,
2 = midline shift to the right, *3* = right frontal
horn of the lateral ventricle.

deficit. The lenticulostriate arteries are prone to **hemorrhagic infarction** due to hypertension or atherosclerotic occlusion. Since these arteries branch at right angles, they are not likely sites for an embolus to lodge causing an embolic infarction. The computed tomography (CT) scan shows a large left MCA territory stroke with edema and mass effect. No visible hemorrhage is apparent since most strokes are caused by thrombosis or embolism.

> **b.** **Outer cortical branches** supply the motor cortex and sensory cortex for the face and arm. Occlusion results in **contralateral paralysis and contralateral anesthesia of the face and arm.**

4. **Anterior communicating artery** connects the two anterior cerebral arteries. It is the most common site of an **aneurysm** (e.g., congenital Berry aneurysm), which, if ruptured, will result in a **subarachnoid hemorrhage** and possibly **bitemporal lower quadrantanopia** due to its close proximity to the optic chiasm.

5. **Posterior communicating artery** connects the anterior circulation of the circle of Willis with the posterior circulation of the circle of Willis. It is the second most common site of an **aneurysm** (e.g., congenital Berry aneurysm), which, if ruptured, will result in a **subarachnoid hemorrhage** and possibly **oculomotor nerve (CN III) paralysis** (droopy upper eyelid, eye "looks down and out," diplopia, fixed and dilated pupil, and lack of accommodation).

D. The **right vertebral artery** (a branch of the right subclavian artery) and the **left vertebral artery** (a branch of the left subclavian artery) both pass through the transverse foramina of C1–6 vertebrae (and foramen magnum) and form the posterior circulation of the circle of Willis.

1. The **basilar artery** is formed by the union of the right and left vertebral arteries. The basilar artery gives off a number of branches, which include the posterior cerebral artery.

2. The **posterior cerebral artery (PCA)** supplies the midbrain, thalamus, and occipital lobe with visual cortex. Occlusion results in **contralateral sensory loss of all modalities with concomitant severe pain (i.e., thalamic syndrome of Dejerine and Roussy)** due to damage to the thalamus and **contralateral hemianopia with macular sparing.**

VI Venous Drainage (Figure 22-7)

A. FACIAL AND SCALP AREAS

1. The **facial vein** (no valves) provides the major venous drainage of the face and drains into the **internal jugular vein.** The facial vein makes clinically important connections with the **cavernous sinus** via the **superior ophthalmic vein, inferior ophthalmic vein,** and **pterygoid plexus of veins.** This connection with the cavernous sinus provides a potential route of infection from the superficial face ("danger zone of the face") to the dural venous sinuses within the cranium.

2. **Diploic veins** (no valves) run within the flat bones of the skull.

3. **Emissary veins** (no valves) form an anastomosis between the superficial veins on the outside of the skull and the dural venous sinuses.

B. DURAL VENOUS SINUSES (no valves) form between the external periosteal layer and the internal meningeal layer of the dura mater. The dural venous sinuses consist of the following:

1. **Superior sagittal sinus** is located along the superior aspect of the falx cerebri. Arachnoid granulations, which transmit CSF from the subarachnoid space to the dural venous sinuses, protrude into its wall.

2. **Inferior sagittal sinus** is located along the inferior aspect (free edge) of the falx cerebri.

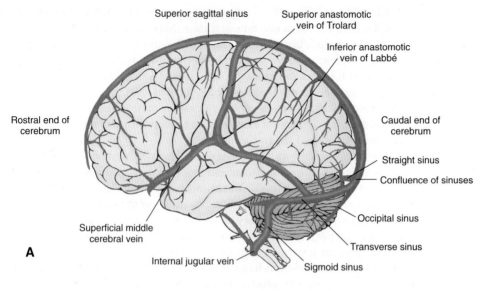

Superior sagittal sinus

Superior anastomotic
vein of Trolard

Inferior anastomotic
vein of Labbé

Rostral end of
cerebrum

Caudal end of
cerebrum

Straight sinus

Confluence of sinuses

Occipital sinus

Superficial middle
cerebral vein

Transverse sinus

A

Internal jugular vein

Sigmoid sinus

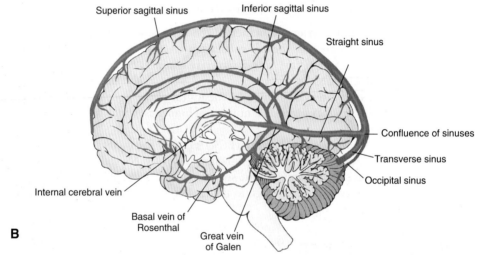

Superior sagittal sinus

Inferior sagittal sinus

Straight sinus

Confluence of sinuses

Transverse sinus

Occipital sinus

Internal cerebral vein

B

Basal vein of
Rosenthal

Great vein
of Galen

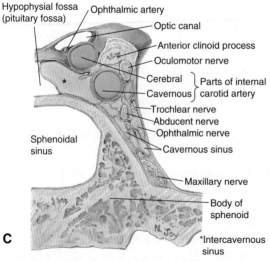

Hypophysial fossa
(pituitary fossa)

Ophthalmic artery

Optic canal

Anterior clinoid process

Oculomotor nerve

Cerebral ⎱ Parts of internal
Cavernous ⎰ carotid artery

Trochlear nerve

Abducent nerve

Ophthalmic nerve

Cavernous sinus

Sphenoidal
sinus

Maxillary nerve

Body of
sphenoid

N. Joy

C

*Intercavernous
sinus

Posterior view of coronal section of right cavernous sinus

● **Figure 22-7 Venous drainage. A, B:** Major dural sinuses and veins. **C:** Cavernous sinus. Coronal section.

3. **Straight sinus** is formed by the union of the inferior sagittal sinus and the **great vein of Galen** (drains venous blood from deep areas of the brain).

4. **Occipital sinus** is located in the attached border of the tentorium cerebelli.

5. **Confluence of sinuses** is formed by the union of the superior sagittal sinus, straight sinus, and occipital sinus.

6. **Transverse sinus** drains venous blood from the confluence of sinuses to the sigmoid sinus.

7. **Sigmoid sinus** drains into the internal jugular vein.

8. **Cavernous sinuses**
   a. The cavernous sinuses are located on either side of the sphenoid bone and receive venous blood from the facial vein, superior ophthalmic vein, inferior ophthalmic vein, pterygoid plexus of veins, central vein of the retina, and each other via the **intercavernous sinuses** that pass anterior and posterior to the hypophyseal stalk.
   b. They drain venous blood into the superior petrosal sinus $\rightarrow$ transverse sinus and the inferior petrosal sinus $\rightarrow$ internal jugular vein.
   c. They are anatomically related to the **internal carotid artery** (carotid siphon), postganglionic sympathetic nerves, and **CN III, CN IV, CN VI, CN V_1, and CN V_2.**
   d. **The cavernous sinuses** are the most clinically significant sinuses, for example:
      i. In infections of the superficial face
      ii. **Thrombophlebitis** can result in poor drainage and enlargement involving CN III, CN IV, CN VI, CN V_1, and CN V_2, thereby producing ocular signs.
      iii. Infections can spread from one side to the other through the intercavernous sinuses.
      iv. Poor drainage may result in exophthalmus and edema of the eyelids and conjunctiva.
      v. Carotid artery–cavernous sinus fistula can result in a headache, orbital pain, diplopia, arterialization of the conjunctiva, and ocular bruit.

VII **Clinical Consideration.** Hemorrhages (**Figure 22-8**) within the head area include an **epidural hemorrhage, subdural hemorrhage, subarachnoid hemorrhage,** and **extracranial hemorrhage,** the important aspects of which are shown in Figure 22-8.

VIII # Lymph Drainage of the Head and Neck Region

A. Lymph from the head and neck drains as follows: perivascular collar of superficial lymph nodes $\rightarrow$ retropharyngeal nodes, paratracheal nodes, infrahyoid nodes, prelaryngeal nodes, pretracheal nodes, and lingual nodes $\rightarrow$ deep cervical lymph nodes $\rightarrow$ right jugular lymph trunk (for the right side of the head and neck) or left jugular lymph trunk (for the left side of the head and neck).

B. The head and neck have a **pericervical collar of superficial lymph nodes** and **deep cervical lymph nodes.**

1. **Pericervical Collar of Superficial Lymph Nodes**
   a. The pericervical collar of superficial lymph nodes includes **submental nodes, submandibular nodes, parotid nodes, buccal nodes, mastoid nodes, occipital nodes, superficial cervical nodes,** and **anterior cervical nodes.**
   b. The lymph from these superficial nodes drains into the deep cervical lymph nodes: **perivascular collar of superficial lymph nodes** $\rightarrow$ **deep cervical lymph nodes.**

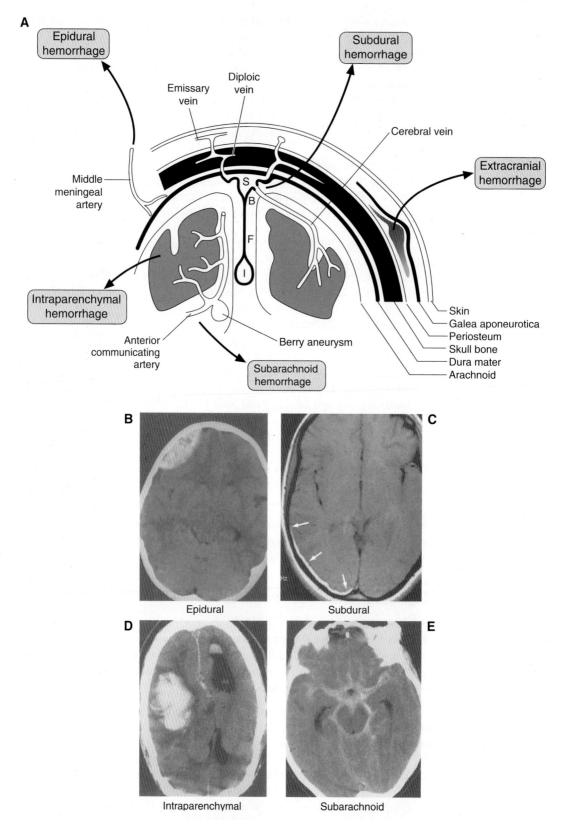

● **Figure 22-8 Hemorrhages. A:** Diagram depicting an epidural, subdural, subarachnoid, intraparenchymal, and extracranial hemorrhage. B = bridging vein, F = falx cerebri, I = inferior sagittal sinus, S = superior sagittal sinus. **B:** Epidural hemorrhage. **C:** Subdural hemorrhage. **D:** Intraparenchymal hemorrhage. **E:** Subarachnoid hemorrhage.

EPIDURAL, EPIDURAL SUBDURAL, SUBARACHNOID, INTRAPARENCHYMAL, AND EXTRACRANIAL HEMORRHAGES

Hemorrhage	Injury	Blood Vessel	Clinical Features
Epidural (B)	Skull fracture near pterion or greater wing of sphenoid Middle cranial fossa A medical emergency	Middle meningeal artery	CT scan shows lens-shaped (biconvex) hyperdensity adjacent to bone; arterial blood located between skull and dura Lucid interval for a few hours followed by death ("talk and die syndrome") May cause a transtentorial herniation that compresses (1) CN III, causing ipsilateral dilated pupil, and (2) cerebral peduncles, causing contralateral hemiparesis No blood in CSF after lumbar puncture
Subdural (C)	Violent shaking of head (e.g., child abuse or car accident) Common in alcoholics and elderly	Superior cerebral veins ("bridging veins")	CT scan shows a thin, crescent-shaped hyperdensity that hugs contours of brain; venous blood located between dura and arachnoid (*arrows*) Blood accumulates slowly (days to weeks after trauma) No blood in CSF after lumbar puncture
Intraparenchymal (D)	Hemorrhage Trauma	Intraparenchymal cerebral artery	CT scan shows hyperdensity within the substance of the brain
Subarachnoid (E)	Contusion or laceration injury to brain Berry aneurysm	Cerebral artery Anterior or posterior communicating artery	CT scan shows hyperdensity in the cisterns, fissures and sulci; thickening of falx cerebri; arterial blood within subarachnoid space Irritation of meninges causes sudden onset of the "worst headache of my life," stiff neck, nausea, vomiting and decreased mentation; earlier "herald headaches" may occur Blood within the CSF after lumbar puncture
Extracranial	Depressed cranial fracture Normal childbirth	Emissary veins Branches of superficial temporal and occipital arteries	Venous and arterial blood located between galea aponeurotica and skull (subaponeurotic space) Lumpy clot, "black eye" No blood in CSF after lumbar puncture

CSF = cerebrospinal fluid, CT = computed tomography.

2. **Deep Cervical Lymph Nodes**
    a. The deep cervical lymph nodes are found alongside of the carotid sheath associated with the internal jugular vein and deep to the sternocleidomastoid muscle.
    b. The deep cervical nodes receive lymph not only from the pericervical collar of superficial lymph nodes, but also from the **retropharyngeal nodes, paratracheal nodes, infrahyoid nodes, prelaryngeal nodes, pretracheal nodes**, and **lingual nodes.**

IX Cranial Nerves (Figure 22-9)

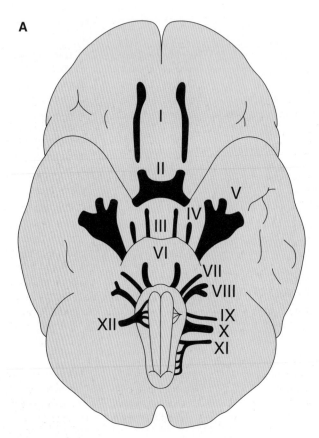

● **Figure 22-9 Cranial nerves. A:** Diagram of the base of the brain showing the location of the cranial nerves. (*continued*)

Cranial Nerve	Clinical Aspects
I Olfactory	Mediates the sense of smell (olfaction)
II Optic	Mediates the sense of sight (vision)
III Oculomotor	CN III lesion (e.g., **transtentorial [uncal] herniation**) results in droopy upper eyelid as a result of paralysis of levator palpebrae muscle; eye "looks down and out" as a result of paralysis of superior rectus muscle, medial rectus muscle, inferior rectus muscle, and inferior oblique muscle and the unopposed action of the superior oblique muscle (CN IV) and lateral rectus muscle (CN VI); double vision (diplopia) when patient looks in direction of paretic muscle; fixed and dilated pupil as a result of paralysis of sphincter pupillae muscle; lack of accommodation (cycloplegia) as a result of paralysis of the ciliary muscle; CN III lesions are associated with diabetes; an aneurysm of the posterior cerebral artery or superior cerebellar artery may exert pressure on CN III as it passes between these vessels
IV Trochlear	Innervates the superior oblique muscle CN IV lesion results in extortion of the eye, vertical diplopia that increases when looking down (e.g., reading a book), head tilting to compensate for extortion
V Trigeminal	Provides sensory innervation to the face and motor innervation to the muscles of mastication CN V lesion results in hemianesthesia of the face, loss of afferent limb of corneal reflex, loss of afferent limb of oculocardiac reflex, paralysis of muscle of mastication, deviation of jaw to the injured side, hypoacusis as a result of paralysis of tensor tympani muscle, and tic douloureux (recurrent, stabbing pain)
VI Abducens	Innervates the lateral rectus muscle CN VI lesion results in convergent strabismus, inability to abduct the eye, horizontal diplopia when patient looks toward paretic muscle; an aneurysm of the labyrinthine artery or anterior inferior cerebellar artery may exert pressure on CN VI as it passes between these vessels
VII Facial	Provides motor innervation to the muscles of facial expression, mediates taste, salivation, and lacrimation CN VII lesion results in paralysis of muscle of facial expression (upper and lower face; called Bell palsy), loss of efferent limb of corneal reflex, hyperacusis as a result of paralysis of stapedius muscle, and **crocodile tears syndrome** (tearing during eating) as a result of aberrant regeneration after trauma
VIII Vestibulocochlear	Mediates equilibrium and balance (vestibular) and hearing (cochlear) CN VIII (vestibular) lesion results in disequilibrium, vertigo, and nystagmus CN VIII (cochlear) lesion (e.g., **acoustic neuroma**) results in hearing loss and tinnitus
IX Glossopharyngeal	Mediates taste, salivation, swallowing, and input from the carotid sinus and carotid body CN IX lesion results in loss of afferent limb of gag reflex, loss of taste from posterior one third of tongue, loss of sensation from pharynx, tonsils, fauces, and back of tongue
X Vagus	Mediates speech and swallowing; innervates viscera in thorax and abdomen CN X lesion results in paralysis of pharynx and larynx, uvula deviates to **opposite side** of injured nerve, loss of efferent limb of gag reflex, and loss of efferent limb of oculocardiac reflex
XI Spinal Accessory	Innervates the sternocleidomastoid and trapezius muscle CN XI lesion results in inability to turn head to **opposite side** of injured nerve, inability to shrug shoulder
XII Hypoglossal	Innervates intrinsic and extrinsic muscles of the tongue CN XII lesion results in tongue deviation to the **same side** of injured nerve

CN = cranial nerve.

B

● **Figure 22-9** *Continued* **B:** Table of important clinical aspects of the cranial nerves.

Case Study 22-1

A 75-year-old man is brought into the emergency room by paramedics after experiencing an acute left-sided weakness approximately 40 minutes before arrival. His wife informs you that her husband has a medical history of hypertension and atrial fibrillation. He is presently being treated with atenolol and warfarin. She also tells you that her husband underwent successful electrical conversion 2 days ago after his international normalized ratio (INR) was confirmed to be within therapeutic range. What is the most likely diagnosis?

Relevant Physical Exam Findings

- Blood pressure 170/80 mm Hg
- Heart rate 55 bpm
- Respiratory rate 12 breaths/min
- Heart rhythm is normal.
- Significant expressive aphasia is present.
- Eyes are deviated to the right.
- Visual fields are absent on the left.
- Pupils are round and reactive to light bilaterally.
- There is flaccid muscle tone on the left upper and lower extremities.
- Muscle strength is 0/5 on the left side.
- Light touch sensation is decreased on the left side.

Relevant Lab Findings

- Electrocardiogram (ECG) shows sinus bradycardia with no acute ST-T–wave changes.
- Serum glucose = 110 mg/dL
- INR = 2.2 (1.0 normal clotting time)
- Computed tomography (CT) scan of the head showed a right-sided hyperdense middle cerebral artery sign.
- Magnetic resonance image (MRI) showed a large area of acute ischemia involving the right middle cerebral artery territory.

Diagnosis

Classic Middle Cerebral Artery Territory Stroke

- Middle cerebral artery (MCA) territory stroke demonstrates contralateral hemiparesis and hypoesthesia.
- MCA territory stroke also demonstrates ipsilateral gaze preference and hemianopia.
- Aphasia (speech disorder) typically occurs if the lesion is located on the dominant hemisphere (usually the left hemisphere).

Case Study 22-2

An unidentified 32-year-old man is brought into the emergency room by paramedics after being found unresponsive on the street by bystanders. The paramedics administered 2 mg naloxone on site without any effect and subsequently intubated orotracheally and transported the man to the emergency room. On arrival, the patient is agitated and pulls at the endotracheal tube. The patient is sedated with benzodiazepine to prevent injury and to allow further examination. What is the most likely diagnosis?

Relevant Physical Exam Findings

- Blood pressure 120/85 mm Hg
- Heart rate 95 bpm
- Respiratory rate 12 breaths/min with assistance of a ventilator
- A 4-inch bleeding scalp laceration is found in the occipital region.
- Pupils are 2 mm and symmetric with sluggish reflexes.
- Corneal reflex is intact.
- Patient periodically moves all four extremities.

Relevant Lab Findings

- Blood alcohol level = 420 mg/dL
- Normal complete blood count (CBC)
- Normal electrolyte panel
- Normal liver function panel
- Normal coagulation profile
- Computed tomography (CT) scan of the head showed a thin, crescent-shaped hyperdensity that hugged the contours of the brain, a 10-mm midline shift, and uncal herniation.

Diagnosis

Subdural Hematoma

- A subdural hematoma usually occurs due to violent shaking of the head (e.g., child abuse or car accident) and is also very common among alcoholics and the elderly, who are prone to uncontrolled falls.
- A subdural hematoma is an accumulation of venous blood located between the dura and arachnoid.
- A subdural hematoma involves injury to the superior cerebral veins ("bridging veins").

Chapter 23

Neck

I. Muscles of the Neck

A. MUSCLES OF THE SUPERFICIAL NECK include the platysma, sternocleidomastoid, and trapezius.

B. MUSCLES OF THE ANTERIOR CERVICAL REGION include the mylohyoid, geniohyoid, stylohyoid, digastric, sternohyoid, omohyoid, sternothyroid, and thyrohyoid.

C. MUSCLES OF THE PREVERTEBRAL AREA include the longus colli, longus capitis, rectus capitis anterior, anterior scalene, rectus capitis lateralis, splenius capitis, levator scapulae, middle scalene, and posterior scalene.

D. MUSCLES OF THE LARYNX include the cricothyroid, thyroarytenoid, posterior cricoarytenoid, lateral cricoarytenoid, transverse and oblique arytenoids, and vocalis.

E. MUSCLES OF THE PHARYNX include the superior constrictor, middle constrictor, inferior constrictor, palatopharyngeus, salpingopharyngeus, and stylopharyngeus.

II. Cervical Plexus (Figure 23-1).
The cervical plexus is formed by the ventral primary rami of C1–4 and has both sensory and motor branches.

A. SENSORY NERVES
1. **Lesser occipital nerve** ascends along the sternocleidomastoid muscle and innervates the skin of the scalp behind the auricle.
2. **Great auricular nerve** ascends on the sternocleidomastoid muscle and innervates the skin behind the auricle and on the parotid gland.
3. **Transverse cervical nerve** turns around the posterior border of the sternocleidomastoid muscle and innervates the skin of the anterior cervical triangle.
4. **Supraclavicular nerve** emerges as a common trunk from under the sternocleidomastoid muscle and divides into the anterior branch, middle branch, and lateral branch to innervate the skin over the clavicle and the shoulder.

B. MOTOR NERVES
1. **Ansa cervicalis** is a nerve loop formed by the descendens hypoglossi (C1) and the descendens cervicalis (C2 and C3). The ansa cervicalis innervates the infrahyoid muscles (except the thyrohyoid).
2. **Phrenic nerve** if formed by C3, C4, and C5 cervical nerves and innervates the diaphragm (motor and sensory).

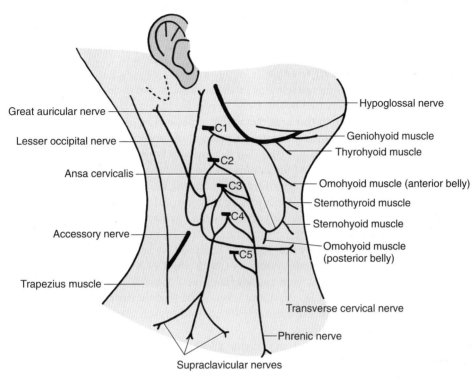

● **Figure 23-1 Cervical plexus.**

Ⅲ Cervical Triangles of the Neck (Figure 23-2)

A. GENERAL FEATURES

1. The sternocleidomastoid muscle divides the neck into the **anterior triangle** and **posterior triangle**, both of which are further subdivided into the **carotid triangle** and **occipital triangle**, respectively.

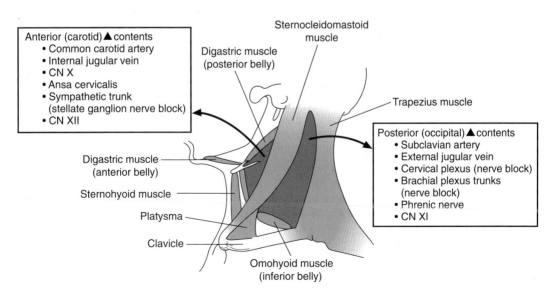

● **Figure 23-2 Cervical triangles of the neck.** A diagram of the lateral aspect of the neck shows the cervical triangles and their contents. Note that the common carotid artery, internal jugular vein, and cranial nerve X all lie within the carotid sheath.

2. The carotid triangle and occipital triangle contain important anatomic structures as indicated in Figure 23-2.

B. CLINICAL CONSIDERATIONS

1. Anterior (Carotid) Triangle

a. The **platysma muscle** lies in the superficial fascia above the anterior triangle and is innervated by the **facial nerve**. Accidental damage during surgery of the facial nerve in this area can result in **distortion of the shape of the mouth**.

b. The **carotid pulse** is easily palpated at the anterior border of the sternocleido-mastoid muscle at the level of the superior border of the thyroid cartilage (C5).

c. The **bifurcation of the common carotid artery** into the internal carotid artery and external carotid artery occurs in the anterior triangle of the neck at the level of C4. At the bifurcation, the **carotid body** and **carotid sinus** can be found. The carotid body is an **oxygen chemoreceptor**. Its sensory information is carried to the central nervous system by **cranial nerve (CN) IX** and **CN X**. The carotid sinus is a **pressure receptor**. Its sensory information is carried to the central nervous system by **CN IX** and **CN X**.

d. **Internal jugular vein catheterization**. The most commonly used approach is on the right side, above the level of the thyroid cartilage (C5; high approach) and medial to the sternocleidomastoid muscle within the anterior (carotid) triangle (see Figure 5-4).

e. **Carotid endarterectomy (Figure 23-3).** Carotid endarterectomy is a surgical procedure to remove blockages of the internal carotid artery and is performed in the anterior (carotid) triangle. This procedure can reduce the risk of stroke in patients who have emboli or plaques that cause **transient monocular blindness (amaurosis fugax)**. Transient monocular blindness is the classic ocular symptom of a **transient ischemic attack (TIA)** that should not be ignored since it involves emboli to the **central artery of the retina**, a terminal branch of the internal carotid artery (internal carotid artery → ophthalmic artery → central artery of retina). **Hollenhorst cholesterol plaques** are observed during a retinal exam. Contralateral hemiplegia and contralateral hemi-anesthesia may also occur due to insufficient blood flow to the middle cerebral artery (MCA). The diagram of the surgical exposure used in a carotid endarterectomy within the anterior (carotid) triangle of the neck is shown. Note the anatomic structures in this area that may be put in jeopardy during this procedure.

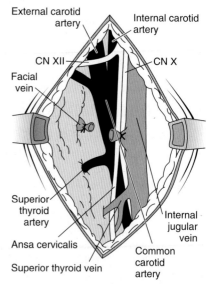

● **Figure 23-3 Carotid endarterectomy.**

f. **Stellate ganglion nerve block (Figure 23-4).** The stellate ganglion is the lowest of the three ganglia of the cervical sympathetic trunk. The term "stellate ganglion nerve block" is not strictly correct since injection of anesthetic is made above the stellate ganglion and enough anesthetic is injected to spread up and down. The needle is inserted between the trachea medially and the sternocleidomastoid muscle and the common carotid artery laterally using the **cricoid cartilage (C6)** and the **transverse process of C6 vertebra** as landmarks. A successful block results in **vasodilation** of the blood vessels of the head, neck, and upper limb and **Horner syndrome**, in which **miosis** (constriction of the pupil due to paralysis of the dilator pupillae muscle), **ptosis** (drooping of the eyelid due to paralysis of superior tarsal muscle), and **hemianhydrosis** (loss of sweating on one side) occurs. A stellate ganglion nerve block is used in Raynaud phenomenon and to relieve vasoconstriction after frostbite or microsurgery of the hand. The photograph shows a stellate ganglion nerve block procedure.

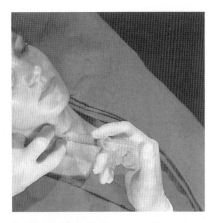

● **Figure 23-4 Stellate ganglion nerve block.**

2. **Posterior (Occipital) Triangle**
   a. **Injury to CN XI** within the posterior (occipital) triangle due to surgery or a penetrating wound will cause paralysis of the **trapezius muscle** so that **abduction of the arm** *past* **the horizontal position** is compromised.
   b. **Injuries to the trunks of the brachial plexus,** which lie in the posterior (occipital) triangle, will result in **Erb-Duchenne** or **Klumpke** syndromes (see Chapter 20).
   c. **Severe upper limb hemorrhage** may be stopped by compressing the subclavian artery against the first rib by applying downward and posterior pressure. The brachial plexus and subclavian artery enter the posterior (occipital) triangle in an area bounded anteriorly by the **anterior scalene muscle,** posteriorly by the **middle scalene muscle,** and inferiorly by the **first rib.**
   d. **Enlarged supraclavicular lymph nodes** due to upper gastrointestinal (GI) or lung cancer may be palpated in the posterior (occipital) triangle.

e. **Cervical plexus nerve block (Figure 23-5).** The needle is inserted at **vertebral level C3** along a landmark line connecting the mastoid process to the transverse process of C6 and enough anesthetic is injected to spread up and down. A cervical plexus nerve block is used in superficial surgery on the neck or thyroid gland and for pain management. The photograph shows a cervical plexus nerve block procedure.

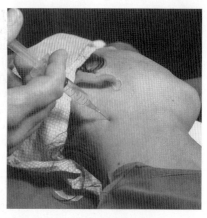

● Figure 23-5 Cervical plexus nerve block.

f. **Brachial plexus nerve block (Figure 23-6).** The needle is inserted at **vertebral level C6** into the interscalene groove (between the anterior and middle scalene muscles) using the cricoid cartilage (C6) and sternocleidomastoid muscle as landmarks. The photograph shows a brachial plexus nerve block procedure.

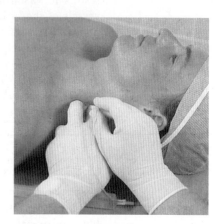

● Figure 23-6 Brachial plexus nerve block.

IV Larynx (Figure 23-7)

A. GENERAL FEATURES

1. The larynx consists of five major cartilages, which include the **cricoid** (1), **thyroid** (1), **epiglottis** (1), and **arytenoid** (2) cartilages.
2. The **ventricle** of the larynx is bounded superiorly by the **vestibular folds** (false vocal cords) and inferiorly by the **vocal folds** (true vocal cords).
3. All intrinsic muscles of the larynx are innervated by the **inferior laryngeal nerve of CN X** (a continuation of the **recurrent laryngeal nerve**), except the **cricothyroid muscle**, which is innervated by the **external branch of the superior laryngeal nerve of CN X**.
4. The intrinsic muscles of the larynx include the:
    a. **Posterior cricoarytenoid muscle:** *abducts* the vocal folds and opens airway during respiration. This is the *only* muscle that abducts the vocal folds.
    b. **Lateral cricoarytenoid muscle:** adducts the vocal folds
    c. **Arytenoideus muscle:** adducts the vocal folds
    d. **Thyroarytenoid muscle:** relaxes the vocal folds
    e. **Vocalis muscle:** alters the vocal folds for speaking and singing
    f. **Transverse and oblique arytenoid muscles:** close the laryngeal aditus (sphincter function)
    g. **Cricothyroid muscle:** stretches and tenses the vocal folds

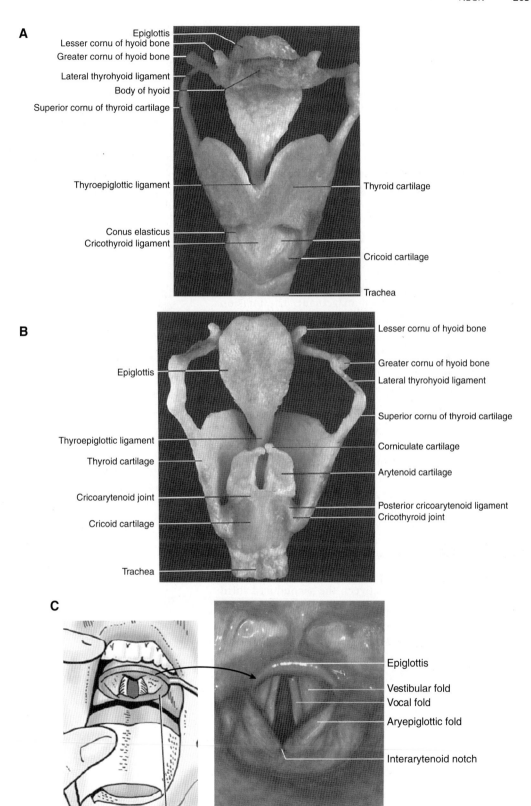

● **Figure 23-7 Larynx. A:** Anterior view of the laryngeal cartilages. **B:** Posterior view of the laryngeal cartilages. **C:** Diagram depicting the anatomic structures observed during inspection of the vocal folds using a laryngeal mirror.

B. CLINICAL CONSIDERATIONS

1. **Unilateral damage to the recurrent laryngeal nerve** can result from dissection around the ligament of Berry or ligation of the inferior thyroid artery during thyroidectomy. It will result in a hoarse voice, inability to speak for long periods, and movement of the vocal fold on the affected side toward the midline.

2. **Bilateral damage to the recurrent laryngeal nerve** can result from dissection around the ligament of Berry or ligation of the inferior thyroid artery during thyroidectomy. It will result in acute breathlessness (dyspnea) since both vocal folds move toward the midline and close off the air passage.

3. **Damage to the superior laryngeal nerve** can result when ligating the superior thyroid artery during thyroidectomy. This can be avoided by ligating the superior thyroid artery at its entrance into the thyroid gland. It will result in a weak voice with loss of projection, and the vocal cord on the affected side appears flaccid.

4. **Cricothyroidotomy (Figure 23-8).** A cricothyroidotomy is a procedure in which a tube is inserted between the cricoid and thyroid cartilages for emergency airway management. The incision made for this procedure will pass through the following structures: skin → superficial fascia and platysma muscle (avoiding the anterior jugular veins) → deep cervical fascia → pretracheal fascia (avoiding the sternohyoid muscle) → cricothyroid ligament (avoiding the cricothyroid muscle). This procedure may be complicated by a **pyramidal lobe** of the thyroid gland in the midline (present in 75% of the population). The diagram shows a lateral view of the laryngeal cartilages indicating the location for a cricothyroidotomy (*arrow*) and the anatomic layers that must be penetrated.

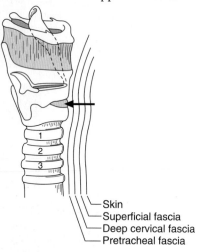

Skin
Superficial fascia
Deep cervical fascia
Pretracheal fascia

● **Figure 23-8 Cricothyroidotomy.**

5. **Tracheotomy (Figure 23-9).** A tracheotomy is a procedure in which a tube is inserted between the **second and third tracheal cartilage rings** when long-term ventilation support is necessary since it reduces the incidence of vocal cord paralysis or subglottic stenosis. The incision made for this procedure will pass through the following structures: skin → superficial fascia and platysma muscle (avoiding the anterior jugular veins) → deep cervical fascia → pretracheal fascia → cartilage rings. The following structures are put in jeopardy of injury: **inferior thyroid veins,** which form a plexus anterior to the trachea; the **thyroid ima artery** (present in 10% of people), which supplies the inferior border of the isthmus of the thyroid gland; and the **thymus gland** in infants. This procedure can be complicated by

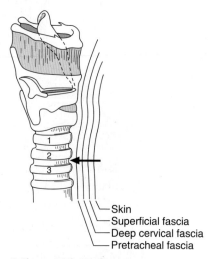

Skin
Superficial fascia
Deep cervical fascia
Pretracheal fascia

● **Figure 23-9 Tracheotomy.**

massive hemorrhage 1 to 2 weeks after placement of the tube due to erosion of the **brachiocephalic (innominate) artery**. The diagram shows a lateral view of the laryngeal cartilages indicating the location for a tracheotomy (*arrow*) and the anatomic layers that must be penetrated.

(V) Thyroid Gland

A. GENERAL FEATURES

1. The arterial supply of the thyroid gland is from the external carotid artery via the **superior thyroid artery** and the subclavian artery/thyrocervical trunk via the **inferior thyroid artery**, and sometimes the arch of the aorta via the **thyroid ima artery** (present in about 10% of the population).

2. The venous drainage is to the **superior thyroid veins, middle thyroid veins**, and **inferior thyroid veins**, all of which empty into the internal jugular vein.

3. The **right recurrent laryngeal nerve** (which recurs around the subclavian artery) and **left recurrent laryngeal nerve** (which recurs around the arch of the aorta at the ligamentum arteriosum) run in the tracheoesophageal groove along the posterior surface of the thyroid gland.

4. **The ligament of Berry** is the superior suspensory ligament of the thyroid gland located adjacent to the cricoid cartilage on the posterior surface of the thyroid gland.

B. CLINICAL CONSIDERATIONS

1. **Complications of a thyroidectomy** include thyroid storm (hyperpyrexia and tachyarrhythmias), hypoparathyroidism (develops within 24 hours due to low serum calcium), and recurrent laryngeal nerve or superior laryngeal nerve damage.

2. **Aberrant thyroid tissue** may occur anywhere along the path of its embryologic descent, that is, from the base of the tongue (foramen cecum), where it is called a **lingual cyst**, to the superior mediastinum.

3. **Thyroglossal duct cyst** is a cystic remnant of the descent of the thyroid during embryologic development.

(VI) Parathyroid Gland

A. GENERAL FEATURES

1. The parathyroid glands are yellowish-brown masses ($2 \times 3 \times 5$ mm in size; 40 g in weight).

2. Most people have four parathyroid glands, but five, six, or seven glands are possible. These glands are rarely embedded within the thyroid gland.

3. The **superior parathyroid glands** are consistently located on the posterior surface of the upper thyroid lobes near the inferior thyroid artery.

4. The **inferior parathyroid glands** are usually located on the lateral surface of the lower thyroid lobes (are more variable in location than the superior parathyroid glands).

5. The arterial supply of the superior and inferior parathyroid glands is from the **inferior thyroid artery**.

B. CLINICAL CONSIDERATIONS

1. **Primary hyperparathyroidism** results from autonomous secretion of parathyroid hormone (PTH) due to glandular hyperplasia, adenoma, or rarely carcinoma. The clinical sign is persistent hypercalcemia. The surgical removal of the hyperfunctioning glands results in a 90% cure rate.

2. **Injury to the parathyroid glands** most commonly results during a thyroidectomy due to disruption of the blood supply from the inferior thyroid artery.

Parotid Gland

A. GENERAL FEATURES

1. The parotid gland secretes a serous saliva that enters the mouth via the **parotid duct of Stensen.**

2. The facial nerve (CN VII) enters the substance of the parotid gland after emerging from the stylomastoid foramen and branches into the **temporal, zygomatic, buccal, mandibular,** and **cervical branches,** which innervate the muscles of facial expression. (Note: CN VII has no function in the parotid gland.)

3. The arterial supply is from branches of the **external carotid artery.**

4. The venous drainage is to the **retromandibular vein** → **external jugular vein.**

5. The parotid gland is innervated by preganglionic parasympathetic axons with cell bodies in the **inferior salivatory nucleus of the glossopharyngeal nerve (CN IX).** These axons travel within the **tympanic nerve** and the **lesser petrosal nerve** to synapse on cell bodies within the **otic ganglion.** Postganglionic parasympathetic axons leave the otic ganglion and are distributed with the **auriculotemporal nerve of trigeminal nerve (CN V)** to the parotid gland to stimulate saliva secretion.

6. The parotid gland is also innervated by postganglionic sympathetic axons from the superior cervical ganglion that reach the parotid gland with the arteries to stimulate saliva secretion.

B. CLINICAL CONSIDERATIONS

1. **Surgery on the parotid gland** may damage the **auriculotemporal nerve of CN V** and cause loss of sensation in the auriculotemporal area of the head. Since the auriculotemporal nerve also carries postganglionic sympathetic nerve fibers to the sweat glands of the head and postganglionic parasympathetic nerve fibers to the parotid gland for salivation, if this nerve is severed, aberrant regeneration may result in a person sweating during eating (**Frey syndrome**).

2. **Surgery on the Parotid Gland or Bell Palsy.** Both of these conditions cause a lower motor neuron lesion of the facial nerve (CN VIII). This results in an **ipsilateral paralysis** of muscles of facial expression of the **upper and lower face,** loss of corneal reflex (efferent limb), loss of taste from the anterior two thirds of the tongue, and hyperacusis (increased acuity to sound). Clinical signs include inability to blink the eye or raise the eyebrow (upper face deficit involving orbicularis oculi and frontalis muscles, respectively) and inability to seal the lips or smile properly (lower face deficit involving orbicularis oris muscle) on the affected side.

3. **Stroke.** A stroke within the internal capsule affecting the corticobulbar tract causes an upper motor neuron lesion of the facial nerve (CN VII). This results in a **contralateral paralysis** of the **lower face** but spares the upper face. Clinical signs include inability to seal the lips or smile properly (lower face deficit involving orbicularis oris muscle) on the contralateral side.

4. **Facial Laceration.** A facial laceration near the anterior border of the masseter muscle will cut the **parotid duct of Stensen** and the **buccal branch of CN VII.**

Chapter 24

Eye

I. Bony Orbit

A. GENERAL FEATURES

1. The bony orbit is a pyramid-shaped cavity surrounded by a shell of bone to protect the eyeball.
2. The roof of the orbit is formed by the **frontal bone** and the **lesser wing of the sphenoid bone.**
3. The medial wall is formed by the **ethmoid bone** and **lacrimal bone.**
4. The lateral wall is formed by the **zygomatic bone** and the **greater wing of the sphenoid bone.**
5. The floor of the orbit is formed by the **maxilla bone** and the **palatine bone.**

B. FISSURES, FORAMINA, AND CANALS

1. **Superior Orbital Fissure**
   a. The superior orbital fissure is formed by a gap between the greater and lesser wings of the sphenoid bone and communicates with the middle cranial fossa.
   b. This fissure transmits the following: **oculomotor nerve (cranial nerve [CN] III), trochlear nerve (CN IV), ophthalmic nerves (branches of the ophthalmic division of the trigeminal nerve [CN V$_1$]), abducens nerve (CN VI), and ophthalmic vein.**

2. **Inferior Orbital Fissure**
   a. The inferior orbital fissure is formed by a gap between the greater wing of the sphenoid bone and maxillary bone and communicates with the infratemporal fossa and pterygopalatine fossa.
   b. This fissure transmits the following: **infraorbital nerve (a branch of the maxillary division of CN V$_2$), infraorbital artery, and inferior ophthalmic vein.**

3. **Infraorbital Foramen and Groove**
   a. This foramen and groove transmits the following: **infraorbital nerve, infraorbital artery, and inferior ophthalmic vein.**

4. **Supraorbital Foramen (or Notch)**
   a. This foramen transmits the following: **supraorbital nerve, supraorbital artery, and superior ophthalmic vein.**

5. **Anterior Ethmoidal Foramen**
   a. This foramen transmits the following: **anterior ethmoidal nerve** and **anterior ethmoidal artery.**

6. **Posterior Ethmoidal Foramen**
   a. This foramen transmits the following: **posterior ethmoidal nerve** and **posterior ethmoidal artery.**

7. **Optic Canal**
   a. This canal is formed by an opening through the lesser wing of the sphenoid bone and communicates with the middle cranial fossa.
   b. This canal transmits the following: **optic nerve (CN II)** and **ophthalmic artery (a branch of the internal carotid artery).**

8. **Nasolacrimal Canal**
   a. This canal is formed by the maxilla bone, lacrimal bone, and inferior nasal concha.
   b. This canal transmits the following: **nasolacrimal duct** from the lacrimal sac to the inferior nasal meatus.

Ⅱ Eyelids and Lacrimal Apparatus (Figure 24-1)

A. EYELIDS

1. The exterior surface of the eyelid is typical **thin skin.**
2. The interior surface of the eyelid is a mucous membrane called the **palpebral conjunctiva.**
3. The palpebral conjunctiva is reflected onto the eyeball, where it is then called the **bulbar conjunctiva.** The bulbar conjunctiva is continuous with the corneal epithelium.
4. The palpebral and bulbar conjunctiva enclose a space called the **conjunctival sac.**
5. Within upper and lower eyelids, there is a dense plate of collagen called the **superior tarsal plate** and **inferior tarsal plate**, respectively. The superior and inferior tarsal plates merge on either side of the eye to form the **medial** and **lateral palpebral ligaments.**
6. The tarsal plates contain **tarsal glands**, which are specialized sebaceous glands opening via a duct onto the edge of the eyelid.
7. The **medial** and **lateral palpebral commissures** are formed where the upper and lower eyelids come together, thus defining the **angles of the eye.** There are three important muscles associated with the eyelid, which include:
   a. **Levator palpebrae superioris muscle.** This **skeletal muscle** is located in the upper eyelid and attaches to the skin of the upper eyelid and anterior surface of the superior tarsal plate. This muscle is innervated by **CN III** and its function is **to keep the eye open (main player).**
   b. **Superior tarsal muscle.** This **smooth muscle** is located in the upper eyelid and attaches to the superior tarsal plate. This muscle is innervated by **postganglionic sympathetic neurons** that follow the carotid arterial system into the head and neck and its function is **to keep the eye open (minor player).**
   c. **Orbicularis oculi muscle (palpebral portion).** This **skeletal muscle** is located in the upper and lower eyelid and lies superficial to the tarsal plates. This muscle in innervated by **CN VII** and its function is **to close the eye.**

B. LACRIMAL APPARATUS

1. **Lacrimal glands** are located in the superior lateral aspect of each orbit and secrete a lacrimal fluid (or tears).
2. Lacrimal fluid, mucus secretion from the palpebral and bulbar conjunctiva, and sebaceous secretions from the tarsal glands all contribute to form the **tear film** on the surface of the eye.
3. Lacrimation is stimulated by the parasympathetic nervous system.
   a. The preganglionic parasympathetic neuronal cell bodies are located in **superior salivatory nucleus and lacrimal nucleus.**
   b. Preganglionic axons from the superior salivatory nucleus and the lacrimal nucleus run with **CN VII** (by way of the **nervus intermedius, greater petrosal nerve, and the nerve of the pterygoid canal**) and enter the **pterygopalatine ganglion**, where they synapse with postganglionic parasympathetic neurons.
   c. Postganglionic axons leave the pterygopalatine ganglion and run with the **zygomaticofacial branch of CN V$_2$** and the **lacrimal branch of CN V$_1$** to innervate the lacrimal gland.

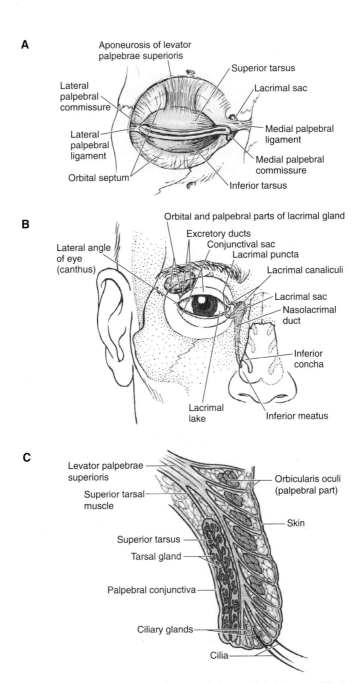

● **Figure 24-1 Eyelids and lacrimal apparatus. A:** Skeleton of the eyelid. **B:** Upper eyelid. **C:** Lacrimal apparatus.

III **The Globe or Eyeball.** The globe of the eye consists of three concentric tunics that make up the wall of the eye, as indicated below.

A. CORNEOSCLERAL TUNIC. This is the outermost fibrous tunic and consists of the cornea, sclera, and **corneoscleral junction (limbus)**.

    1. Cornea. The cornea is an avascular structure, but is highly innervated by **branches of CN V_1 (ophthalmic division of trigeminal nerve)**.

    2. Sclera. The sclera is a white, opaque structure that provides attachments for the extraocular eye muscles.

3. **Corneoscleral Junction (Limbus)**
   a. The limbus is the junction of the transparent cornea and the opaque sclera.
   b. The limbus contains a **trabecular network** and the **canal of Schlemm**, which are involved in the flow of aqueous humor.
   c. The flow of aqueous humor follows this route: **posterior chamber → anterior chamber → trabecular network → canal of Schlemm → aqueous veins → episcleral veins.**
   d. An obstruction of aqueous humor flow will increase intraocular pressure, causing a condition called **glaucoma.**

B. **UVEAL TUNIC.** This is the middle vascular tunic and consists of the **choroid, stroma of the ciliary body,** and **stroma of the iris.**
   1. **Choroid.** The choroid is a pigmented vascular bed that lies immediately deep to the corneoscleral tunic. The profound vascularity of the choroid is responsible for the "red eye" that occurs with flash photography.
   2. **Stroma of the Ciliary Body**
      a. The stroma of the ciliary body contains the **ciliary muscle,** which is circularly arranged around the entire circumference of the ciliary body and is innervated by the parasympathetic nervous system.
         i. The preganglionic parasympathetic neuronal cell bodies are located in the **Edinger-Westphal nucleus of CN III.**
         ii. Preganglionic axons from the Edinger-Westphal nucleus travel with CN III and enter the **ciliary ganglion,** where they synapse with postganglionic parasympathetic neurons.
         iii. Postganglionic axons leave the ciliary ganglion, where they travel with the **short ciliary nerves** to innervate the ciliary muscle.
      b. **Accommodation** is the process by which the lens becomes rounder to focus a nearby object or flatter to focus a distant object.
         i. For **close vision** (e.g., reading), the ciliary muscle contracts, which reduces tension on the zonular fibers attached to the lens and thereby allows the lens to take a rounded shape.
         ii. **For distant vision,** the ciliary muscle relaxes, which increases tension on the zonular fibers attached to the lens and thereby allows the lens to take a flattened shape.
   3. **Stroma of the Iris.** The stroma contains the **dilator pupillae muscle** and **sphincter pupillae muscle.**
      a. **Dilator pupillae muscle**
         i. The dilator pupillae muscle is radially arranged around the entire circumference of the iris and is innervated by the sympathetic nervous system.
            (a) The preganglionic sympathetic neuronal cell bodies are located in the gray matter of the T1–L2/L3 spinal cord.
            (b) Preganglionic axons project from this area, enter the paravertebral chain ganglia, and ascend to the **superior cervical ganglion,** where they synapse with postganglionic sympathetic neurons.
            (c) Postganglionic axons leave the superior cervical ganglion and follow the carotid arterial system into the head and neck, where they travel with the **long ciliary nerves** to innervate the dilator pupillae muscle, which **dilates the pupil.**
         ii. Any pathology that compromises this sympathetic pathway will result in **Horner syndrome.**
      b. **Sphincter pupillae muscle**
         i. The sphincter pupillae muscle is circularly arranged around the entire circumference of the iris and is innervated by the parasympathetic nervous system.

(a)   The preganglionic parasympathetic neuronal cell bodies are located in the **Edinger-Westphal nucleus of CN III.**

(b)   Preganglionic axons from the Edinger-Westphal nucleus enter the **ciliary ganglion**, where they synapse with postganglionic parasympathetic neurons.

(c)   Postganglionic axons leave the ciliary ganglion, where they travel with the **short ciliary nerves** to innervate the sphincter pupillae muscle, which **constricts the pupil (or miosis).**

ii.   Lesions involving CN III will result in a **fixed and dilated pupil.**

C.   RETINAL TUNIC. This is the innermost tunic and consists of the **outer pigment epithelium** and the **inner neural retina (posteriorly)**, the **epithelium of the ciliary body (anteriorly)**, and the **epithelium of the iris (anteriorly).**

1.   **Outer Pigment Epithelium and Inner Neural Retina**

a.   The outer pigment epithelium and the inner neural retina together constitute the **retina.**

b.   The **intraretinal space** separates the outer pigment epithelium from the inner neural retina. Although the intraretinal space is obliterated in the adult, it remains a weakened area prone to **retinal detachment.**

c.   The posterior two thirds of the retina is a light-sensitive area (**pars optica**) and the anterior one third is a light-insensitive area (**par ciliaris and iridis**). These two areas are separated by the **ora serrata.**

d.   The retina has a number of specialized areas, which include the following:

i.   **Optic disc**

(a)   The optic disc is the site where axons of the ganglion cells converge to form the optic nerve (CN II) by penetrating the sclera, forming the **lamina cribrosa.**

(b)   The optic disc lacks rods and cones and is therefore a **blind spot.**

(c)   The **central artery and vein of the retina** pass through the optic disc.

ii.   **Fovea**

(a)   The fovea is a shallow depression of the retina located 3 mm lateral (temporal side) to the optic disc along the visual axis.

(b)   The **fovea centralis** is located at the center of the fovea and is the area of highest visual acuity and color vision.

(c)   The fovea centralis contains **only cones (no rods or capillaries)** that are arranged **at an angle** so that light directly impinges on the cones without passing through other layers of the retina and is linked to a single ganglion, both of which contribute to visual acuity.

(d)   The **macula lutea** is a yellowish area (due to xanthophyll pigment accumulation in ganglion cells) surrounding the fovea centralis.

2.   **Epithelium of the Ciliary Body.** The ciliary epithelium **secretes aqueous humor** and **produces the zonular fibers** that attach to the lens.

3.   **Epithelium of the Iris**

D.   CONTENTS OF THE GLOBE. The globe is divided into two cavities by the lens: the **anterior cavity** and **posterior cavity.**

1.   The **anterior cavity** consists of the **anterior chamber** (the area between the cornea and iris) and the **posterior chamber** (the area between the iris and lens). These chambers are filled with the watery **aqueous humor** that is secreted by the epithelium of the ciliary body.

2.   The **posterior cavity** consists of the **vitreous chamber** (the area between the lens and retina). The vitreous chamber is filled with the **vitreous body** (a jellylike substance) and **vitreous humor** (a watery fluid), which hold the retina in place and support the lens.

Ⅳ Extraocular Musculature (Figure 24-2)

A. There are seven extraocular muscles of the orbit: **levator palpebrae superioris muscle, superior rectus muscle, inferior rectus muscle, lateral rectus muscle, medial rectus muscle, superior oblique muscle,** and **inferior oblique muscle.**

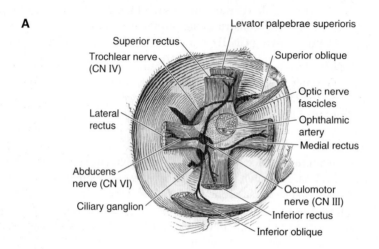

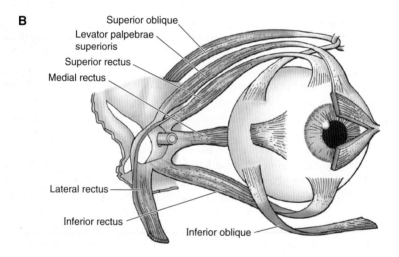

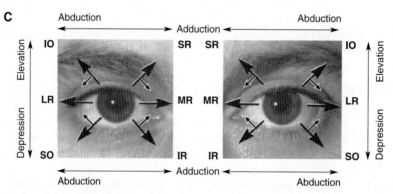

● **Figure 24-2 Nerves of the orbit and eye movements. A:** Nerves of the orbit after excision of the eyeball. **B:** Extraocular muscles (anterolateral view from the right). **C:** Eye movements. Large arrows indicate the direction of eye movements caused by the various extraocular muscles. Small arrows indicate either intorsion or extorsion. IO = inferior oblique, IR = inferior rectus, LR = lateral rectus, MR = medial rectus, SO = superior oblique, SR = superior rectus. (*continued*)

Muscle	Nerve	Function and Clinical Test[a]
Extraocular Muscles		
Levator palpebrae	CN III	Elevates upper eyelid (keeps eye open)
Superior rectus	CN III	Elevates, adducts, and intorts[b] Patient is asked first to look to the side, and then to look up
Medial rectus	CN III	Adducts Patient is asked to look to the nose (medially)
Inferior rectus	CN III	Depresses, adducts, and extorts[c] Patient is asked first to look to the side, and then to look down
Inferior oblique	CN III	Elevates, abducts, and extorts Patient is asked first to look to the nose, and then to look up **"up and in" toward bridge of the nose**
Superior oblique	CN IV	Depresses, abducts, and intorts Patient is asked first to look to the nose, and then to look down **"down and in" toward the tip of the nose**
Lateral rectus	CN VI	Abducts Patient is asked to look to the side (laterally)
Other Eye Muscles		
Orbicularis oculi	CN VII	Closes the eye; efferent limb of corneal reflex
Dilator pupillae	Postganglionic sympathetic	Dilates the pupil
Superior tarsal	Postganglionic sympathetic	Keeps eye open
Sphincter pupillae	Postganglionic parasympathetic	Constricts the pupil
Ciliary muscle	Postganglionic parasympathetic	Performs accommodation

[a]Since the actions of the superior rectus, inferior rectus, superior oblique, and inferior oblique are complicated, the physician tests eye movements with the eye placed in a position where a single action of the muscle predominates.
[b]Intorsion is the medial rotation of the superior pole of the eyeball.
[c]Extorsion is the lateral rotation of the superior pole of the eyeball.

D

● **Figure 24-2** *Continued* **D:** Summary table.

B. The four rectus muscles arise from the **common tendinous ring** at the apex of the orbit and extend anteriorly to insert in the sclera of the globe.

C. Although individual rectus muscles exert unique forces on the globe, they rarely act independently.

D. The innervation to the extraocular muscles can be remembered by the chemical formula $LR_6SO_4AO_3$ (lateral rectus, CN VI; superior oblique, CN IV; all others, CN III).

V **Arterial Supply of the Orbit.** The **ophthalmic artery** is a branch of the **internal carotid artery** and is the primary arterial supply to the orbit. The ophthalmic artery enters the orbit via the optic canal and gives off various branches, which include the **central artery of the retina, supraorbital artery, supratrochlear artery, lacrimal artery, dorsal nasal artery, short posterior ciliary arteries, long posterior ciliary arteries, anterior ethmoidal artery, posterior ethmoidal artery, anterior ciliary artery,** and **infraorbital artery.**

VI # Venous Drainage of the Orbit

A. SUPERIOR OPHTHALMIC VEIN. The superior ophthalmic vein is formed by the union of the supraorbital vein, supratrochlear vein, and angular vein. The superior ophthalmic vein ultimately drains into the **cavernous sinus. Superior ophthalmic vein →
cavernous sinus.**

B. INFERIOR OPHTHALMIC VEIN. The inferior ophthalmic vein is formed by the union of small veins in the floor of the orbit. The inferior ophthalmic vein communicates with the pterygoid venous plexus and empties into the superior ophthalmic vein. **Inferior
ophthalmic vein → superior ophthalmic vein → cavernous sinus.**

C. CENTRAL VEIN OF THE RETINA. The central vein of the retina most often drains into the cavernous sinus directly, but may join the superior or inferior ophthalmic vein. **Central vein of the retina → cavernous sinus.**

D. VORTICOSE VEINS. The vorticose veins from the choroid layer of the globe drain into the inferior ophthalmic vein. **Vorticose veins → inferior ophthalmic vein → superior
ophthalmic vein → cavernous sinus.**

 VII # Clinical Considerations

A. ORBITAL FRACTURES. A direct impact to the face (e.g., being punched in the eye) is transmitted to the walls of the bony orbit. A portion of the **ethmoid bone** known as the **lamina papyracea** (as the name implies, a "paper thin" bone) is the weakest segment of the medial wall and is thus prone to fracture. This fracture results in direct communication between the orbit and the nasal cavity by way of the ethmoid sinuses. The **infraorbital canal** (which contains the infraorbital branch of the maxillary nerve) is the weakest portion of the orbital floor and is also prone to fracture. This fracture results in direct communication between the orbit and nasal cavity by way of the maxillary sinus.

B. STY. A sty is a painful, erythematous, suppurative swelling of the eyelid that results from an obstructed and infected ciliary gland found at the margin of the eyelid. When a sebaceous gland of the eyelid becomes obstructed and forms a cyst, this is known as a **chalazion.** Obstruction of the tarsal glands produces inflammation known as a **tarsal
chalazion.**

C. DRY EYE is caused by a disruption in the production of tears or damage to the eyelid. This may lead to ulceration, perforation, loss of aqueous humor, and blindness.

D. RED EYE is caused most commonly by conjunctivitis (i.e., inflammation of the conjunctiva). A purulent discharge indicates bacterial infection. A watery discharge indicates a viral infection.

E. BOGORAD SYNDROME (CROCODILE TEARS). This syndrome is the spontaneous lacrimation during eating caused by a lesion of CN VII proximal to the geniculate ganglion. This syndrome occurs after facial paralysis and is due to the misdirection of regenerating preganglionic parasympathetic axons (that formerly innervated the salivary glands) to the lacrimal glands.

F. GLAUCOMA is the obstruction of aqueous humor flow that results in an increased intraocular pressure. This increased pressure causes impaired retinal blood flow producing retinal ischemia; degeneration retinal cells, particularly at the optic disc; defects in the visual field; and blindness. There are two types of glaucoma:
1. **Open-angle glaucoma** (most common) occurs when the trabecular network is open but the canal of Schlemm is obstructed.

2. **Closed-angle glaucoma** occurs when the trabecular network is closed usually due to an inflammatory process of the uvea (uveitis; e.g., infection by cytomegalovirus).

G. **OBSTRUCTION OF THE CENTRAL ARTERY OF THE RETINA** is generally caused by an embolus and leads to retinal ischemia with instantaneous complete blindness. The blindness is often described as a dark curtain coming down over the eye and when the attack is brief it is called **amaurosis fugax.** These events are most often monocular and may last only a few seconds or can result in permanent blindness.

H. **CAVERNOUS SINUS THROMBOSIS.** The anastomoses between the angular vein of the face and the inferior ophthalmic vein can result in spread of infectious agents from periorbital and perinasal areas to the cavernous sinus, resulting in thrombosis. This thrombosis prevents retinal drainage, eventually leading to retinal ischemia and blindness.

I. **PAPILLEDEMA (CHOKED DISC)** is a noninflammatory edema of the optic disc (papilla) due to increased intracranial pressure usually caused by brain tumors, subdural hematoma, or hydrocephalus. It usually does not alter visual acuity, but may cause bilateral **enlarged blind spots.**

J. **RETINAL DETACHMENT** may result from head trauma or may be congenital. The site of detachment is between the outer pigment epithelium and the inner neural retina (i.e., outer segment layer of the rods and cones of the neural retina).

K. **STRABISMUS (CROSSED EYE)** is caused by damage to CN III, which results in weakness or paralysis of the extraocular eye muscles. Strabismus is a visual disorder in which the visual axes do not meet the desired objective point (or the eyes are misaligned and point in different directions) due to the uncoordinated action of the extraocular eye muscles. The affected eye may turn inward, outward, upward, or downward, leading to decreased vision and misaligned eyes.

L. **DIPLOPIA (DOUBLE VISION)** is caused by paralysis of one or more extraocular muscles resulting from injury of the nerves supplying them.

M. **HORNER SYNDROME** is caused by injury to the cervical sympathetic nerves and results in **miosis** (constriction of pupil due to paralysis of dilator pupillae muscle), **ptosis** (drooping of eyelid due to paralysis of superior tarsal muscle), **hemianhidrosis** (loss of sweating on one side), **enophthalmos** (retraction of the eyeball into the orbit due to paralysis of the orbitalis muscle), and **flushing** (vasodilation and increased blood flow to the head and neck). Horner syndrome may be caused by brainstem stroke, tuberculosis, Pancoast tumor, trauma, and injury to the carotid arteries.

N. **COMMON OPHTHALMOSCOPIC PATHOLOGIES (FIGURE 24-3)**

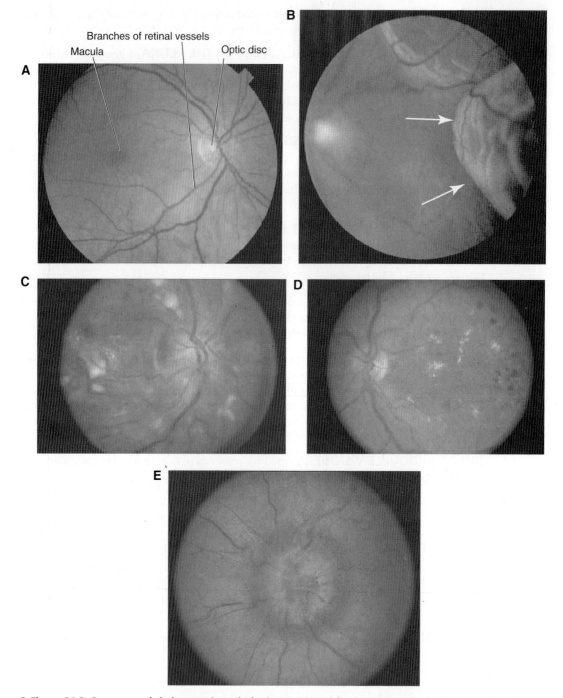

Branches of retinal vessels
Macula
Optic disc

A

B

C

D

E

● **Figure 24-3 Common ophthalmoscopic pathologies. A:** Normal fundus. **B:** Detached retina (*arrows*). **C:** Hypertensive retinopathy. The optic nerve head is edematous and the retina contains numerous exudates and "cotton-wool" spots. **D:** Diabetic retinopathy. The retina contains several yellowish "hard" exudates, which are rich in lipids, and several relatively small retinal hemorrhages. **E:** Chronic papilledema. The optic nerve head is congested and protrudes anteriorly toward the interior of the eye. The optic disc has blurred margins and the blood vessels within it are poorly seen.

Chapter 25

Ear

I **General Features.** The ear is the organ of hearing and balance. The ear consists of the external ear, middle ear, and inner ear.

II **External Ear (Figure 25-1)** consists of the following:

A. AURICLE

1. The auricle (known as "the ear" by laypeople) is supported by elastic cartilage and covered by skin.
2. The auricle develops from **six auricular hillocks** that surround pharyngeal groove 1.
3. The auricle is innervated by **cranial nerve (CN) V_3, CN VII, CN IX, and CN X,** and **cervical nerves C2 and C3.**

B. EXTERNAL AUDITORY MEATUS

1. The external auditory meatus is an air-filled tubular space.
2. The external auditory meatus develops from the **pharyngeal groove 1**, which becomes filled with ectodermal cells, forming a temporary **meatal plug** that disappears before birth.
3. The external auditory meatus is innervated by CN V_3 and CN IX.

C. TYMPANIC MEMBRANE (EARDRUM)

1. The tympanic membrane separates the middle ear from the external auditory meatus of the external ear.
2. The tympanic membrane develops from **pharyngeal membrane 1.**
3. The **pars flaccida** is a small triangular portion between the anterior and posterior malleolar folds; the remainder of the tympanic membrane is called the **pars tensa.**
4. The **cone of light** is a triangular reflection of light seen in the anterior-inferior quadrant.
5. The **umbo** is the most depressed center point of the tympanic membrane concavity.
6. The external (lateral) concave surface is innervated (sensory) by the **auriculotemporal branch of CN V_3** and the **auricular branch of CN X.**
7. The internal (medial) surface is innervated by the **tympanic branch of CN IX.**

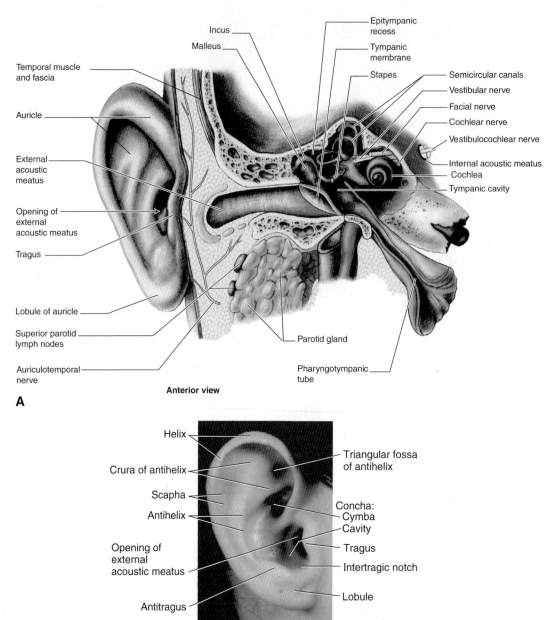

Incus
Malleus
Epitympanic recess
Tympanic membrane
Stapes
Semicircular canals
Vestibular nerve
Facial nerve
Cochlear nerve
Vestibulocochlear nerve
Internal acoustic meatus
Cochlea
Tympanic cavity
Temporal muscle and fascia
Auricle
External acoustic meatus
Opening of external acoustic meatus
Tragus
Lobule of auricle
Superior parotid lymph nodes
Auriculotemporal nerve
Parotid gland
Pharyngotympanic tube

Anterior view

A

Helix
Crura of antihelix
Scapha
Antihelix
Opening of external acoustic meatus
Antitragus
Triangular fossa of antihelix
Concha: Cymba Cavity
Tragus
Intertragic notch
Lobule

B

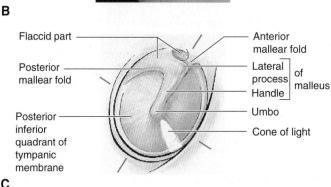

Flaccid part
Posterior mallear fold
Posterior inferior quadrant of tympanic membrane
Anterior mallear fold
Lateral process
Handle
of malleus
Umbo
Cone of light

C

● **Figure 25-1 Anatomy of the external ear. A:** Coronal section of the ear. **B:** External ear. **C:** An otoscopic view of the right tympanic membrane.

Middle Ear (Figure 25-2). The middle ear is an air-filled chamber lined by a mucosa that is innervated (sensory) by the **tympanic nerve of CN IX**, which forms the **tympanic plexus** with caroticotympanic nerves from the arterial carotid sympathetic plexus. The middle ear consists of the following:

A. TYMPANIC (MIDDLE EAR) CAVITY

1. The **tympanic cavity proper** is a space internal to the tympanic membrane.
2. The **epitympanic recess** is a space superior to the tympanic membrane that contains the head of the malleus and body of the incus.
3. The tympanic cavity communicates with the nasopharynx via the auditory (eustachian) tube and the mastoid air cells and mastoid antrum.

B. OSSICLES. The ossicles function as amplifiers to overcome the impedance mismatch at the air–fluid interface between the tympanic cavity (air) and the inner ear (fluid).

1. **Malleus (Hammer)**
   a. The malleus develops from cartilage of **pharyngeal arch 1** (Meckel cartilage).
   b. The malleus consists of a **head, neck,** and **handle** along with **anterior** and **lateral processes.**
   c. The head of the malleus articulates with the body of the incus in the epitympanic recess.
   d. The handle of the malleus is fused to the internal (medial) surface of the tympanic membrane and is moved by the **tensor tympani muscle,** which is innervated by CN V_3.

2. **Incus (Anvil)**
   a. The anvil develops from the cartilage of **pharyngeal arch 1** (Meckel cartilage).
   b. The incus consists of a **body, short process,** and **long process.**
   c. The body of the incus articulates with the head of the malleus.
   d. The short process of the incus extends horizontally backward and attaches to the ligament of the incus.
   e. The long process of the incus descends vertically and articulates with the stapes.

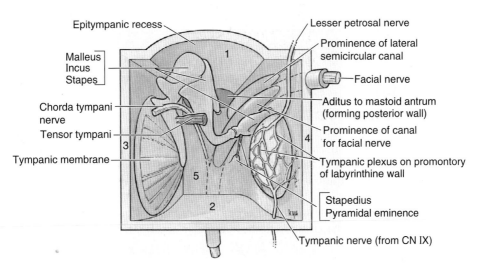

● **Figure 25-2 View of the tympanic cavity.** The anterior wall (carotid canal) has been removed. *1* = tegmen tympani, *2* = jugular fossa, *3* = lateral wall (tympanic membrane), *4* = medial wall (promontory, oval window, round window), *5* = posterior wall (mastoid air cells and mastoid antrum).

3. **Stapes (Stirrup)**
   a. The stapes develops from the cartilage of **pharyngeal arch 2** (Reichert cartilage).
   b. The stapes consists of a **head**, a **neck**, **two processes**, and a **footplate**.
   c. The stapes is moved by the **stapedius** muscle, which is innervated by CN VII.
   d. The footplate is attached to the **oval window** of the vestibule.

C. **MUSCLES**
   1. **Tensor Tympani Muscle**
      a. The tensor tympani muscle inserts on the handle of the malleus.
      b. The tensor tympani muscle draws the tympani membrane medially and tightens it in response to a loud noise, thereby reducing the vibration of the tympanic membrane.
      c. The tensor tympani muscle is innervated by **CN V_3**.
   2. **Stapedius Muscle**
      a. The stapedius muscle inserts on the neck of the stapes.
      b. The stapedius muscle pulls the stapes posteriorly and reduces excessive oscillation, thereby protecting the inner ear from injury from a loud noise.
      c. The stapedius muscle is innervated by **CN VII**.

D. **OVAL WINDOW (FENESTRA VESTIBULI)**
   1. The oval window is pushed back and forth by the footplate of the stapes and transmits sonic vibrations of the ossicles to the perilymph of the scala vestibuli of the inner ear.

E. **ROUND WINDOW (FENESTRA COCHLEA)**
   1. The round window is closed by the mucous membrane of the middle ear and accommodates pressure waves transmitted to the perilymph of the scala tympani.

F. **AUDITORY (EUSTACHIAN) TUBE**
   1. The auditory tube connects the middle ear to the nasopharynx.
   2. The auditory tube allows air to enter or leave the tympanic cavity, thereby balancing the air pressure of the tympanic cavity with the atmospheric pressure. This allows free movement of the tympanic membrane.
   3. The auditory tube can be opened by the contraction of the tensor veli palatini and the salpingopharyngeus muscles.

 Inner Ear (Figure 25-3). The inner ear consists of the **semicircular ducts, utricle, saccule,** and **cochlear duct,** all of which are referred to as the **membranous labyrinth** containing **endolymph.**

A. **SEMICIRCULAR DUCTS (KINETIC LABYRINTH)**
   1. The semicircular ducts consist of the **anterior (superior)**, lateral, and **posterior ducts** along with their dilated ends, called **ampullae.**
   2. Type I and **type II hair cells** that cover the **cristae ampullaris** (a prominent ridge within the ampulla) have numerous stereocilia and a single **kinocilium** on their apical border.
   3. These cells synapse with bipolar neurons of the **vestibular ganglion of CN VIII.**
   4. The semicircular ducts respond to **angular acceleration** and **deceleration of the head.**

B. **UTRICLE AND SACCULE (STATIC LABYRINTH)**
   1. The utricle and saccule are dilated membranous sacs that contain specialized receptors called **maculae.**
   2. Type I and **type II hair cells** within **maculae** have stereocilia and a single **kinocilium** on their apical border.

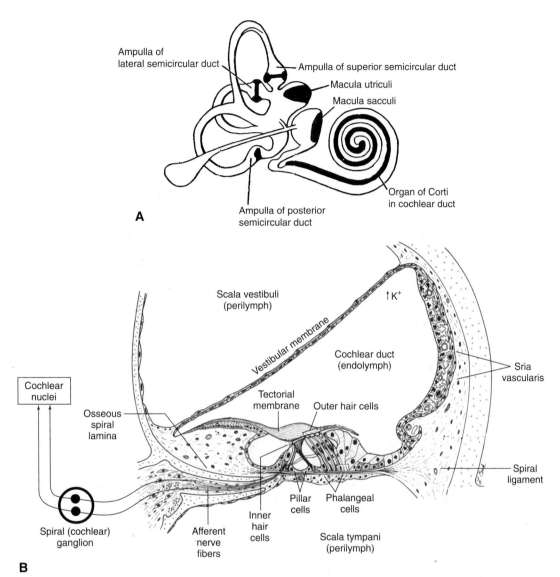

● **Figure 25-3 Anatomy of the inner ear. A:** Diagram of the membranous labyrinth. Note the location of the specialized sensory areas (*black color*) for angular acceleration (cristae ampullaris), linear acceleration (maculae), and hearing (organ of Corti). **B:** Organ of Corti of the cochlear duct. The organ of Corti responds to sound. The hearing process begins when airborne sound waves cause vibration of the tympanic membrane, which moves the stapes against the oval window. This produces waves of perilymph within the scala vestibuli and scala tympani. The waves of perilymph cause an upward displacement of the basilar membrane such that the stereocilia of the hair cells hit the tectorial membrane. As a result, K^+ ion channels open, hair cells are depolarized, and afferent nerve fibers are stimulated. Note that the endolymph has a high K^+ concentration, which is maintained by the stria vascularis. The basilar membrane extends between the osseous spiral lamina and the spiral ligament.

   3. These cells synapse with bipolar neurons of the **vestibular ganglion of CN VIII.**
   4. The utricle and saccule respond to the position of the head with respect to **linear acceleration** and **pull of gravity.**

C. COCHLEAR DUCT
   1. The cochlear duct is a triangular duct wedged between the scala vestibuli and scala tympani.
   2. The cochlear duct consists of a **vestibular membrane** (roof), **basilar membrane** (floor), and **stria vascularis** (lateral wall).

3. The stria vascularis participates in the formation of endolymph.
4. The cochlear duct contains the **organ of Corti**.
5. The organ of Corti contains a single row of **inner hair cells** and three rows of **outer hair cells**.
6. The hair cells synapse with bipolar neurons of the **cochlear (spiral) ganglion of CN VIII** (90% of these bipolar neurons synapse with inner hair cells).
7. The organ of Corti responds to **sound**. High-frequency sounds cause maximum displacement of the basilar membrane and stimulation of hair cells at the **base of the cochlea**. Low-frequency sounds cause maximum displacement of the basilar membrane and stimulation of hair cells at the **apex of the cochlea**.

Ⓥ Clinical Considerations

A. **RUBELLA VIRUS.** The organ of Corti may be damaged by exposure to rubella virus during week 7 and week 8 of embryologic development.

B. **MÉNIÈRE DISEASE** is caused by an increase in endolymph. Clinical findings include vertigo (the illusion of rotational movement), nausea, positional nystagmus (involuntary rhythmic oscillations of the eye), vomiting, and tinnitus (ringing of the ears).

C. **WAARDENBURG SYNDROME** is an autosomal dominant congenital deafness associated with pigment abnormalities resulting from abnormal neural crest cell migration. These patients classically have a white forelock.

D. **OTITIS MEDIA** is a middle ear infection that may spread from the nasopharynx through the auditory tube. This may cause temporary or permanent deafness. Acute otitis media could be further complicated by **mastoiditis** since the mastoid air cells are continuous with the middle ear.

E. **HYPERACUSIS** is caused by a lesion of CN VII, which results in the paralysis of the stapedius muscle. This results in increased sensitivity to loud sounds due to the uninhibited movements of the stapes.

F. **CONDUCTIVE HEARING LOSS** results from an interference of sound transmission through the external ear or middle ear. This is most commonly caused by **otitis media** in children or **otosclerosis** (abnormal bone formation around the stapes) in adults.

G. **SENSORINEURAL HEARING LOSS** results from a loss of hair cells in the organ of Corti or a lesion of the cochlear part of CN VIII or to a central nervous system auditory pathway.

H. **PRESBYCUSIS** is caused by a progressive loss of hair cells at the base of the organ of Corti, which results in high-frequency hearing loss in the elderly.

Appendix **1**

Muscles of the Arm

MUSCLES OF THE ANTERIOR AXIOAPPENDICULAR REGION

Muscle	Innervation	Action
Pectoralis major	Lateral and medial pectoral nerves (C5, C6)	Adducts and medially rotates the arm
Pectoralis minor	Medial pectoral nerve (C8, T1)	Stabilizes the scapula
Subclavius	Nerve to subclavius (C5, C6)	Anchors and depresses the clavicle
Serratus anterior	Long thoracic nerve (C5, C6, C7)	Protracts the scapula and holds it against thoracic wall Rotates the scapula

MUSCLES OF THE POSTERIOR AXIOAPPENDICULAR AND SCAPULOHUMERAL REGION

Muscles	Innervation	Action
Trapezius	Accessory nerve (cranial nerve [CN] XI)	Elevates, depresses, and retracts the scapula Rotates glenoid cavity superiorly
Latissimus dorsi	Thoracodorsal nerve (C6, C7, C8)	Extends, adducts, and medially rotates the arm
Levator scapulae	Dorsal scapular nerve (C5) Cervical nerves (C3, C4)	Elevates the scapula Rotates glenoid cavity inferiorly
Rhomboid major and minor	Dorsal scapular nerve (C4, C5)	Retracts and rotates the scapula to depress glenoid cavity
Deltoid	Axillary nerve (C5, C6)	Flexes and medially rotates the arm Abducts the arm Extends and laterally rotates the arm
Supraspinatus	Suprascapular nerve (C4, C5, C6)	Abducts the arm Acts with rotator cuff muscles
Infraspinatus	Suprascapular nerve (C4, C5)	Laterally rotates the arm Holds humeral head in glenoid cavity
Teres minor	Axillary nerve (C5, C6)	Laterally rotates the arm Holds humeral head in glenoid cavity
Teres major	Lower subscapular nerve (C5, C6)	Adducts and medially rotates the arm
Subscapularis	Upper and lower subscapular nerve (C5, C6, C7)	Adducts and medially rotates the arm Holds humeral head in glenoid cavity

MUSCLES OF THE ANTERIOR (FLEXOR) COMPARTMENT OF THE ARM

Muscle	Innervation	Action
Biceps brachii	Musculocutaneous nerve (C5, C6)	Flexes and supinates the forearm
Brachialis	Musculocutaneous nerve (C5, C6)	Flexes forearm in all positions
Coracobrachialis	Musculocutaneous nerve (C5, C6)	Flexes and adducts the arm

MUSCLES OF THE POSTERIOR (EXTENSOR) COMPARTMENT OF THE ARM

Muscle	Innervation	Action
Triceps brachii	Radial nerve (C6, C7, C8)	Extends the forearm
Anconeus	Radial nerve (C7, C8, T1)	Extends the forearm Stabilizes the elbow joint

MUSCLES OF THE ANTERIOR (FLEXOR) COMPARTMENT OF THE FOREARM

Muscle	Innervation	Action
Pronator teres	Median nerve (C6, C7)	Flexes and pronates the forearm
Flexor carpi radialis	Median nerve (C6, C7)	Flexes and abducts the hand
Palmaris longus	Median nerve (C7, C8)	Flexes the hand
Flexor carpi ulnaris	Median nerve (C7, C8)	Flexes and adducts the hand
Flexor digitorum superficialis	Median nerve (C7, C8, T1)	Flexes proximal phalanges at metacarpophalangeal joints Flexes middle phalanges at proximal interphalangeal joints
Flexor digitorum profundus 　Medial part 　Lateral pat	 Ulnar nerve (C8, T1) Anterior interosseous nerve (C8, T1)	 Flexes distal phalanges at distal interphalangeal joint
Flexor pollicis longus	Anterior interosseous nerve (C8, T1)	Flexes phalanges of the thumb
Pronator quadratus	Anterior interosseous nerve (C8, T1)	Pronates the forearm

MUSCLES OF THE POSTERIOR (EXTENSOR) COMPARTMENT OF THE FOREARM

Muscle	Innervation	Action
Brachioradialis	Radial nerve (C5, C6, C7)	Flexes the forearm when in midpronated position
Extensor carpi radialis longus	Radial nerve (C6, C7)	Extends and abducts the hand
Extensor carpi radialis brevis	Deep branch of radial nerve (C7, C8)	Extends and abducts the hand
Extensor digitorum	Posterior interosseous nerve (C7, C8)	Extends the fingers at the metacarpophalangeal joints
Extensor digiti minimi	Posterior interosseous nerve (C7, C8)	Extends the fifth finger at the metacarpophalangeal joints
Extensor carpi ulnaris	Posterior interosseous nerve (C7, C8)	Extends and adducts the hand
Supinator	Deep branch of radial nerve (C7, C8)	Supinates the forearm
Extensor indicis	Posterior interosseous nerve (C7, C8)	Extends the second finger
Abductor pollicis longus	Posterior interosseous nerve (C7, C8)	Abducts the thumb Extends the thumb at the carpometacarpal joint
Extensor pollicis longus	Posterior interosseous nerve (C7, C8)	Extends the thumb at the carpometacarpal, metacarpophalangeal, and interphalangeal joints
Extensor pollicis brevis	Posterior interosseous nerve (C7, C8)	Extends the thumb at the carpometacarpal and metacarpophalangeal joints

INTRINSIC MUSCLES OF THE HAND

Muscle	Innervation	Action
Opponens pollicis	Recurrent branch of median nerve (C8, T1)	Opposes the thumb
Abductor pollicis brevis	Recurrent branch of median nerve (C8, T1)	Abducts the thumb
Flexor pollicis brevis	Recurrent branch of median nerve (C8, T1)	Flexes the thumb
Adductor pollicis	Deep branch of ulnar nerve (C8, T1)	Adducts the thumb
Abductor digiti minimi	Deep branch of ulnar nerve (C8, T1)	Abducts the fifth finger
Flexor digiti minimi brevis	Deep branch of ulnar nerve (C8, T1)	Flexes the fifth finger at the metacarpophalangeal joint
Opponens digiti minimi	Deep branch of ulnar nerve (C8, T1)	Opposes the fifth finger
First and second lumbricals	Median nerve (C8, T1)	Flex fingers at the metacarpophalangeal joints
Third and fourth lumbricals	Deep branch of ulnar nerve (C8, T1)	Flex fingers at the metacarpophalangeal joints
First through fourth dorsal interossei	Deep branch of ulnar nerve (C8, T1)	Abduct second through fourth fingers
First through third palmar interossei	Deep branch of ulnar nerve (C8, T1)	Adduct second, fourth, and fifth fingers

Muscles of the Leg

MUSCLES OF THE GLUTEAL REGION (ABDUCTORS AND ROTATORS OF THE THIGH)

Muscle	Innervation	Action
Gluteus maximus	Inferior gluteal nerve (L5, S1, S2)	Extends the thigh (especially from a flexed position) Assists in lateral rotation of the thigh Assists in rising from a sitting position
Gluteus medius	Superior gluteal nerve (L5, S1)	Adducts and medially rotates the thigh
Gluteus minimus	Superior gluteal nerve (L5, S1)	Adducts and medially rotates the thigh
Tensor of fascia lata	Superior gluteal nerve (L5, S1)	Adducts and medially rotates the thigh
Piriformis	Branches of anterior rami (S1, S2)	Laterally rotates extended thigh Abducts flexed thigh
Obturator internus	Nerve to obturator internus (L5, S1)	Laterally rotates extended thigh Abducts flexed thigh
Superior gemellus	Nerve to obturator internus (L5, S1)	Laterally rotates extended thigh Abducts flexed thigh
Inferior gemellus	Nerve to quadratus femoris (L5, S1)	Laterally rotates extended thigh Abducts flexed thigh
Quadratus femoris	Nerve to quadratus femoris (L5, S1)	Laterally rotates the thigh

MUSCLES OF THE ANTERIOR COMPARTMENT OF THE THIGH
(FLEXORS OF THE HIP JOINT AND EXTENSORS OF THE KNEE JOINT)

Muscle	Innervation	Action
Pectineus	Femoral nerve (L2, L3)	Adducts and flexes the thigh Assists in medial rotation of the thigh
Psoas major	Anterior rami of L1, L2, L3	Flexes the thigh
Psoas minor	Anterior rami of L1, L2	Flexes the thigh
Iliacus	Femoral nerve (L2, L3)	Flexes the thigh
Sartorius	Femoral nerve (L2, L3)	Flexes, abducts, and laterally rotates the thigh Flexes the leg
Rectus femoris	Femoral nerve (L2, L3, L4)	Extends the leg
Vastus lateralis	Femoral nerve (L2, L3, L4)	Extends the leg
Vastus medialis	Femoral nerve (L2, L3, L4)	Extends the leg
Vastus intermedius	Femoral nerve (L2, L3, L4)	Extends the leg

MUSCLES OF THE MEDIAL COMPARTMENT OF THE THIGH
(ADDUCTORS OF THE THIGH)

Muscle	Innervation	Action
Adductor longus	Obturator nerve (L2, L3, L4)	Adducts the thigh
Adductor brevis	Obturator nerve (L2, L3, L4)	Adducts the thigh
Adductor magnus	Obturator nerve (L2, L3, L4) Tibial part of sciatic nerve (L4)	Adducts the thigh Adductor part: flexes the thigh Hamstring part: extends the thigh
Gracilis	Obturator nerve (L2, L3)	Adducts the thigh Flexes the leg
Obturator externus	Obturator nerve (L3, L4)	Laterally rotates the thigh

MUSCLES OF THE POSTERIOR COMPARTMENT OF THE THIGH
(EXTENSORS OF THE HIP JOINT AND FLEXORS OF THE KNEE JOINT)

Muscles	Innervation	Action
Semitendinosus	Tibial part of sciatic nerve (L5, S1, S2)	Extends the thigh Flexes the leg Medially rotates the flexed leg
Semimembranosus	Tibial part of sciatic nerve (L5, S1, S2)	Extends the thigh Flexes the leg Medially rotates the flexed leg
Biceps femoris	Long head: tibial part of sciatic nerve (L5, S1, S2) Short head: common fibular part of sciatic nerve (L5, S1, S2)	Flexes the leg Laterally rotates the flexed leg

MUSCLES OF THE ANTERIOR AND LATERAL COMPARTMENTS OF THE LEG

Muscles	Innervation	Action
Anterior Compartment		
Tibialis anterior	Deep fibular nerve (L4, L5)	Dorsiflexes the foot Inverts the foot
Extensor digitorum longus	Deep fibular nerve (L5, S1)	Dorsiflexes the foot Extends lateral four toes
Extensor hallucis longus	Deep fibular nerve (L5, S1)	Dorsiflexes the foot Extends the big toe
Fibularis tertius	Deep fibular nerve (L5, S1)	Dorsiflexes the foot Assists in inversion of the foot
Lateral Compartment		
Fibularis longus	Superficial fibular nerve (L5, S1, S2)	Everts the foot
Fibularis brevis	Superficial fibular nerve (L5, S1, S2)	Everts the foot

MUSCLES OF THE POSTERIOR COMPARTMENT OF THE LEG

Muscle	Innervation	Action
Gastrocnemius	Tibial nerve (S1, S2)	Plantar flexes the ankle when leg is extended Flexes the leg
Soleus	Tibial nerve (S1, S2)	Plantar flexes the ankle independent of leg position
Plantaris	Tibial nerve (S1, S2)	Assists the gastrocnemius
Popliteus	Tibial nerve (L4, L5, S1)	Flexes the knee weakly Medially rotates the unplanted leg
Flexor hallucis longus	Tibial nerve (S2, S3)	Flexes the big toe at all joints Plantar flexes the ankle weakly
Flexor digitorum longus	Tibial nerve (S2, S3)	Flexes the lateral four toes Plantar flexes the ankle
Tibialis posterior	Tibial nerve (L4, L5)	Plantar flexes the ankle Inverts the foot

MUSCLES OF THE FOOT

Muscle	Innervation	Action
Abductor hallucis	Medial plantar nerve (S2, S3)	Abducts and flexes the big toe
Flexor digitorum brevis	Medial plantar nerve (S2, S3)	Flexes lateral four toes
Abductor digiti minimi	Lateral plantar nerve (S2, S3)	Abducts and flexes the little toe
Quadratus plantae	Lateral plantar nerve (S2, S3)	Flexes lateral four toes
Lumbricals	Medial 1: medial plantar nerve (S2, S3) Lateral 3: lateral plantar nerve (S2, S3)	Flex proximal phalanges of lateral four toes Extend middle and distal phalanges of lateral four toes
Flexor hallucis brevis	Medial plantar nerve (S2, S3)	Flexes proximal phalanx of the big toe
Adductor hallucis	Lateral plantar nerve (S2, S3)	Adducts the big toe
Flexor digiti minimi brevis	Lateral plantar nerve (S2, S3)	Flexes proximal phalanx of the little toe
Plantar interossei (3)	Lateral plantar nerve (S2, S3)	Adduct toes 2–4 Flex metatarsophalangeal joints
Dorsal interossei (4)	Lateral plantar nerve (S2, S3)	Abduct toes 2–4 Flex metatarsophalangeal joints
Extensor digitorum brevis	Deep fibular nerve (L5, S1)	Extends the four medial toes at the metatarsophalangeal and interphalangeal joints
Extensor hallucis brevis	Deep fibular nerve (L5, S1)	Extends the big toe at the metatarsophalangeal joint

Credits

Figure 1-1: A1, A2: From Moore KL, Dalley AF. *Clinically Oriented Anatomy*. 5th Ed. Philadelphia: Lippincott Williams & Wilkins, 2006:479. **A3:** From Moore KL, Dalley AF. *Clinically Oriented Anatomy*. 5th Ed. Philadelphia: Lippincott Williams & Wilkins, 2006:479. **Original source:** Olson TR. *Student Atlas of Anatomy*. Baltimore: Williams & Wilkins, 1996. **B:** From Moore KL, Dalley AF. *Clinically Oriented Anatomy*. 5th Ed. Philadelphia: Lippincott Williams & Wilkins, 2006:514. **C:** From Moore KL, Dalley AF. *Clinically Oriented Anatomy*. 5th Ed. Philadelphia: Lippincott Williams & Wilkins, 2006:515.

Figure 1-2: From Moore KL, Dalley AF. *Clinically Oriented Anatomy*. 5th Ed. Philadelphia: Lippincott Williams & Wilkins, 2006:507.

Figure 1-3: A: From Daffner RH. *Clinical Radiology: The Essentials*. 2nd Ed. Philadelphia: Lippincott Williams & Wilkins, 1999:544.

Figure 1-4: From Chew FS. *Musculoskeletal Imaging: The Core Curriculum*. Philadelphia: Lippincott Williams & Wilkins, 2003:163.

Figure 1-5: From Chew FS. *Musculoskeletal Imaging: The Core Curriculum*. Philadelphia: Lippincott Williams & Wilkins, 2003:163.

Figure 1-6: From Osborn AG. Head trauma. In: Eisenberg RL, Amberg JR, eds. *Critical Diagnostic Pathways in Radiology: An Algorithmic Approach*. Philadelphia: J.B. Lippincott Company, 1981.

Figure 1-7: From Daffner RH. *Clinical Radiology: The Essentials*. 2nd Ed. Philadelphia: Lippincott Williams & Wilkins, 1999:556.

Figure 1-8: From Chew FS. *Musculoskeletal Imaging: The Core Curriculum*. Philadelphia: Lippincott Williams & Wilkins, 2003:88.

Figure 1-9: From Harris JH, Harris WH. *The Radiology of Emergency Medicine*. 4th Ed. Philadelphia: Lippincott Williams & Wilkins, 2000:253.

Figure 1-10: From Chew FS. *Musculoskeletal Imaging: The Core Curriculum*. Philadelphia: Lippincott Williams & Wilkins, 2003:90.

Figure 1-11: From Daffner RH. *Clinical Radiology: The Essentials*. 2nd Ed. Philadelphia: Lippincott Williams & Wilkins, 1999:563.

Figure 1-12: A: From Agur AMR, Dalley AF. *Grant's Atlas of Anatomy*. 11th Ed. Philadelphia: Lippincott Williams & Wilkins, 2005:789. Courtesy of D. Salonen, University of Toronto, Canada. **B:** From Esses SI. *Textbook of Spinal Disorders*. Philadelphia: J.B. Lippincott Company, 1995:85.

Figure 1-13: A, B: From Esses SI. *Textbook of Spinal Disorders*. Philadelphia: J.B. Lippincott Company, 1995:88–89.

Figure 2-1: A: From Moore KL, Dalley AF. *Clinically Oriented Anatomy*. 5th Ed. Philadelphia: Lippincott Williams & Wilkins, 2006:523. **B:** From Dudek RW. *High-Yield Gross Anatomy*. 2nd Ed. Philadelphia: Lippincott Williams & Wilkins, 2002:14. Originally adapted from Chung KW. *BRS Gross Anatomy*. 4th Ed. Philadelphia: Lippincott Williams & Wilkins, 2000:289.

Figure 2-2: A: From Moore KL, Dalley AF. *Clinically Oriented Anatomy*. 5th Ed. Philadelphia: Lippincott Williams & Wilkins, 2006:49.

Figure 2-4: Adapted from Moore KL. *Clinically Oriented Anatomy*. 3rd Ed. Baltimore: Williams & Wilkins, 1992:368. **Inset from:** Scott DB. *Techniques of Regional Anaesthesia*. East Norwalk, CT: Appleton and Lange, 1989:169.

Figure 2-5: From Eisenberg RL. *Clinical Imaging: An Atlas of Differential Diagnosis*. 4th Ed. Philadelphia: Lippincott Williams & Wilkins, 2003:1007. Courtesy of M. Smith, MD, Nashville, TN. With permission from Neuroradiology Companion, M. Castillo, Lippincott-Raven.

Figure 2-6, 2-7: From Runge VM. *Contrast Media in Magnetic Resonance Imaging: A Clinical Approach*. Philadelphia: J.B. Lippincott Company, 1992:97–99.

Figure 5-1: From Moore KL, Dalley AF. *Clinically Oriented Anatomy*. 5th Ed. Philadelphia: Lippincott Williams & Wilkins, 2006:75. **Original source:** Agur AMR. *Grant's Atlas of Anatomy*. 9th Ed. Baltimore: Williams & Wilkins, 1991.

Figure 5-2: A: From Moore KL. *Clinically Oriented Anatomy*. 3rd Ed. Baltimore: Williams & Wilkins, 1992:46. **B:** From Brandt WE, Helms CA. *Fundamentals of Diagnostic Radiology*. 2nd Ed. Philadelphia: Lippincott Williams & Wilkins, 1999:496. **C:** From RH Daffner. *Clinical Radiology: The Essentials*. 2nd Ed. Philadelphia: Lippincott Williams & Wilkins, 1999:243. **D:** From RH Daffner. *Clinical Radiology: The Essentials*. 2nd Ed. Philadelphia: Lippincott Williams & Wilkins, 1999:245.

Figure 5-6: From Swischuck LE. *Imaging of the Newborn, Infant, and Young Child*. 5th Ed. Philadelphia: Lippincott Williams & Wilkins, 2004:288.

Figure 5-7: From Brandt WE, Helms CA. *Fundamentals of Diagnostic Radiology*. 2nd Ed. Philadelphia: Lippincott Williams & Wilkins, 1999:579.

Figure 5-8: From Brandt WE, Helms CA. *Fundamentals of Diagnostic Radiology*. 2nd Ed. Philadelphia: Lippincott Williams & Wilkins, 1999:582.

Figure 5-9: From Brandt WE, Helms CA. *Fundamentals of Diagnostic Radiology*. 2nd Ed. Philadelphia: Lippincott Williams & Wilkins, 1999:596.

Figure 5-10: Diagram: Adapted from Moore KL. *Clinically Oriented Anatomy*. 3rd Ed. Baltimore: Williams & Wilkins, 1992:57.

Figure 5-11: From Chen H, Sonneday CJ, Lillemoe KD, eds. *Manual of Common Bedside Surgical Procedures*. 2nd Ed. Philadelphia: Lippincott Williams & Wilkins, 2000:123.

Figure 5-12: From Scott DB. *Techniques of Regional Anaesthesia*. East Norwalk, CT: Appleton and Lange, 1989:147.

Figure 5-13: Diagram: Adapted from Moore KL. *Clinically Oriented Anatomy*. 3rd Ed. Baltimore: Williams & Wilkins, 1992:213.

Figure 5-14: From Freundlich IM, Bragg DG. *A Radiologic Approach to Diseases of the Chest*. 2nd Ed. Baltimore: Williams & Wilkins, 1997:610.

Figure 5-15: From Collins J, Stern EJ. *Chest Radiology: The Essentials.* Philadelphia: Lippincott Williams & Wilkins, 1999:3.

Figure 5-16: From Collins J, Stern EJ. *Chest Radiology: The Essentials.* Philadelphia: Lippincott Williams & Wilkins, 1999:4.

Figure 5-17: From Moore KL, Dalley AF. *Clinically Oriented Anatomy.* 5th Ed. Philadelphia: Lippincott Williams & Wilkins, 2006:187. Courtesy of Dr. E.L. Lansdown, Professor of Medical Imaging, University of Toronto, Ontario, Canada.

Figure 6-1: From Brandt, WE, Helms CA. *Fundamentals of Diagnostic Radiology.* 2nd Ed. Philadelphia: Lippincott Williams & Wilkins, 1999:463.

Figure 6-2: Adapted from Freundlich IM, Bragg DG. *A Radiologic Approach to Diseases of the Chest.* 2nd Ed. Baltimore: Williams & Wilkins, 1997:270.

Figure 6-3: A: Modified from Rohen JW. *Color Atlas of Anatomy.* 4th Ed. Baltimore: Williams & Wilkins, 1998:235. B, C: Adapted from Collins J, Stern EJ. *Chest Radiology: The Essentials.* Philadelphia: Lippincott Williams & Wilkins, 1999:10–11. D: From Moore KL and Dalley AF. *Clinically Oriented Anatomy.* 5th Ed. Philadelphia: Lippincott Williams & Wilkins, 2006:127.

Figure 6-4: From Collins J, Stern EJ. *Chest Radiology: The Essentials.* Philadelphia: Lippincott Williams & Wilkins, 1999:50.

Figure 6-5: From Daffner RH. *Clinical Radiology: The Essentials.* 2nd Ed. Philadelphia: Lippincott Williams & Wilkins, 1999:158.

Figure 6-6: From Freundlich IM, Bragg DG. *A Radiologic Approach to Diseases of the Chest.* 2nd Ed. Baltimore: Williams & Wilkins, 1997:716.

Figure 6-7: From Daffner RH. *Clinical Radiology: The Essentials.* 2nd Ed. Philadelphia: Lippincott Williams & Wilkins, 1999:146.

Figure 6-8: From Daffner RH. *Clinical Radiology: The Essentials.* 2nd Ed. Philadelphia: Lippincott Williams & Wilkins, 1999:152.

Figure 6-9: From Freundlich IM, Bragg DG. *A Radiologic Approach to Diseases of the Chest.* 2nd Ed. Baltimore: Williams & Wilkins, 1997:309.

Figure 6-10: From Collins J, Stern EJ. *Chest Radiology: The Essentials.* Philadelphia: Lippincott Williams & Wilkins, 1999:94.

Figure 6-11–6-14: From Slaby F, Jacobs ER. *NMS Radiographic Anatomy.* Philadelphia: Harwal, 1990:108, 110, 112, 116.

Figure 7-1: A, B: From Allen HD, Clark EB, Clark AG, et al., eds. *Moss and Adam's Heart Disease in Infants, Children, and Adolescents,* vol 1. 6th Ed. Philadelphia: Lippincott Williams & Wilkins, 2001:86.

Figure 7-2: From Moore KL, Dalley AF. *Clinically Oriented Anatomy.* 5th Ed. Philadelphia: Lippincott Williams & Wilkins, 2006:168.

Figure 7-3: A–D: From GM Pohost, O'Rourke RA, Berman DS, et al., eds. *Imaging in Cardiovascular Disease.* Philadelphia: Lippincott Williams & Wilkins, 2000:354–355.

Figure 7-4: Modified from Dudek RW. *High-Yield Heart.* Philadelphia: Lippincott Williams & Wilkins, 2006:42.

Figure 7-5: Modified from Dudek RW. *High-Yield Histology.* 3rd Ed. Philadelphia: Lippincott Williams & Wilkins, 2004:99. A: Modified from Dudek RW. *High-Yield Gross Anatomy.* 2nd Ed. Philadelphia: Lippincott Williams & Wilkins, 2002:46 and Damjanov I. *High-Yield Pathology.* Philadelphia: Lippincott Williams & Wilkins, 2000:38. D: From Damjanov I. *Histopathology: A Color Atlas and Textbook.* Baltimore: Williams & Wilkins, 1996:101.

Figure 7-6: A, C, D: From Brandt WE, Helms CA. *Fundamentals of Diagnostic Radiology.* 2nd Ed. Philadelphia: Lippincott Williams & Wilkins, 1999:374, 364, 532. B: From Collins J, Stern EJ. *Chest Radiology: The Essentials.* Philadelphia: Lippincott Williams & Wilkins, 1999:263.

Figures 7-7–7-9: Barrett CP, Poliakoff SJ, Holder LE. *Primer of Sectional Anatomy with MRI and CT Correlation.* 2nd Ed. Baltimore: Williams & Wilkins, 1994:53–54, 59–60, 69–70.

Figure 10-1: A: Adapted from Moore KL. *Clinically Oriented Anatomy.* 3rd Ed. Baltimore, Williams & Wilkins, 1992:168. B: From Erkonen WE, Smith WL. *Radiology 101: The Basics and Fundamentals of Imaging.* 2nd Ed. Philadelphia: Lippincott Williams & Wilkins, 2005:111.

Figure 10-2: From Moore KL, Dalley AF. *Clinically Oriented Anatomy.* 5th Ed. Philadelphia: Lippincott Williams & Wilkins, 2006:338.

Figure 10-3: From Moore KL, Dalley AF. *Clinically Oriented Anatomy.* 5th Ed. Philadelphia: Lippincott Williams & Wilkins, 2006:182.

Figure 10-5: From Moore KL, Dalley AF. *Clinically Oriented Anatomy.* 5th Ed. Philadelphia: Lippincott Williams & Wilkins, 2006:212. **Original source:** Lockhart RD, Hamilton GF, Fyfe FW. *Anatomy of the Human Body.* Philadelphia: Lippincott, 1959.

Figure 11-1: From Sternberg SS. *Histology for Pathologists.* 2nd Ed. Philadelphia: Lippincott-Raven, 1997:463.

Figure 11-2: A: From Sternberg SS. *Histology for Pathologists.* 2nd Ed. Philadelphia: Lippincott-Raven, 1997:467. B: From Gartner LP, Hiatt JL. *Color Atlas of Histology.* 4th Ed. Philadelphia: Lippincott Williams & Wilkins, 2006:291.

Figure 11-3: A: From Cormack DH. *Clinically Integrated Histology.* Philadelphia: Lippincott Williams & Wilkins, 1998:193. B: From Erkonen WE, Smith WL. *Radiology 101: The Basics and Fundamentals of Imaging.* 2nd Ed. Philadelphia: Lippincott Williams & Wilkins, 2005:91.

Figure 11-4: A,B: From Erkonen WE, Smith WL. *Radiology 101: The Basics and Fundamentals of Imaging.* 2nd Ed. Philadelphia: Lippincott Williams & Wilkins, 2005:113.

Figure 11-5: A: From Yamada T, Alpers DH, Laine L, et al., eds. *Atlas of Gastroenterology.* 2nd Ed. Philadelphia: Lippincott Williams & Wilkins, 1999:334. B: From Erkonen WE, Smith WL. *Radiology 101: The Basics and Fundamentals of Imaging.* 2nd Ed. Philadelphia: Lippincott Williams & Wilkins, 2005:114. C: From Rubin E, Farber JL. *Pathology.* 3rd Ed. Philadelphia: Lippincott Williams & Wilkins, 1999:731. **Original source:** Mitros FA. *Atlas of Gastrointestinal Pathology.* New York: Gower Medical Publishing, 1988. D: From Erkonen WE, Smith WL. *Radiology 101: The Basics and Fundamentals of Imaging.* 2nd Ed. Philadelphia: Lippincott Williams & Wilkins, 2005:115.

Figure 11-6: B: From Agur AMR, Dalley AF. *Grant's Atlas of Anatomy.* 11th Ed. Philadelphia: Lippincott Williams & Wilkins, 2004:151. Courtesy of G.B. Haber, University of Toronto, Canada. C: From Erkonen WE, Smith WL. *Radiology 101: The Basics and Fundamentals of Imaging.* 2nd Ed. Philadelphia: Lippincott Williams & Wilkins, 2005:138. D: From Erkonen WE, Smith WL. *Radiology 101: The Basics and Fundamentals of Imaging.* 2nd Ed. Philadelphia: Lippincott Williams & Wilkins, 2005:150. E: From Erkonen WE, Smith WL. *Radiology 101: The Basics and Fundamentals of Imaging.* 2nd Ed. Philadelphia: Lippincott Williams & Wilkins, 2005:352.

Figure 11-7, 11-8: From Yamada T, Alpers DH, Laine L, et al., eds. *Atlas of Gastroenterology.* 2nd Ed. Philadelphia: Lippincott Williams & Wilkins, 1999:474–475.

Figure 11-10: From Sternberg SS. *Histology for Pathologists.* 2nd Ed. Philadelphia: Lippincott-Raven, 1997:614. B: From Cubilla AL, Fitzgerald PJ. Tumors of the exocrine pancreas. In: Hartmann WH, Sobin LH, eds. *Atlas of Tumor Pathology,* 2nd series, fascile 19. Washington, DC: Armed Forces Institute of Pathology, 1984.

Figure 11-11: From Misiewicz JJ, Bartram CI. *Atlas of Clinical Gastroenterology.* London: Gower Medical Publishing, 1987.

Figures 11-12–11-14: From Barrett CP, Poliakoff SJ, Holder LE. *Primer of Sectional Anatomy with MRI and CR Correlation.* 2nd Ed. Baltimore: Williams & Wilkins, 1994:75–76, 79–80, 81–82.

Figure 11-15: A: From Erkonen WE, Smith WL. *Radiology 101: The Basics and Fundamentals of Imaging.* 2nd Ed. Philadelphia: Lippincott Williams & Wilkins, 2005:92. B: From Agur AMR, Dalley AF. *Grant's Atlas of Anatomy.* 11th

Ed. Philadelphia: Lippincott Williams & Wilkins, 2005:134. Courtesy of C.S. Ho, University of Toronto, Canada.

Figure 12-2: From Brandt WE, Helms CA. *Fundamentals of Diagnostic Radiology.* 2nd Ed. Philadelphia: Lippincott Williams & Wilkins, 1999:1168.

Figure 12-3: From Daffner RH. *Clinical Radiology: The Essentials.* Philadelphia: Lippincott Williams & Wilkins, 1999:309.

Figure 12-5: A,B: From Yamada T, Alpers DH, Laine L, et al., eds. *Textbook of Gastroenterology*, vol 2. 3rd Ed. Philadelphia: Lippincott Williams & Wilkins, 1999: 2959, 2956.

Figure 13-2: From Harris JH, Harris WH. *The Radiology of Emergency Medicine.* 4th Ed. Philadelphia: Lippincott Williams & Wilkins, 2000:616.

Figure 14-1: A: From Dudek RW. *High-Yield Kidney.* Philadelphia: Lippincott Williams & Wilkins, 2005:12. B: Adapted from Moore KL. *Clinically Oriented Anatomy.* 3rd Ed. Baltimore: Williams & Wilkins, 1992:174.

Figure 14-2: From Dudek RW. *High-Yield Kidney.* Philadelphia: Lippincott Williams & Wilkins, 2005:14.

Figure 14-3: A: From Wicke L. *Atlas of Radiologic Anatomy.* 6th Ed. Baltimore: Williams & Wilkins, 1998:133. B: From Daffner RH. *Clinical Radiology: The Essentials.* 2nd Ed. Philadelphia: Lippincott Williams & Wilkins, 1999:349.

Figure 14-9: From Dudek RW. *High-Yield Kidney.* Philadelphia: Lippincott Williams & Wilkins, 2005:31.

Figure 14-10: A: From Wicke L. *Atlas of Radiologic Anatomy.* 6th Ed. Baltimore: Williams & Wilkins, 1998:135. B: From Brandt WE, Helms CA. *Fundamentals of Diagnostic Radiology.* 2nd Ed. Philadelphia: Lippincott Williams & Wilkins, 1999:778. C: From Eisenberg RL. *Clinical Imaging: An Atlas of Differential Diagnosis.* 4th Ed. Philadelphia: Lippincott Williams & Wilkins, 2003:601.

Figure 14-11: A: From Taveras JM, Ferrucci JT, eds. *Radiology: Diagnosis, Imaging, Intervention,* Vol 4. Philadelphia: Lippincott, 1988:5. B: From Brandt WE, Helms CA. *Fundamentals of Diagnostic Radiology.* 2nd Ed. Philadelphia: Lippincott Williams & Wilkins, 1999:809.

Figure 14-12: A–C: From Barrett CP, Poliakoff SJ, Holder LE. *Primer of Sectional Anatomy with MRI and CT Correlation.* 2nd Ed. Baltimore: Williams & Wilkins, 1994:80, 82, 84.

Figure 15-1: A: From Premkumar K. *The Massage Connection Anatomy and Physiology.* Philadelphia: Lippincott Williams & Wilkins, 2004. B: Asset provided by the Anatomical Chart Co.

Figure 15-2: From Brandt WE, Helms CA. *Fundamentals of Diagnostic Radiology.* 2nd Ed. Philadelphia: Lippincott Williams & Wilkins, 1999:771.

Figure 15-3: A: From Rubin E, Farber JL. *Pathology.* 3rd Ed. Philadelphia: Lippincott Williams & Wilkins, 1999:1193. B: From Brandt WE, Helms CA. *Fundamentals of Diagnostic Radiology.* 2nd Ed. Philadelphia: Lippincott Williams & Wilkins, 1999:770.

Figure 15-4: Courtesy of Dr. J. Kitchin, Department of Obstetrics and Gynecology, University of Virginia.

Figure 15-5: From Sternberg SS. *Diagnostic Surgical Pathology,* Vol 1. 3rd Ed. Philadelphia: Lippincott Williams & Wilkins, 1999:614.

Figure 15-6: From Dudek RW. *High-Yield Histology.* 3rd Ed. Philadelphia: Lippincott Williams & Wilkins, 2004:212.

Figure 16-1: A: From Snell RS. *Clinical Anatomy.* 7th Ed. Philadelphia: Lippincott Williams & Wilkins, 2004:409.

Figure 17-1: From Rohen JW, Yokochi C, Lutjen-Drecoll E. *Color Atlas of Anatomy.* 4th Ed. Philadelphia: Lippincott Williams & Wilkins, 1998:316.

Figure 17-2: A,B: From Moore KL, Dalley AF. *Clinically Oriented Anatomy.* 5th Ed. Philadelphia: Lippincott Williams & Wilkins, 2006:223, 403.

Figure 17-7: A: From Dudek RW. *High-Yield Histology.* 3rd Ed. Philadelphia: Lippincott Williams & Wilkins, 2004:240. **Original source:** From Cormack DH. *Clinically Integrated Histology.* 2nd Ed. Philadelphia: Lippincott-Raven, 1998:272. B: From Brandt WE, Helms CA. *Fundamentals of*

Diagnostic Radiology. 2nd Ed. Philadelphia: Lippincott Williams & Wilkins, 1999:825.

Figure 17-8: From Daffner RH. *Clinical Radiology: The Essentials.* 2nd Ed. Philadelphia: Lippincott Williams & Wilkins, 1999:352.

Figure 17-9: From Brandt WE, Helms CA. *Fundamentals of Diagnostic Radiology.* 2nd Ed. Philadelphia: Lippincott Williams & Wilkins, 1999:826.

Figure 17-10: A,B: From Moore KL, Dalley AF. *Clinically Oriented Anatomy.* 5th Ed. Philadelphia: Lippincott Williams & Wilkins, 2006:454–455. C: From Dudek RW. *High-Yield Gross Anatomy.* 2nd Ed. Philadelphia: Lippincott Williams & Wilkins, 2002:119. Adapted from Chung KW. *BRS Gross Anatomy.* 2nd Ed. Baltimore: Williams & Wilkins, 1991.

Figure 18-1: A–D: From Rohen JW, Yokochi C, Lutjen-Drecoll E. *Color Atlas of Anatomy.* 4th Ed. Philadelphia: Lippincott Williams & Wilkins, 1998:409–412.

Figures 18-2, 18-3: A: From Moore KL, Dalley AF. *Clinically Oriented Anatomy.* 5th Ed. Philadelphia: Lippincott Williams & Wilkins, 2006:368, 370.

Figure 18-5: A: From Scott DB. Techniques of Regional Anaesthesia. East Norwalk, CT: Appleton & Lange, 1989:159. B: From Olson TR, Pawlina W. *ADAM Student Atlas of Anatomy.* Baltimore: Williams & Wilkins, 1996:169.

Figure 18-6: A, B: From Snell RS. *Clinical Anatomy.* 7th Ed. Philadelphia: Lippincott Williams & Wilkins, 2004: 407–408.

Figure 19-1: A: From Moore KL, Dalley AF. *Clinically Oriented Anatomy.* 5th Ed. Philadelphia: Lippincott Williams & Wilkins, 2006:435. **Original source:** Bassett DL. A Stereoscopic Atlas of Anatomy. Portland, OR: Sawyers, 1961:153–154. B: From Moore KL, Dalley AF. *Clinically Oriented Anatomy.* 5th Ed. Philadelphia: Lippincott Williams & Wilkins, 2006:435.

Figure 19-2: A, B: From Chung KW. *BRS Gross Anatomy.* 5th Ed. Philadelphia: Lippincott Williams & Wilkins, 2005:263–264.

Figure 20-1: C: Modified from Chung KW. *BRS Gross Anatomy.* 5th Ed. Philadelphia: Lippincott Williams & Wilkins, 2005:57.

Figure 20-3: A: Adapted from April EW. *NMS Clinical Anatomy.* 2nd Ed. Baltimore: Williams & Wilkins, 1990:58.

Figure 20-4: A–C: From Agur AMR, Ming JL. *Grant's Atlas of Anatomy.* 10th Ed. Philadelphia: Lippincott Williams & Wilkins, 1999:424–425.

Figure 20-5: From Agur AMR, Dalley AF. *Grant's Atlas of Anatomy.* 11th Ed. Philadelphia: Lippincott Williams & Wilkins, 2005:516. Courtesy of E. Becker, University of Toronto, Canada.

Figure 20-6: From Harris JH, Harris WH. *The Radiology of Emergency Medicine.* 4th Ed. Philadelphia: Lippincott Williams & Wilkins, 2000:310.

Figure 20-7: From Daffner RH. *Clinical Radiology: The Essentials.* 2nd Ed. Philadelphia: Lippincott Williams & Wilkins, 1999:417.

Figure 20-8: From Harris JH, Harris WH. *The Radiology of Emergency Medicine.* 4th Ed. Philadelphia: Lippincott Williams & Wilkins, 2000:304.

Figure 20-9: From Harris JH, Harris WH. *The Radiology of Emergency Medicine.* 4th Ed. Philadelphia: Lippincott Williams & Wilkins, 2000:306.

Figure 20-10: From Harris JH, Harris WH. *The Radiology of Emergency Medicine.* 4th Ed. Philadelphia: Lippincott Williams & Wilkins, 2000:327.

Figure 20-11: A: From April EW. *NMS Clinical Anatomy.* 2nd Ed. Baltimore: Williams & Wilkins, 1990:62. B, C: From Slaby F, Jacobs ER. *NMS Radiographic Anatomy.* Media, PA: Harwal, 1990:10, 12.

Figure 20-12: From Harris JH, Harris WH. *The Radiology of Emergency Medicine.* 4th Ed. Philadelphia: Lippincott Williams & Wilkins, 2000:357.

Figure 20-13: From Harris JH, Harris WH. *The Radiology of Emergency Medicine.* 4th Ed. Philadelphia: Lippincott Williams & Wilkins, 2000:352.

Figure 20-14: From Harris JH, Harris WH. *The Radiology of Emergency Medicine.* 4th Ed. Philadelphia: Lippincott Williams & Wilkins, 2000:350.

Figure 20-15: From Harris JH, Harris WH. *The Radiology of Emergency Medicine.* 4th Ed. Philadelphia: Lippincott Williams & Wilkins, 2000:364.

Figure 20-16: From Slaby F, Jacobs ER. *NMS Radiographic Anatomy.* Media, PA: Harwal, 1990:22.

Figure 20-17: From Chew FS. *Musculoskeletal Imaging: The Core Curriculum.* Philadelphia: Lippincott Williams & Wilkins, 2002:41.

Figure 20-18: From Harris JH, Harris WH. *The Radiology of Emergency Medicine.* 4th Ed. Philadelphia: Lippincott Williams & Wilkins, 2000:382.

Figure 20-19: From Chew FS. *Musculoskeletal Imaging: The Core Curriculum.* Philadelphia: Lippincott Williams & Wilkins, 2002:38.

Figure 20-20: From Harris JH, Harris WH. *The Radiology of Emergency Medicine.* 4th Ed. Philadelphia: Lippincott Williams & Wilkins, 2000:418.

Figure 20-21: A, B: Modified from Moore KL, Dalley AF. *Clinically Oriented Anatomy.* 5th Ed. Philadelphia: Lippincott Williams & Wilkins, 2006:670. **Original source:** Basmajian JV, Slonecker CE. *Grant's Method of Anatomy: A Clinical Problem-Solving Approach.* 11th Ed. Baltimore: Williams & Wilkins, 1989.

Figure 21-1: A: Adapted from Fleckenstein P, Tranum-Jensen J. *Anatomy in Diagnostic Imaging.* Philadelphia: WB Saunders, 1993:109. B, C: Modified from Moore KL, Dalley AF. *Clinically Oriented Anatomy.* 5th Ed. Philadelphia: Lippincott Williams & Wilkins, 2006:670. **Original source:** Basmajian JV, Slonecker CE. *Grant's Method of Anatomy: A Clinical Problem-Solving Approach.* 11th Ed. Baltimore: Williams & Wilkins, 1989.

Figure 21-2: From Chung KW. *BRS Gross Anatomy.* 5th Ed. Philadelphia: Lippincott Williams & Wilkins, 2005:96.

Figure 21-4: From Agur AMR, Dalley AF. *Grant's Atlas of Anatomy.* 11th Ed. Philadelphia: Lippincott Williams & Wilkins, 2005:348.

Figure 21-5: A: From Agur AMR, Dalley AF. *Grant's Atlas of Anatomy.* 11th Ed. Philadelphia: Lippincott Williams & Wilkins, 2005:380.

Figure 21-6: A: From Harris JH, Harris WH. *The Radiology of Emergency Medicine.* 4th Ed. Philadelphia: Lippincott Williams & Wilkins, 2000:780. B: From Chew FS. *Musculoskeletal Imaging.* Philadelphia: Lippincott Williams & Wilkins, 2003:107.

Figure 21-7: From Chew FS. *Musculoskeletal Imaging.* Philadelphia: Lippincott Williams & Wilkins, 2003:109.

Figure 21-8: From Harris JH, Harris WH. *The Radiology of Emergency Medicine.* 4th Ed. Philadelphia: Lippincott Williams & Wilkins, 2000:795.

Figure 21-9: From Harris JH, Harris WH. *The Radiology of Emergency Medicine.* 4th Ed. Philadelphia: Lippincott Williams & Wilkins, 2000:801.

Figure 21-10: From Eisenberg RL. *Clinical Imaging: An Atlas of Differential Diagnosis.* 4th Ed. Philadelphia: Lippincott Williams & Wilkins, 2003:889.

Figure 21-11: From Agur AMR, Dalley AF. *Grant's Atlas of Anatomy.* 11th Ed. Philadelphia: Lippincott Williams & Wilkins, 2005:393.

Figure 21-13: A(2), B: From Agur AMR, Dalley AF. *Grant's Atlas of Anatomy.* 11th Ed. Philadelphia: Lippincott Williams & Wilkins, 2005:432, 427. Courtesy of E. Becker, University of Toronto, Canada.

Figure 21-14: A (2): From Harris JH, Harris WH. *The Radiology of Emergency Medicine.* 4th Ed. Philadelphia: Lippincott

Williams & Wilkins, 2000:861. A (3): From Chew FS. *Musculoskeletal Imaging.* Philadelphia: Lippincott Williams & Wilkins, 2003:140.

Figure 21-15: B (2): From Harris JH, Harris WH. *The Radiology of Emergency Medicine.* 4th Ed. Philadelphia: Lippincott Williams & Wilkins, 2000:857.

Figure 22-1: Adapted from Moore KL. *Clinically Oriented Anatomy.* 3rd Ed. Baltimore: Williams & Wilkins, 1992.

Figure 22-2: From McMillan JA, DeAngelis CD, Feigin RD, et al., eds. *Oski's Pediatrics: Principles and Practice.* 3rd Ed. Philadelphia: Lippincott Williams & Wilkins, 1999:396.

Figure 22-3: From Swischuk LE. *Imaging of the Newborn Infant and Young Child.* 5th Ed. Philadelphia: Lippincott Williams & Wilkins, 2004:1016.

Figure 22-4: A: Adapted from Moore KL, Dalley AF. *Clinically Oriented Anatomy.* 3rd Ed. Baltimore: Williams & Wilkins, 1992:666. B: From Harwood-Nuss A, Wolfson AB, Linden CH, et al., eds. *The Clinical Practice of Emergency Medicine.* 3rd Ed. Philadelphia: Lippincott Williams & Wilkins, 2001.

Figure 22-5: From Siegel A, Sapru HN. *Essential Neuroscience.* Philadelphia: Lippincott Williams & Wilkins, 2006:47.

Figure 22-6: From Jenkins JR, da Costa Lite C. *Neurodiagnostic Imaging: Pattern Analysis and Differential Diagnosis.* Philadelphia: Lippincott-Raven, 1998:789.

Figure 22-7: A, B: From Siegel A, Sapru HN. *Essential Neuroscience.* Philadelphia: Lippincott Williams & Wilkins, 2006:53. C: From Agur AMR, Dalley AF. *Grant's Atlas of Anatomy.* 11th Ed. Philadelphia: Lippincott Williams & Wilkins, 2005:620.

Figure 22-8: A: From Dudek RW. *High-Yield Gross Anatomy.* 2nd Ed. Philadelphia: Lippincott Williams & Wilkins, 2002:170. B–E: From Haines DE. *Neuroanatomy: An Atlas of Structure, Sections, and Systems.* 6th Ed. Philadelphia: Lippincott Williams & Wilkins, 2004:48, 49, 51.

Figure 23-1: From Chung KW. *BRS Gross Anatomy.* 5th Ed. Philadelphia: Lippincott Williams & Wilkins, 2005:350.

Figure 23-3: From Blackbourne LH. *Advanced Surgical Recall.* Baltimore: Williams & Wilkins, 1997:787.

Figures 23-4–23-6: From Scott BD. *Techniques of Regional Anaesthesia.* Stamford, CT: Appleton and Lange, 1989:77, 93, 209.

Figure 23-7: A, B, C (right): From Rohen JW, Yokochi, Lutjen-Drecoll E. *Color Atlas of Anatomy.* 4th Ed. Philadelphia: Lippincott Williams & Wilkins, 1998:154, 157. C (left): Adapted from Moore KL, Dalley AF. *Clinically Oriented Anatomy.* 5th Ed. Philadelphia: Lippincott Williams & Wilkins, 2006:1095. **Original source:** Rohen JW, Yokochi C, Lutjen-Drecoll E. *Color Atlas of Anatomy.* 5th Ed. Philadelphia: Lippincott Williams & Wilkins, 2002.

Figure 23-8, 23-9: Redrawn from Agur AMR, Ming JL. *Grant's Atlas of Anatomy.* 10th Ed. Baltimore: Williams & Wilkins, 1999.

Figure 24-1: A: From Moore KL, Dalley AF. *Clinically Oriented Anatomy.* 5th Ed. Baltimore: Lippincott Williams & Wilkins, 2006:962. B: From Ross MH, Kaye GI, Pawlina W. *Histology: A Text and Atlas.* 4th Ed. Baltimore: Lippincott Williams & Wilkins, 2003:809. C: From Moore KL, Dalley AF. *Clinically Oriented Anatomy.* 5th Ed. Baltimore: Lippincott Williams & Wilkins, 2006:962.

Figure 24-2: A: From Moore KL, Dalley AF. *Clinically Oriented Anatomy.* 5th Ed. Philadelphia: Lippincott Williams & Wilkins, 2006:970. B, C: From Moore KL, Dalley AF. *Clinically Oriented Anatomy.* 5th Ed. Philadelphia: Lippincott Williams & Wilkins, 2006:971.

Figure 24-3: From Fix J. *High-Yield Neuroanatomy.* 3rd Ed. Philadelphia: Lippincott Williams & Wilkins, 2005:124.

Index

Page numbers followed by f indicate figure; those followed by t indicate table.

M